Dynamics and Applications of Photon-Nanostructured Systems

Dynamics and Applications of Photon-Nanostructured Systems

Editor

Evangelia Sarantopoulou

MDPI • Basel • Beijing • Wuhan • Barcelona • Belgrade • Manchester • Tokyo • Cluj • Tianjin

Editor
Evangelia Sarantopoulou
Theoretical and Physical Chemistry Institute, National Hellenic Research Foundation
Greece

Editorial Office
MDPI
St. Alban-Anlage 66
4052 Basel, Switzerland

This is a reprint of articles from the Special Issue published online in the open access journal *Nanomaterials* (ISSN 2079-4991) (available at: https://www.mdpi.com/journal/nanomaterials/special_issues/photon).

For citation purposes, cite each article independently as indicated on the article page online and as indicated below:

LastName, A.A.; LastName, B.B.; LastName, C.C. Article Title. *Journal Name* **Year**, *Article Number*, Page Range.

ISBN 978-3-03943-328-5 (Hbk)
ISBN 978-3-03943-329-2 (PDF)

Cover image courtesy of Zoran Samardžija and Alkiviadis-Constantinos Cefalas.

Contents

About the Editor

Evangelia Sarantopoulou is a Senior Researcher at National Hellenic Research Foundation, Theoretical and Physical Chemistry Institute, Athens, Greece. She is the author or co-author of 117 peer-reviewed articles and books, and has participated in more than 130 international conferences. Her current research interests include complexity issues in physics and materials, photonic surface engineering and physics of interphases, nanomechanics, biophotonics and nanomedicine. As a scientist at the European Space Agency, she was responsible for European and National research projects, and she also serves as an expert in the European Commission.

nanomaterials

Editorial

Dynamics and Applications of Photon-Nanostructured Systems

Evangelia Sarantopoulou

Theoretical and Physical Chemistry Institute, National Hellenic Research Foundation, 48 Vassileos Constantinou Avenue, 11635 Athens, Greece; esarant@eie.gr; Tel.:+ 30-210-7273840

Received: 13 August 2020; Accepted: 29 August 2020; Published: 3 September 2020

In a speedy and complicated word, only a small number of book readers have the time to dig out the hidden "gemstones" between the text lines. The ambition of the present Special Issue titled "Dynamics and Applications of Photon-Nanostructured Systems" is to offer the readers the opportunity to look at nanosystems differently. Besides photon surface engineering, topics, such as non-equilibrium nano thermodynamics, nonlinear dynamic evolution of nanosystems, quantum, size-effects, and photonic states from photon–nanosystem interactions are also discussed. Naturally, the description "photonic–nanostructured systems" outline photonic-crafted nanodevices with specific functionalities [1,2]. However, because electromagnetic waves are carriers of information, photons inscribe not only nanostructured domains in engineering materials but, also, frequency, phase, noise, and the state of quantum coherence are carriers of information of the structure and shape of nano-entities. The present Special Issue leverages the above topics and compiles twelve original theoretical and applied research articles and three review papers. The articles cover a broad range of thematic areas in physics, engineering, and biology, including photonic interaction with nanocavities and resonators. The collection of articles underlines not only a connection between light, nanodimensionality, surface, and dynamics, but it also accentuates essential issues such as topology, hierarchy, time, information, and irreversibility.

Contrary to common conceptions for time, light, surface, and dynamics, topology, hierarchy, information, and irreversibility are notions not readily discussed in regular nanotechnology research topics. However, these "hidden treasures" are in a position to shape tomorrow's world by inspiring an implausible set of applications. Topology is the science of geometrical aspects of objects, aiming to identify geometrical invariants. In that sense topology and physics are inherently interconnecting because both scientific domains trace "*immutability*". Time, on the other hand, in the classical conception of dynamics and quantum mechanics, flows continuously in a homogeneous way, following only a universal constraint that the equations of motion are invariant in time reversal. In this *static* and *reversible* world, dynamical systems are traced not only back, but their states are also projected in future. In the non-relativistic view, the physical observables, such as momentum, energy, and angular momentum, stand for the eternal, immutable landmarks of the physical–geometrical space. The topology is Euclidian, and the flow of time and the structure of space are both homogeneous. However, even in this immovable world, some hidden gems are waiting to be discovered.

Let us clarify the above statement by giving an example. Two cars climb a mountain from opposite sides (Figure 1a). They both aim to reach the top of the hill. At first glance, the situation might well describe two similar processes. However, the two events might have diverging endings, and, therefore, the two events represent two different states. If the slopes and the morphology of the two hillsides are alike or even very similar, we expect the physical observables, associated with the two cars, to attain comparable values in both directions during the up-hilling stage. The two vehicles are in the position to reach the top of the hill because of a similar hillside topology. In the opposite case, where the slopes attain different values, the two vehicles require diverging energy, power, and torque to achieve the goal of reaching the top of the mountain. In the extreme case, where the inclination of one side exceeds, say, 50°, the associated vehicle might reach the top of the hill, but it might overturn because of the high speed required to climb the mountain (Figure 1b). We have here an example of instability in the "mountain–car" system during the dynamical evolution of the system imposed by the Euclidean topology. The excess power required to climb the steeper side of the mountain might be responsible for a car overturned at the top.

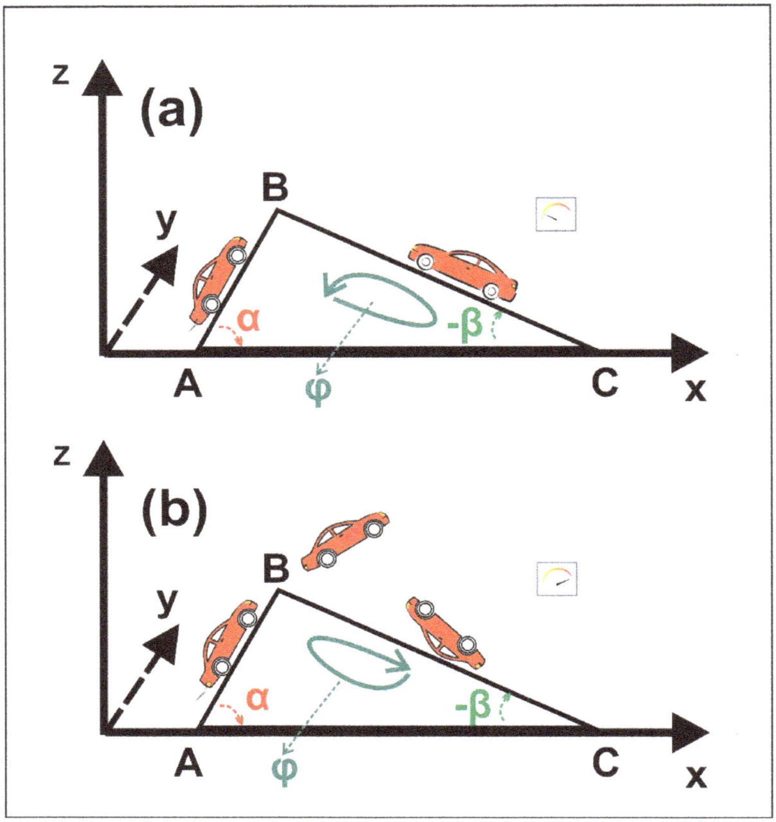

Figure 1. The connection between topology and irreversibility in classical dynamics: (**a**) Two cars climb two asymmetric sides of a mountain in opposite directions. Because of different side slopes, the engine power and the torque during the climbing stage are different; (**b**) the left car requires a high engine power and torque to reach the top of the mountain, which might lead to an over-flip (instability) and, thus, to an irreversible situation.

The final state of the system "mountain–car" depends on the topology of the mountain surface. In the case of vehicle demolition during the climbing stage, ascending the top of a hill with a vehicle might be irreversibly dangerous, because topology might lead to an irreversible damaging state.

We have, here, a simplified example where surface topology entails different endings (states) of similar physical processes and also imply irreversibility during the evolution of a system in the same steps following classical dynamics. The two states and the topology imply a hidden degeneracy of physical observables, e.g., different values of vehicles' power along opposite directions to obtain the objective of reaching the top of the mountain. With a similar way in 2D nanosystems, Euclidean topology is directly correlated with the topology of the electric components of a surface (local charge, electric field, potential). Because of hidden degeneracies and space inhomogeneity, we expect different physical phenomena to appear in tracing paths along opposite directions. It is plausible therefore to ask ourselves the question: "could one identify similar hidden degeneracies ('hidden gems') during light interaction with surfaces, and in the case of an affirmative answer, what kind of 'gems' one could be in a position of digging out?" Following the example of the two cars, we are in the position to conclude that irreversibility, time, and topology are connected inherently in the melting pot of classical notions of physics and because of everyday experience, humans realize that irreversibility is a rather ordinary state of Nature. The perceptions of evolution and irreversibility appear to be central in our understanding of the Cosmos and life. Irreversibility and evolution emerged at full speed during the nineteenth century in almost every scientific field and physics through the second law of thermodynamics, the important principle of the increase of entropy. In this classical view of the word, the second law of thermodynamics describes molecular disorder, and Boltzmann's thermodynamic equilibrium corresponds to a state of maximum probability. However, in the physics of tiny systems, as well as in the evolution of life, entropy and irreversibility imply transformations to *higher levels of complexity* (hierarchical levels) and "information" which, contrary to the static view of the world (e.g., the planetary motions), follow unidirectional evolution pathways (the mortality is an irreversible evolution process). Let us clarify the above points with another physical example: irradiation of a polymeric surface with low-energy density photons in the visible region of the spectrum leads to the *reversible* fast dynamic transient response of electron and vibrational states. After some time, the system retains its original state with a similar hierarchical level before irradiation.

On the contrary, irradiation of the same system with vacuum ultraviolet photons (VUV, 110–180 nm) modifies the surface in *a non-reversible* way putting the system at a *higher hierarchical level*. The topologies of the surface prior and after irradiation are different. The two physical states have different fractal dimensionality and structure, and transition from the non-irradiated state to the radiative one is unidirectional (irreversible).

The entropy (and thus the transfer of information) of a system might follow non-thermodynamic pathways, and thus the transfer of physical information in tiny topological spaces could be chaotic. It is plausible therefore to ask another question: "how one could relate thermodynamic irreversibility with time (characteristic time of a central physical process), and hierarchy, or even to ask for the correlation between hierarchy and non-thermodynamic chaotic (random) motions, where the physical rates of systems follow completely random pathways". The key to this answer is again topology. Indeed, confinement and escape of molecules from tiny spaces follow either thermodynamic or chaotic behavior. The size and topology of physical entities and space outline a set of time-space boundaries between thermodynamic equilibrium and chaotic motions, between physical laws and chaos, where physical rates vary randomly. This is the result of entropic variations from molecular confinement in tiny spaces, emerging from an irreversible surface restructuring at a high hierarchical level [3].

In the applications domain now, nanotechnologies trail diverging steps of innovative technological applications and the bet, in this case, is the successful integration of molecular functionalities with the macro-world [4]. Photons, besides their use in all practical aspects of modern life, convey a vast amount of *quantum information*, which, when joining nanosciences and nanotechnological tools, allow one to visualize new technological breakthroughs such as quantum computing (stages of higher hierarchical levels) [5].

Along the above lines, an extensive scientific and technological effort has been devoted to designing and integrating micro-nano sensors with improved (molecular) sensitivity, quick responses, high stabilities, and robustness in lab-on-a-chip devices. Among applications, mechanical nanosensors are detecting mechanical frequency variations, wave velocity, pressure, and strain [6]. Therefore, it would not be possible in this Special Issue to leave out nanomechanics and nanoresonators. The latter belongs to a new class of nanoelectromechanical systems enabling applications such as atomic and molecular sensing and separation, molecular transportation, high-frequency signal processing, and bioimaging [6]. Here, we have a *reversible state* of matter, but an *irreversible state of photons* which occupy a higher hierarchical quantum state after the interaction of photons with matter.

Along the above lines, the mechanical and quantum characteristics of nanomechanical resonators coupled to a superconducting resonator were theoretically studied by Choi et al. [7] and thus is possible to predict phenomena that could lead to the development of novel technologies for quantum information processing.

Whispering-gallery-mode microresonator-based sensors with high local field intensities also configure novel platforms for enhancing the interactions between light and matter in both *reversible* (matter) and *non-reversible* (photons) states. Such states could bear low detection limits, down to a single molecule and nanoparticles. An "open" sensing configuration with the whispering-gallery-mode microresonator-based sensor, to monitor chemical reaction progress in the water droplet, is discussed and supported by a proof-of-a concept demonstration by Lu et al. [8]. This "open" configuration arrangement provides a real-time accuracy and sensitivity chemical/biochemical reaction kinetics platform.

Furthermore, optofluidic microcavity laser systems bear a wide span of potential applications in tunable single-mode on-chip lasers, biosensors in photobiology and photomedicine. For the first time in aqueous media, Guo et al. [9] investigated theoretically and experimentally single-frequency laser and mode splitting phenomena in optofluidic microdisk device that combines solid-state dye-doped polymer microdisks with a microfluidic channel device.

Guided mode resonance (GMR) structures allow obtaining complete bioreaction information. Zhou et al. [10] systematically presented a parametric analysis elucidating the influence of structural design factors (i.e., grating period and groove depth) at the nanoscale for "grating–waveguide" GMR sensors performance to achieve higher angular sensitivity and optimized wavelength figure of merit. By combining the analytical model and numerical simulations, higher performance sensors with lower detection limits in biosensing can be designed.

In the applications domain, nanostructures have attracted considerable research interest for their many advantages in photonic sensors applications. The response of the passive-type visible-blind ultraviolet photodetectors of ZnO nanorod, with different structure morphologies, was investigated by Khan et al. [11]. The fabricated ultraviolet sensors based on the ZnO NR-gated AlGaN/GaN high electron mobility transistor structure with nanoscale fin isolation demonstrate high-responsivity. Moreover, the sensing mechanism upon UV illumination was revealed.

Furthermore, porous Si–SiO_2 UV microcavities with the thickness of a few tenths nm were also applied as filters by Jimenéz-Vivanco et al. [12] to modulate a broad responsivity photodetector with a detection range from 300 to 510 nm. The photodetectors had a broad, but porous Si–SiO_2 UV microcavities improved the broad response silicon photodetector inside specific UV range of wavelengths, and they can be applied as UV-heated mirrors or UV bandpass filters.

In the *information* domain, numerical simulations were applied by Cao et al. [13] to prove that the electromagnetic field formed in the localized region of the mesoscale dielectric sphere can be modulated by introducing a nanohole structure at its shadow surface. Thus the authors were able to improve the spatial resolution of the information transfer up to $\lambda/40$, well beyond the stable immersion diffraction limit. This finding is essential for advancing the particle-lens-based super-resolution technologies, including sub-diffraction imaging, interferometry, surface fabrication, enhanced Raman scattering, and optical tweezer.

Purtov et al. [14] reported on the fabrication of defect-free arrays of pillars with diameters down to 184 nm. The two-dimensional photonic structures compared to theoretical predictions from Monte Carlo simulations and the optical reflectivities of the nanopillar gratings were analyzed by optical microscopy and verified by coupled-wave simulations. Ding et al. [15] reported on near and deep-subwavelength ripples on stainless-steel surfaces. A qualitative description based on the surface plasmon polariton modulated periodic Coulomb explosion is proposed for the interpretation of their formation mechanism. In the work of Wei et al. [16], a miniaturized nanowire laser with high end-facet reflection was realized by integrating an Ag grating between the nanowire and the substrate. The proposed nanowire laser with a lowered threshold and reduced dimensions is significant in on-chip information systems and networks.

Nano/micro-scale native random levels of roughness are responsible for the field enhancement effect unwanted in current electrical and optical systems. To design and optimize the networks, including the selection of materials, structures, and operating conditions, the plasmonic local energy enhancement effect around the metal surfaces, is required. Fukuoka et al. [17] investigated, numerically, the plasmonic enhancement of the electromagnetic field energy density at the sharp tips of nanoparticles or nanoscale surface levels of the roughness of hydrogen-absorbing transition metals, Pd, Ti, and Ni. Last but not least, three review articles are included in this particular issue. A brief review of enhancing the photoelectric performance of nanostructured semiconductor-based photodetectors is presenting by Ding et al. [18]. The authors give the latest research surface/interface engineering. The key factors and the challenges for improving nanostructured photodetectors are also pointed out.

A comprehensive review by Tian et al. [19] focuses on the synthetic methods and optoelectronic properties of inorganic boron-based nanostructures. Also, the optoelectronic behaviors of known inorganic boron-based nanostructures and future applications are presenting.

Finally, an extensive review of materials design strategies and performance of high sensitivity resists for extreme ultraviolet (EUV) lithography at 13.5 nm by Manouras and Argitis [20] is presented in this Special Issue. The last review article entails both fundamental information on the radiation-induced processes in this spectral region and a large number of new ideas targeting at the design of new highly sensitive and top-performing EUV resists.

All authors are confident that this current Special Issue entitled "Dynamics and Applications of Photon-Nanostructured Systems" will not only provide experts in the field but also curious readers with overarching insights into this complex and high cross-disciplinary field.

Funding: This research was funded under the frame of the projects "ELI–LASERLAB Europe Synergy, HiPER and IPERION-CH.gr" (MIS 5002735) which is implemented under the Action "Reinforcement of the Research and Innovation Infrastructure", funded by the Operational Programme "Competitiveness, Entrepreneurship, and Innovation" (NSRF 2014–2020), co-financed by Greece and the European Union (European Regional Development Fund, and "Advanced Materials and Devices" (MIS 5002409) which is implemented under the "Action for the Strategic Development on the Research and Technological Sector" funded by the Operational Programme "Competitiveness, Entrepreneurship and Innovation" (NSRF 2014–2020) and co-financed by Greece and the European Union (European Regional Development Fund).

Acknowledgments: A deep acknowledgement to all authors who submitted their research work to this Special Issue "Dynamics and Applications of Photon-Nanostructured Systems" and to the *Nanomaterials* reviewers participating in the peer-review process for improving the quality and effect of the submitted papers. I also wish to thank the Editor-in-Chief, Shirley Chiang, and the assistant editors, Erika Zhao and Frances Yuan, for their dedication to the creation of a smooth and efficient process.

Conflicts of Interest: The author declares no conflict of interest.

References

1. Konstantatos, G.; Sargent, E.H. Nanostructured materials for photon detection. *Nat. Nanotechnol.* **2010**, *5*, 391–400. [CrossRef] [PubMed]
2. Pergolesi, D.; Grojo, D.; Rebollar, E.; Dinescu, M. (Eds.) Photon-Assisted Synthesis and Processing of Materials in Nano-Microscale. In *Special Issue of Advanced Surface Science Dedicated to the Proceedings of Symposium X of the European Materials Research Society Spring Meeting, Strasbourg, France, 18–22 June 2018*; Applied Surface Science; Elsevier: Amsterdam, The Netherlands, 2018. Available online: https://www.sciencedirect.com/journal/applied-surface-science/special-issue/1097BQJDV9Q (accessed on 21 February 2019).
3. Gavriil, V.; Chatzichristidi, M.; Christofilos, D.; Kourouklis, G.A.; Kollia, Z.; Bakalis, E.; Cefalas, A.C.; Sarantopoulou, E. Entropy and random walk trails water confinement and non-thermal equilibrium in photon-induced nanocavities. *Nanomaterials* **2020**, *10*, 1101. [CrossRef] [PubMed]
4. Fecht, H.J.; Werner, M.; Van de Voorde, M. *The Nano-Micro Interface: Bridging Micro and Nano Worlds*, 2nd ed.; Wiley-VCH Verlag GmbH & Co. KGaA Boschstr: Weinheim, Germany, 2015.
5. Mäntynen, H.; Anttu, N.; Sun, Z.; Lipsanen, H.K. Single-Photon sources with quantum dots in III–V nanowires. *Nanophotonics* **2019**, *8*, 747–769. [CrossRef]

6. Arash, B.; Jiang, J.-W.; Rabczuk, T. A review on nanomechanical resonators and their applications in sensors and molecular transportation. *Appl. Phys. Rev.* **2015**, *2*, 21301. [CrossRef]
7. Choi, J.; Ju, S. Quantum characteristics of a nanomechanical resonator coupled to a superconducting LC resonator in quantum computing systems. *Nanomaterials* **2019**, *9*, 20. [CrossRef] [PubMed]
8. Lu, Q.; Chen, X.; Fu, L.; Xie, S.; Wu, X. On-Chip Real-Time chemical sensors based on water-immersion-objective pumped whispering-gallery-mode microdisk laser. *Nanomaterials* **2019**, *9*, 479. [CrossRef] [PubMed]
9. Guo, Z.; Wang, H.; Zhao, C.; Chen, L.; Liu, S.; Hu, J.; Zhou, Y.; Wu, X. Spectral modulation of optofluidic coupled-microdisk lasers in aqueous media. *Nanomaterials* **2019**, *9*, 1439. [CrossRef] [PubMed]
10. Zhou, Y.; Wang, B.; Guo, Z.; Wu, X. Guided mode resonance sensors with optimized figure of merit. *Nanomaterials* **2019**, *9*, 837. [CrossRef] [PubMed]
11. Khan, F.; Khan, W.; Kim, S.-D. High-Performance ultraviolet light detection using nano-scale-fin isolation AlGaN/GaN heterostructures with ZnO nanorods. *Nanomaterials* **2019**, *9*, 440. [CrossRef] [PubMed]
12. Jimenéz-Vivanco, M.R.; García, G.; Carrillo, J.; Morales-Morales, F.; Coyopol, A.; Gracia, M.; Doti, R.; Faubert, J.; Lugo, J.E. Porous Si-SiO_2 UV Microcavities to modulate the responsivity of a broadband photodetector. *Nanomaterials* **2020**, *10*, 222. [CrossRef] [PubMed]
13. Cao, Y.; Liu, Z.; Minin, O.; Minin, I. Deep subwavelength-scale light focusing and confinement in nanohole-structured mesoscale dielectric spheres. *Nanomaterials* **2019**, *9*, 186. [CrossRef] [PubMed]
14. Purtov, J.; Rogin, P.; Verch, A.; Johansen, V.E.; Hensel, R. Nanopillar diffraction gratings by two-photon lithography. *Nanomaterials* **2019**, *9*, 1495. [CrossRef] [PubMed]
15. Ding, S.; Zhu, D.; Xue, W.; Liu, W.; Cao, Y. Picosecond laser-induced hierarchical periodic Near- and Deep-subwavelength ripples on stainless-steel surfaces. *Nanomaterials* **2019**, *10*, 62. [CrossRef] [PubMed]
16. Wei, W.; Yan, X.; Zhang, X. Miniaturized GaAs nanowire laser with a metal grating reflector. *Nanomaterials* **2020**, *10*, 680. [CrossRef] [PubMed]
17. Fukuoka, N.; Tanabe, K. Lightning-Rod effect of plasmonic field enhancement on hydrogen-absorbing transition metals. *Nanomaterials* **2019**, *9*, 1235. [CrossRef] [PubMed]
18. Ding, M.; Guo, Z.; Chen, X.; Ma, X.; Zhou, L. Surface/Interface engineering for constructing advanced nanostructured photodetectors with improved performance: A brief review. *Nanomaterials* **2020**, *10*, 362. [CrossRef] [PubMed]
19. Tian, Y.; Guo, Z.; Zhang, T.; Lin, H.; Li, Z.; Chen, J.; Deng, S.; Liu, F. Inorganic boron-based nanostructures: Synthesis, optoelectronic properties, and prospective applications. *Nanomaterials* **2019**, *9*, 538. [CrossRef] [PubMed]
20. Manouras, T.; Argitis, P. High sensitivity resists for EUV lithography: A review of material design strategies and performance results. *Nanomaterials* **2020**, *10*, 1593. [CrossRef] [PubMed]

nanomaterials

MDPI

Article

Entropy and Random Walk Trails Water Confinement and Non-Thermal Equilibrium in Photon-Induced Nanocavities

Vassilios Gavriil [1,2], Margarita Chatzichristidi [3], Dimitrios Christofilos [2], Gerasimos A. Kourouklis [2], Zoe Kollia [1], Evangelos Bakalis [1,4], Alkiviadis-Constantinos Cefalas [1] and Evangelia Sarantopoulou [1,*]

1 National Hellenic Research Foundation, Theoretical and Physical Chemistry Institute, 48 Vassileos Constantinou Avenue, 11635 Athens, Greece; vgavriil@eie.gr (V.G.); zkollia@eie.gr (Z.K.); evangelos.bakalis2@unibo.it (E.B.); ccefalas@eie.gr (A.-C.C.)
2 School of Chemical Engineering and Physics Laboratory, Faculty of Engineering, Aristotle University of Thessaloniki, University Campus, 54124 Thessaloniki, Greece; christof@eng.auth.gr (D.C.); gak@auth.gr (G.A.K.)
3 Department of Chemistry, Laboratory of Industrial Chemistry, Panepistimiopolis Zografou, National and Kapodistrian University of Athens, 15771 Athens, Greece; mchatzi@chem.uoa.gr
4 Dipartimento di Chimica "G. Giamician" University di Bologna, Via F. Selmi 2, 40126 Bologna, Italy
* Correspondence: esarant@eie.gr; Tel.: +30-210-727-3840

Received: 29 April 2020; Accepted: 22 May 2020; Published: 2 June 2020

Abstract: Molecules near surfaces are regularly trapped in small cavitations. Molecular confinement, especially water confinement, shows intriguing and unexpected behavior including surface entropy adjustment; nevertheless, observations of entropic variation during molecular confinement are scarce. An experimental assessment of the correlation between surface strain and entropy during molecular confinement in tiny crevices is difficult because strain variances fall in the nanometer scale. In this work, entropic variations during water confinement in 2D nano/micro cavitations were observed. Experimental results and random walk simulations of water molecules inside different size nanocavitations show that the mean escaping time of molecular water from nanocavities largely deviates from the mean collision time of water molecules near surfaces, crafted by 157 nm vacuum ultraviolet laser light on polyacrylamide matrixes. The mean escape time distribution of a few molecules indicates a non-thermal equilibrium state inside the cavity. The time differentiation inside and outside nanocavities reveals an additional state of ordered arrangements between nanocavities and molecular water ensembles of fixed molecular length near the surface. The configured number of microstates correctly counts for the experimental surface entropy deviation during molecular water confinement. The methodology has the potential to identify confined water molecules in nanocavities with life science importance.

Keywords: nanocavities; non-thermal equilibrium; water; entropy; nanothermodynamics; nanoindentation; AFM; electric dipole interactions; VUV irradiation; random walk

1. Introduction

Confined molecular water in nanocavities shows intriguing and unexpected behavior. The dynamic evolution of confined molecular water swings between bulk response, molecular collective actions and interface binding reactions [1]. Translational and rotational motions of confined water point to different stretching dynamics from its bulk counterpart [2]. It is also known that confined water builds tight hydrogen-bonded (H-bonded) networks, and its flow response is diverging by orders of magnitude from macroscopic hydrodynamics [3]. Possible lack of H-bonding of water molecules in small volumes counts for de-wetting, cavity expulsion [4], water self-dissociation [5] and a diverging dielectric constant [6]. It is plausible; therefore, that diverging behaviors of the biological and geological evolution of molecular enclosures in small systems [7–13] also imply a nanothermodynamic approach [14,15].

The central element of any thermodynamic theory of small systems is based on the hypothesis that nanometer-sized configurations pullout an additional physical component to the free energy of the associated macroscopic system from interactions among nanostructure entities. Moreover, the confinement of a relatively large number of molecules in nanocavities, restraints the molecular degrees of freedom (translational, vibration or rotational), and finally the system evolves through different entropic states before equilibration. Most interesting, the confinement of a small number of molecules in a large number of distinguishable tiny spaces might well indicate a thermodynamic entropic collective behavior [13], space and time local heterogeneities, not-extensive fluctuations and intriguing surface-boundary effects. The reduction of the translational degrees of freedom of molecules in tiny spaces and the deviation of the molecular trapping time inside a cavity from the mean molecular collision time outside, highlight the presence of an entropic barrier that separates the molecular motions inside and outside the cavities.

Today, both theoretical [14–19] and experimental advancements [20,21] gradually disclose the intriguing issues of thermodynamics of small systems, with major impacts on colloids, liquids, surfaces, interphases, chemical sensors, micro/nanofluidics, nanoporous media, proteins and DNA folding [10,22–27]. In cell biology, the presence of different nano-sized molecular scaffolds in the extracellular matrix environment implies a vast diversity of cellular activities and responses, including uncorrelated diverging drug delivery efficiencies [28].

Because thermodynamic potential variations and fluctuations allow for volume and surface stressing, any experimental verification of local volume and surface stress might well point to entropic fluctuations during molecular confinement [13,29]. Commonly, bulk and surface stressing go along with self-assembled structures, translational symmetry breaking, non-linearity, bifurcations, chaos, instability and morphological and shape nano configurations [30,31]. In the non-equilibrium state, rapidly changing thermodynamic potentials across phase boundaries usually force tiny systems to pass from different morphological progressions and physical states by tracing minimum energy and maximum entropy production pathways. This universal principle appears everywhere in Nature; from self-assembled bio and macromolecular structures and folding of large protein molecules [32] to nano/micro flower-like artificial structures [33,34].

The confinement of molecules within nano-size cavitations, usually on the surface of a matrix, is linked to system's entropy diversity before and after trapping [13,27,35,36]. It is also known that for the same translational entropy, any confined molecular state attains a small variation of its rotational entropy compared to the non-confined molecular state. Likewise, rotational restriction affects surface molecular bonding and sorption/desorption kinetics [35]. Specific response of nanoentropic potentials from molecular confinement within photon-induced nanocavitations in PDMS matrixes underlines an inherent correlation between internal stressing and 2D entropy diversion [13].

Commonly, photon-processing of surfaces reconfigures their physicochemical properties, including thermodynamic potentials [37–40]. Irradiation of a polymeric matrix with vacuum ultraviolet (VUV) light in the spectral range from 110 to 180 nm entails an extensive modification of topological and thus of physical features, because of bond breaking and formation of new bonds. Any 2D topological transform is accompanied by a diversion of surface characteristics, such as porosity, sensing efficiency, chemical stability and extensive nanocavitation [41–46]. The adsorption of various molecules on 2D nanostructured surfaces [47–49], might well boost a plethora of surfactant effects along with molecular sensing [43,50], gas separation and storage [51–54], and also applications with particular emphasis on nanomedicine [55], bio-engineering [56,57] and drug delivery systems [58,59]. Among other polymeric matrixes, polyacrylamide (PAM) is a hydrophilic low toxic, biocompatible, water-soluble, synthetic linear or cross-linked molecule, modified accordingly for a wide range of applications, including oil recuperation, wastewater treatment, soil conditioner, cosmetics food and biomedical industries [60–62]. A diverging number of physical and chemical methods are currently applied to optimize the biocompatibility level of different polymers (e.g., PDMS, PET, PTFEMA, PEG), for biomedical applications, biosensors, tissue engineering and artificial organs [46,63,64]. Well established methods of surface functionalization through photon irradiation with UV, VUV and EUV (extreme ultraviolet) light sources and plasma treatment at various wavelengths and electron energies, aim to optimize chemical instability and surface modification for controlling a plethora of surface functionalities [65].

Today, several methods exist to improve the strength and the physicochemical properties of PAM matrixes by blending the matrix with chitosan, starch or other polymers [66]. While functionalization of pure PAM polymeric surfaces is mostly done via sunlight exposure at standard environmental conditions, a limited number of studies include plasma processing [67–71]. However, no data exist for VUV processing of PAM surfaces, preventing thus precise tailoring of PAM's physicochemical surface characteristics (surface roughness, structure size, elasticity, chemical composition, etc.) and the formation of controlled micro/nanopatterns and cavitations for different applications [37,42,43,63,64].

The current work establishes the link between entropy variation and molecular water confinement in small nanocavities fabricated by 157 nm laser photons in polymeric PAM matrixes. The work follows a line of a rational evolution. First, the correlation between 157 nm molecular photodissociation (laser fluence or a number of laser pulses) and surface topological features, including nanocavitations, is established from fractal and surface analysis by using atomic force microscopy (AFM). Next, the correlation between surface strain and 157 nm molecular photodissociation is revealed by applying AFM nanoindentation (AFM-NI), contact angle (CA) wetting and white light reflection spectroscopy (WLRS). Random walk simulations of water molecules inside cavitations differentiate the escape time of confined molecular water and the mean collision time of water molecules near the PAM surface. The different time scales inside and outside the nanocavities point to an additional state of ordered arrangements between nanocavities and the molecular water ensembles of fixed molecular length near the surface. The configured number of microstates properly counts for the experimental surface entropy deviation during molecular water confinement, in agreement with the experimental results. Finally, the mean time distribution for a small number of water molecules for different runs reveals a non-equilibrium state inside tiny cavities. The experimental method has the potential to identify confined water molecules in nanocavities via entropy variation. The proposed roadmap of analysis may be used in applications related to life science.

2. Materials and Methods

2.1. Materials

PAM (typical M_n = 150 K, M_w 400 K) purchased from Sigma-Aldrich (St. Louis, MO, USA) used to prepare solution 5% *w*/*w* in water. Thin layers (426 ± 1 nm) on Si wafer substrates were made by spin-coating for 60 s at 2500 rpm, and finally, cured at 110 °C for 15 min at a temperature rate of 0.37 °C s^{-1} and then left to cool at room temperature. WLRS measures the thickness of PAM films coated on Si wafers.

2.2. 157 nm Laser

PAM layers irradiated with a high power pulse discharged molecular fluorine laser at 157 nm (Lambda Physik 250 (LPF^{TM} 200), Lambda Physik AG (Coherent), Göttingen, Germany), under continuous nitrogen flow (99.999%) at 10^5 Pa and room temperature. The layers were mounted into a computer-controlled X-Y-Z-θ translation-rotation motorized stage, placed inside a 316 stainless-steel chamber. The laser temporal pulse duration at FWHM, the energy of an unfocused laser beam per laser pulse, the photon fluence per laser pulse and laser repetition rate were set up at 15 ns, 28 mJ, 250 J

m^{-2} and 10 Hz. For dipping the amount of oxygen inside the stainless-steel chamber, nitrogen purging of the chamber was applied for 10 min before the irradiating stage.

2.3. AFM Imaging and AFM-NI

An AFM system (*diInnova*, Veeco Instruments Inc. (SPM Bruker), Santa Barbara, CA, USA) used for surface imaging of exposed/non exposed areas and the AFM-NI measurements. The imaging carried out in a tapping mode at a scanning rate of 0.5 Hz, using phosphorus-(n)-doped silicon cantilever (MPP-11123-10), having a spring constant of 40 nN nm^{-1} and tip radius of 8 nm, operating at a resonance frequency of 300 kHz at ambient conditions. The surface parameters of the samples were also evaluated.

The force versus distance (F-D) response from ten different points on each non-exposed and exposed areas was also recorded with the same cantilever. The elastic modulus (Young's modulus) was calculated using the SPIP force curve analysis software by fitting a Hertz model to the force-distance curve. The hysteresis between approach and retract curves were corrected by the same software. Calculations performed with a Poisson's ratio value of 0.3 [72].

2.4. Fractal Analysis

The fractal characteristics of the exposed and non-exposed areas were quantified through the fractal dimensionality D_f that describes the topology and the cavitation of a surface quantitatively. D_f was derived from AFM images by four different algorithms, the cube counting, triangulation, variance and power spectrum methods, besides an algorithm provided by the AFM's "lake pattern" software (diSPMLab Vr.5.01). A detailed description of the concept and the specific methodologies of the different algorithms can be found in [27]. The D_f was calculated for the four different methods using "Gwyddion, SPM data visualisation and analysis tool" [73]. The D_f calculated with the four different algorithms follow the same trend, despite small dimensionality divergences coming up from systematic errors, because of the different converging speed of the fractal analytical approaches.

2.5. Water Contact Angle (CA)

The chemical modification of PAM surfaces following PAM surface laser irradiation was monitored by water CA surface measurements under ambient atmospheric conditions. Distilled water droplets with a volume of 0.5 μL were gently deposited onto the sample surface using a microsyringe. Water CAs on samples before and after irradiation and at different time intervals were measured using a CA measurement system (Digidrop, GBX, Romans sur Isere, Drôme, France) equipped with a CCD camera to capture lateral snapshots of a droplet deposited on top of the preselected area, suitable for both static and dynamic CA measurements. Droplet images captured at a speed of 50 frames/s. CA values were obtained via the Digidrop software analysis, approximating the tangent of the drop profile at the triple point (three-phase contact point). Three different CA measurements were taken from each sample at different sample positions to calculate the average values.

2.6. White Light Reflectance Spectroscopy (WLRS)

The WLRS measurements were performed by an FR-Basic, ThetaMetrisis™ (ThetaMetrisis SA, Athens, Greece) equipped with a VIS–NIR spectrometer (Theta Metrisis SA, Athens, Greece) having 2048 pixels detector and optical resolution of 0.35 nm. The beam of the light source comes from a white light halogen lamp, with a uniquely designed stable power supply and soft-start circuit, ensuring stable operation over time that is necessary for long time duration experiments. Software controls the instrument, performing the data acquisition and film thickness calculations. The PAM films were spin-coated on native oxide Si wafers and SiO_2 layer on the top with a thickness of 2–3 nm.

2.7. Random Walk Model

The mean escape time of a water molecule confined in nanocavities was computed by applying different 3D random walk models with diverging numbers of water molecules, variable spherical size nanocavities, and entrance-escape hole sizes. Two different models of non-interactive and interactive water molecules inside the cavities were used. The first model, the non-interactive random walk model, uses molecular masses of zero volume and elastic collisions of the water molecule with the cavity wall and it records the sequence of positions of water molecules inside the spherical cavity until it gets back to the entrance-escape hole. The collision angle was varied randomly with a uniform distribution. The model calculates the total distance that molecules travel in the cavity before they escape from the entrance-escape hole. The mean escape time was calculated by considering that the molecule attains its kinetic energy after an elastic collision with the walls of the cavity. Therefore the kinetic energy transfer from the wall to the molecule should be equal with the thermal energy of the wall $\frac{3}{2}k_BT$, where k_B is Boltzmann's constant.

The escape time from the entrance-escape hole for a non-interactive water molecule in the cavity is given by the equation:

$$t_e = \frac{\sum_{i=0}^{n}\left(\sqrt{2R^2(1-\sin\theta_{i+1}\sin\theta_i\cos(\varphi_{i+1}-\varphi_i)+\cos\theta_{i+1}\cos\theta_i}\right)-2R+(R_0+R_n)}{\sqrt{\frac{3k_BTN_A}{M_{H_2O}}}} \quad (1)$$

where n is the number of collisions in each run, R is the radius of the spherical cavity, (R,θ_i,φ_i) is the position of the molecule in the ith collision. The entrance and the exit point in the cavity wall are given in spherical coordinates (R_0,θ_0,φ_o) and (R_0,θ_n,φ_n), accordingly, and M_{H_2O} is the molecular mass of water.

The interactive random walk model records the sequence of positions of a specific molecule that enters a spherical cavity through the entrance-escape hole, alongside with the locations of a variable number of neighboring molecules trapped in the cavity, until it gets back to the entrance-escape hole. At first, because of non-thermal equilibrium between water molecules within the cavity, the molecules are placed inside the cavity in random positions with random velocities of uniform distribution between 0 and $\sqrt{\frac{3k_BTN_A}{M_{H_2O}}}$ m/s. The position of each molecule was recorded every 10^{-14}s. The collision of each water molecule with the cavity wall and its neighbouring molecules is considered to be elastic. The collision angle was varied randomly with a uniform distribution. Contrary to the non-interactive model of zero-size molecules, the interactive model uses a spherical molecular diameter of 0.3 nm.

For every pair of the cavity size and entrance-escape hole, the random walk was run 10^2 times and the mean escape time was calculated. In addition, the mean-escape time distribution for different cavities and number of molecules was used to evaluate the thermodynamic state inside the cavity. The model was designed and run in MATLAB. 9.4.0.813654 (R2018a), The MathWorks Inc.; Natick, MA, USA.

3. Results

3.1. Surface Analysis

Commonly, four surface parameters, the surface roughness histogram, the area roughness, the area root mean square (RMS) and the maximum range characterize a surface and mean area values are plotted as functions of the laser pulse number or the laser fluence, Figure 1. The surface parameters are extracted from AFM images, Figure 2a–e.

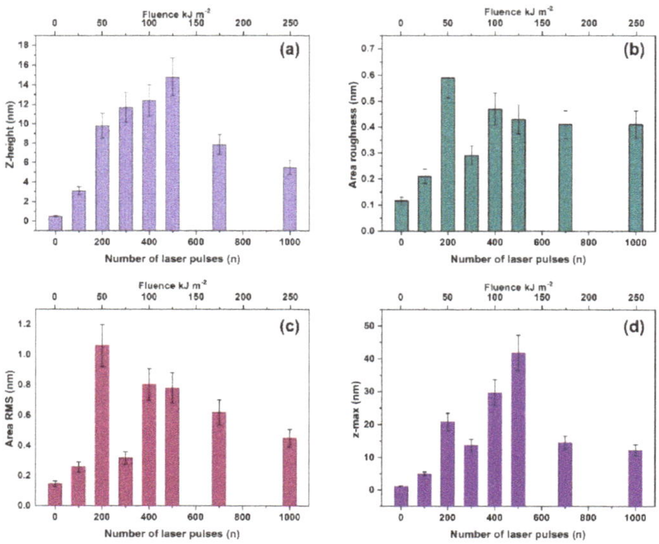

Figure 1. Surface parameters of irradiated polyacrylamide (PAM) layers for a 2 μm × 2 μm area: (**a**) Mean *z*-height; (**b**) area roughness (R_a); (**c**) area RMS; (**d**) maximum range. The area roughness and area RMS parameters show an increment with laser pulses up to ~ 200 lp followed by a dip at 10^3 lp.

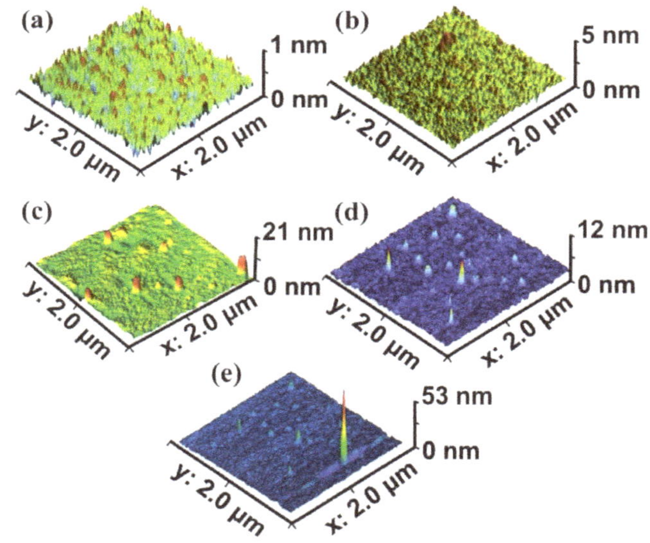

Figure 2. Atomic force microscopy (AFM) surface image of polyacrylamide (PAM) layers. Scan area 2 × 2 μm^2, laser fluence 250 J m^{-2} per pulse: (**a**) non-irradiated PAM layer; (**b**) irradiated PAM layer with 100 laser pulses (lp), 25 kJ m^{-2}; (**c**) 200 lp, 50 kJ m^{-2}; (**d**) 10^3 lp, 250 kJm^{-2}; (**e**) scan area 2.3 × 2.3 μm^2, 10^3 lp, 250 kJ m^{-2}. The surface morphology is area size-dependent.

The surface roughness histogram, or average *z*-height, is the arithmetic mean defined as the sum of all height values divided by the number of data points $|Z| = \frac{1}{N}\sum_{i=1}^{N} Z_i$. Next, the R_a (area roughness or roughness average) is the arithmetic mean of the height deviation from the image's mean value, $R_a = \frac{1}{n}\sum_{i=1}^{n}\left|Z^i - \overline{Z}\right|$. The area RMS ($R_{rms}$) is the value defined as the square root of the mean value of the squares of the distance of the points from the image mean value: $R_{rms} = \sqrt{\frac{1}{N}\sum_{i=1}^{N}(Z - Z_i)^2}$. Finally, the maximum range of Z_{max} is defined as the maximum value of z-heights. The surface parameter values (z-height, area roughness, area RMS, and maximum range) of photon exposed areas were more considerable compared to the non-irradiated ones. However, because surface parameters are area size-dependent (Figure 2d,e), they are utilized only for a comparative qualitative evaluation of area modification under 157 nm laser irradiation.

3.2. Fractal Analysis of 157 nm Photon Processed PAM Polymeric Matrixes

Because of statistical self-similarity between matrix space topology during a scaling-down route, there is a strong correlation between porosity, stage of cavitations and fractal dimensionality. Furthermore, in porous materials, the linear, area and volumetric porosities are alike, and therefore the 3D fractal dimensionality is similar to the area one. The dimensionality of a surface is equal to two for an ideal solid (Euclidean surfaces) and equal to three for completely porous surfaces with a fractal character. Areas with Z_i values above a threshold Z height are known as "islands", while those with Z_i's below the threshold height value are named as "lakes". AFM "island-lake structure" of non-irradiated and VUV irradiated 2 μm × 2 μm areas are shown in Figure 3.

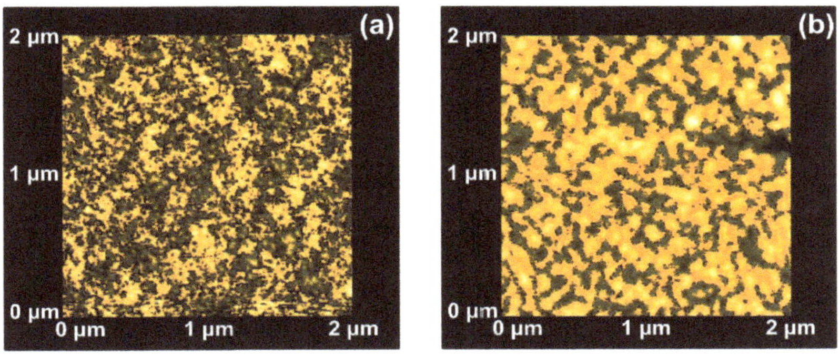

Figure 3. Atomic force microscopy (AFM) image of "lake" (grey) and "island" (orange) for a fractal area of 2×2 μm^2: (**a**) Non-irradiated area; (**b**) irradiated area with 10^3 laser pulses.

The mean Z_i heights of non-irradiated and the irradiated regions (10^3 laser pulses) were set at 0.75 and 1.94 nm respectively, and the irradiated areas show a diverging surface topology, in agreement with previous results [13,16,56,58]. Following a standard procedure, two parameters, the fractal dimensionality D_f (which is a dimensionless number) and the "periphery to the area ratio" (PAR) are used to describe a set of "islands" or "lakes". Both parameters are linked to the surface roughness, cavitations and topological entropy [27,74]. PAR is the ratio of logarithms of the perimeter Π to the area A, where $\Pi = \alpha\left(1 + D_f\right)A^{(1-D_f)/2}$. For assessing the state of cavitations, the fractal dimensionality is calculated by the partitioning, the cube counting, the triangulation, and the power spectrum algorithms [58]. Results are compared with those derived directly from the AFM "lake" pattern software, Figure 4a.

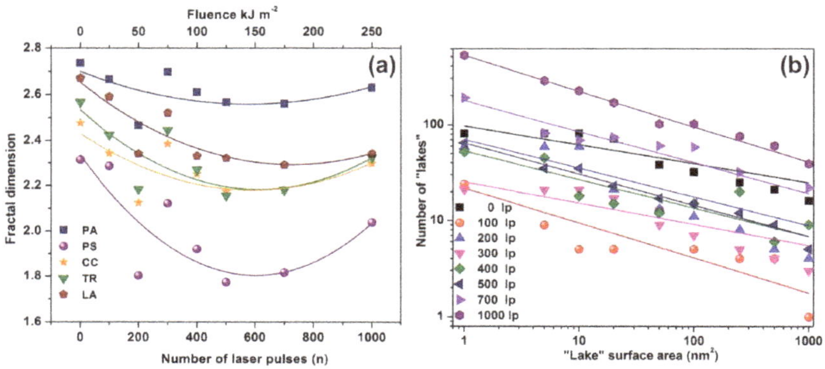

Figure 4. Fractal dimensionality, "lake" surface area and lake number vs. laser fluence. (**a**) Surface -fractal dimensionality calculated with four different fractal analytical methodologies. Colors, symbols and lines to assist the eye: blue squares for partitioning (PA), purple spheres for power spectrum (PS), dark yellow stars for cube counting (CC), green triangles for triangulation (TR), and brown pentagons for atomic force microscopy (AFM) "lake" pattern (LA). The five methods show a similar fractality trend vs. laser fluence; (**b**) "lake" surface area vs. the number of lakes at different number of laser pulses (laser fluence).

AFM images of 2 μm × 2 μm laser-irradiated areas were digitized to a 512 px × 512 px matrix, and then they processed with four different fractal algorithms. It is unveiled that fractal dimensionality, and thus cavitations, are functions of the laser photon fluence. All algorithms exhibit a similar trend of fractal dimensionality with the number of laser pulses (laser fluence), although the fractal dimensionality derived with the power spectra methodology seems slightly different, as expected [13,25]. The fractal dimensionality initially dips, attaining its minimum value around 500 laser pulses and then rises again with a small gradient up to 10^3 laser pulses, Figure 4a. For a constant "lake" surface area the number of "lakes", and thus the number of cavities, is a function of the laser pulses (laser fluence), Figure 4b. The number of "lakes" within a given surface area vs. the number of laser pulses is shown in Figure 5a. The number of "lake" areas rises almost exponentially with the number of laser pulses and small area "lakes" prevail over larger ones The fractal dimensionality vs. laser fluence has a non-monotonous complex structure. Small size features (1–10^2 nm^2) are associated with nanocavity-like structures, Figure 5b. It is also confirmed that below 10^3 laser pulses small size features contribute to a high cavitation state because small size features have a higher dimensionality than large size structures, Figure 5c. On the contrary, large size features are prominent at 10^3 laser pulses, indicating the complexity of the associated processes. In addition, for the same number of laser pulses, small size cavitation prevails over larger ones, Figure 5a. The experimental results indicate that water confinement is rather associated with small cavitations, in agreement with WLRS measurements (vide infra).

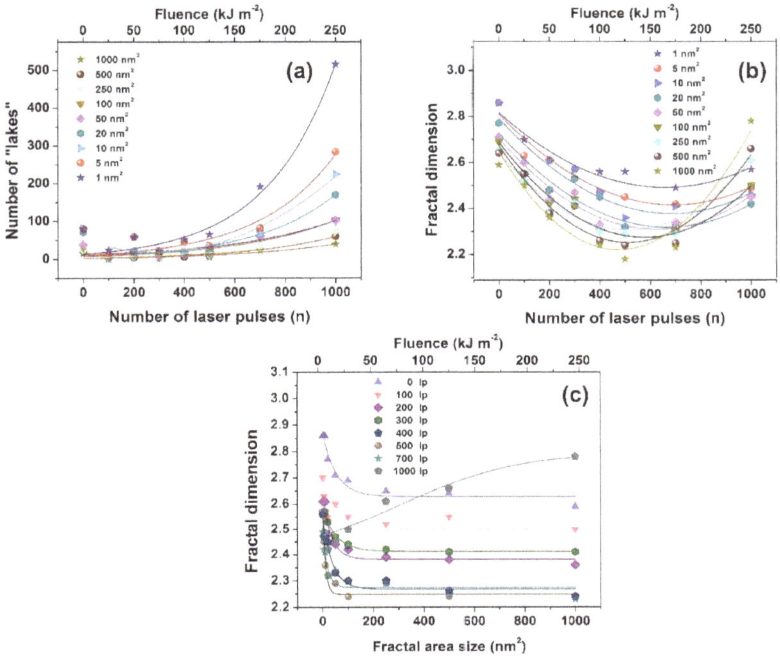

Figure 5. Fractal parameters of polyacrylamide (PAM) at different laser fluence: (**a**) Number of "lakes" for different fractal size vs. the number of laser pulses *(n)*; (**b**) fractal dimensionality vs. the number of laser pulses *(n)* for different fractal size. The concentration of small size nanocavities increases at higher laser fluence; (**c**) fractal dimension vs. fractal size at a different number of laser pulses (lp).

3.3. AFM-NI

The mechanical response of 426 nm-thick PAM polymers was evaluated with nanoscale resolution via the F-D curves at different laser fluence, Figure 6a–d. Young's modulus and adhesion forces were also evaluated. Major non-monotonic modifications were recorded indicating substantial conformational changes of the surface energy of the PAM layers, Figure 6a. A diverging Young's modulus is attributed to accelerated ageing because of molecular bond breaks, accompanied by the formation of new carbon and carbonyl bonds [75–77]. A nonlinear alteration of the elastic modulus of PAM gel formulations during ten days ageing was also reported, revealing substantial changes of PAM's mechanical properties during irradiation [78].

The approach and retract curves follow different paths in all irradiating conditions, describing thus a system evolving out of equilibrium. The elastic modulus of the dry state hydrogels is significantly reduced after immersion to water, e.g., from 18 GPa to 3.3 MPa [79]. The Young's modulus of the non-irradiated PAM hydrogels depends on the hydration conditions, e.g., it decreases from 295 MPa in the dried state to 266 kPa in the fully hydrated state [78,80]. A Young's modulus of 2.84 GPa of uncured PAM hydrogel was recently attributed to the presence of pre-polymerized PAM oligomers [81]. Moreover, enhancement of Young's modulus to 4.84 GPa was predicted via an extension of the 3D polymeric networks at higher cross-linking states [81].

In this work, the non-irradiated PAM surfaces were thermally cured after being spin-coated on a silicon wafer; therefore, their mechanical properties are expected to deviate from those in the gel state. The average Young's modulus prior to and post-irradiation with 500 and 10^3 laser pulses was 2.0 ± 0.8 and 1.6 ± 0.42 and 2.55 ± 1.29 GPa, respectively, Figure 7a. The significant errors of Young's moduli at different points in the same sample are credited to various morphological heterogeneities and a progressive phase transformation to a relatively high carbonized state. Young's moduli follow a similar trend with fractal dimensionality vs. laser fluence, Figures 4a and 5b.

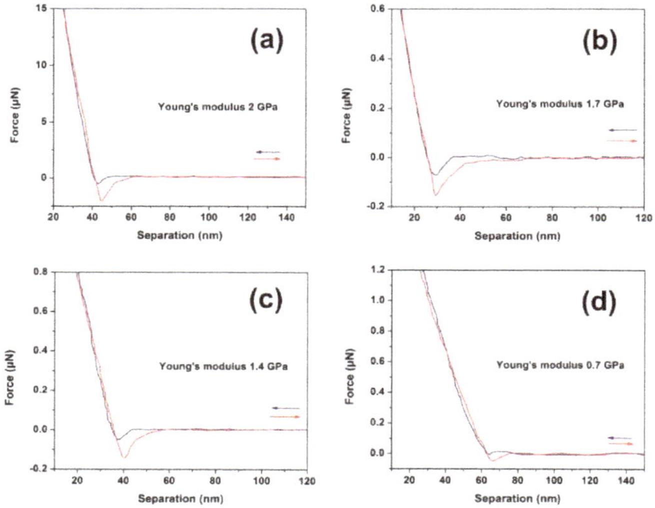

Figure 6. Typical force-distance (F-D) curves of polyacrylamide (PAM) thin layer surfaces (426 nm) irradiated with a different number of laser pulses (lp) (250 J m^{-2} per laser pulse): (**a**) F-D curves of the non-irradiated layer; (**b**) F-D curves with 200 lp; (**c**) F-D curves with 300 lp; (**d**) F-D curves with 400 lp.

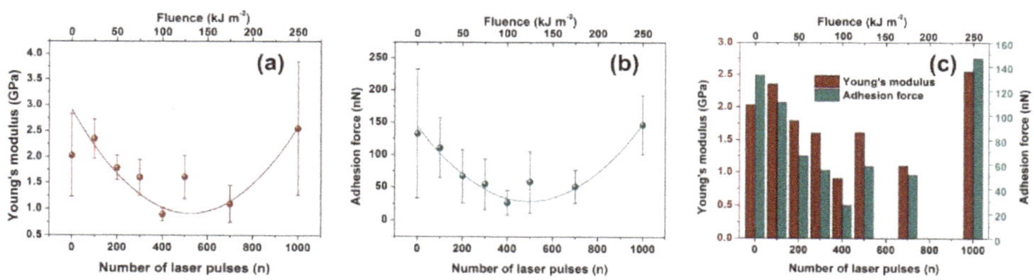

Figure 7. Young's modulus and adhesion force of irradiated polyacrylamide (PAM) surfaces showing enhanced carbonization at higher laser fluence. (**a**) Young's modulus; (**b**) adhesion force of a PAM surface irradiated at different laser fluence up to 250 J m^{-2}; (**c**) Young's modulus and adhesion force column charts of PAM vs. laser fluence.

Additionally, the adhesive force, as it is measured during the penetrating state of AFM's tip, follows a similar trend with Young's modulus, Figure 7b,c. Because of diverging surface carbonization, the adhesive force drops from 130 to 26 nN between 0–400 laser pulses and then it rises again to ~ 150 nN for 10^3 laser pulses.

3.4. Water Contact Angle (CA)

Water CAs of PAM matrixes were recorded for varying photon fluence. The average CAs rise from 20° ± 2° to a saturated "plateau" at ~65° ± 7° after 200 laser pulses, Figure 8a.

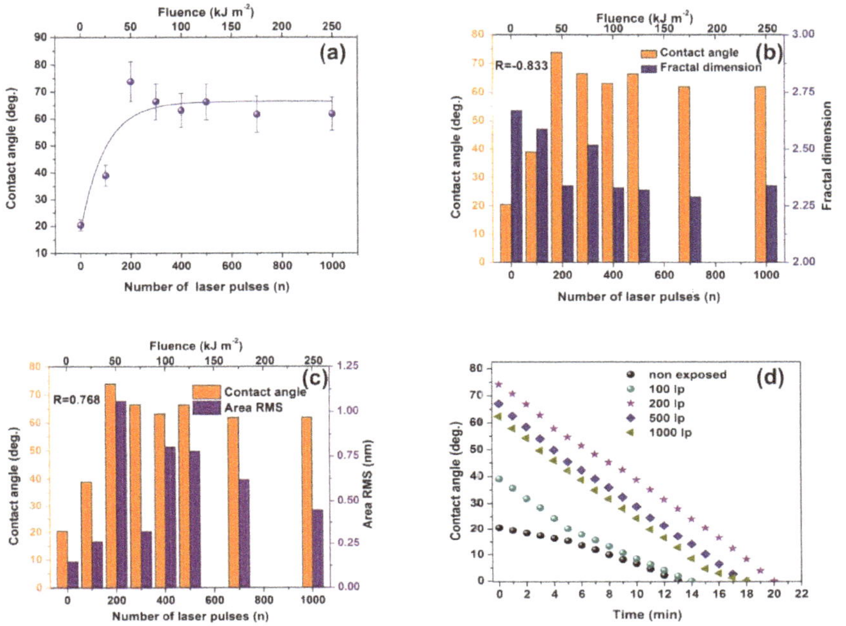

Figure 8. Water contact angle (CA) vs. the number of laser pulses: (**a**) Water CA vs. the number of laser pulses. (**b**) column chart diagram of CA and fractal dimensionality vs. laser fluence. The mean correlation factor is −0.833; (**c**) column chart diagram of CA and area RMS of PAM vs. laser fluence. The mean correlation factor is 0.768 pointing to a strong correlation between fractal dimensionality, CA, and area RMS over a wide range of laser fluence; (**d**) column chart diagram of water CA at different time intervals. The almost similar slopes point to a uniform surface response at different VUV photon fluence. From Figures 4a, 5a,b, 7a,b and 8a the surface chemical modification is saturated at ~500 laser pulses, because of the low penetrating depth of the 157 nm laser photons, indicating the strong correlation between fractal dimensionality, CA, area RMS, Young's modulus and surface modification.

VUV photon processed PAM matrixes attain higher CA values, displaying thus a hydrophobic state, affirming that VUV irradiation has a primary effect on the surface wettability by altering both the material's physicochemical properties and surface nano/micro features, Figure 8a,b. In addition, the mean correlation factors of −0.833 and 0.768 between CA, D_f and area RMS indicate a secure interconnection between surface morphology and D_f, Figure 8b,c.

The wetting behavior was also analyzed with time, Figure 8d. The CAs of non-irradiated and irradiated with 100 laser pulses matrixes decrease consistently for 5 min. The dynamic CA of irradiated samples exhibits similar slope values, suggesting similar diffusion constants for different porosities, a fact that stresses out a picture of molecular water confinement in nanocavitations.

3.5. *White Light Reflectance Spectroscopy (WLRS)*

WLRS uses a broad-band light source and a spectrometer. The white light emitted from the light source is guided to a reflection probe through a number of optical fibers that incident vertically onto a sample. The sample consists of a stack of transparent and semi-transparent films placed over a reflective substrate. A reflection probe collects the reflected light through a fiber, directing it to the spectrometer. The light source beam interacts with the sample and generates a reflectance signal that is constantly recorded by the spectrometer. The number and the shape of interference fringes, registered in the CCD of the spectrometer, depend on the thickness and the refractive index of the film(s). The fitting of the experimental spectrum is performed by using the Levenberg-Marquardt algorithm.

Water confinement is a source of volume strain and the relative surface deformation of the PAM polymeric matrixes caused by molecular water confinement is monitored by WLRS, Figure 9. The layer's thickness during water confinement and the relative surface deformation of the PAM layer prior and after water confinement in the irradiated surfaces is calculated from the phase shift and the superposition of amplitudes of the reflected light beam on the PAM surfaces. The white light beam records the surface strain within a cylindrical volume ~ $V = 4.09 \times 10^{-14}$ m^3, defined by the cross-sectional diameter of the white light beam of 3.5×10^{-4} m and the thickness of the polymeric layer of 426 nm.

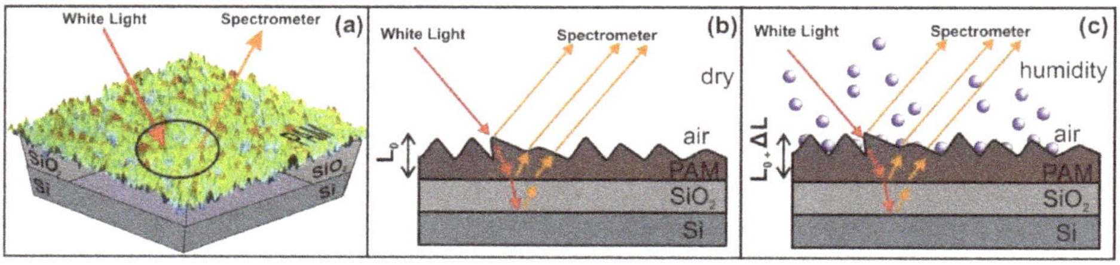

Figure 9. Principle of operation of white light reflectance spectroscopy (WLRS). (**a**) White light beam reflection in PAM surfaces. (**b**) experimental details and geometry of the reflected beams (**c**) surface strain response during water confinement in nanocavities. The contribution of the SiO_2 layer at the interference pattern is negligible.

3.6. Random Walk Model

The comparison between calculations with the diffusion model [82] (water vapour diffusion coefficient in the air at normal pressure at 293 K is ~ 2.42×10^{-5} m^2 s^{-1}) and the current non-interactive random walk model for 10^3 runs is shown in Tables A1 and A2. There is a noticeable difference between the two models for small size nanocavities because the diffusion constant for small size nanocavities is undetermined. The mean escape time from random walk models with the interactive model for the different number of confined molecules, cavity and the entrance-escape hole size is given in Figure 10 and Appendix A, Tables A3–A6.

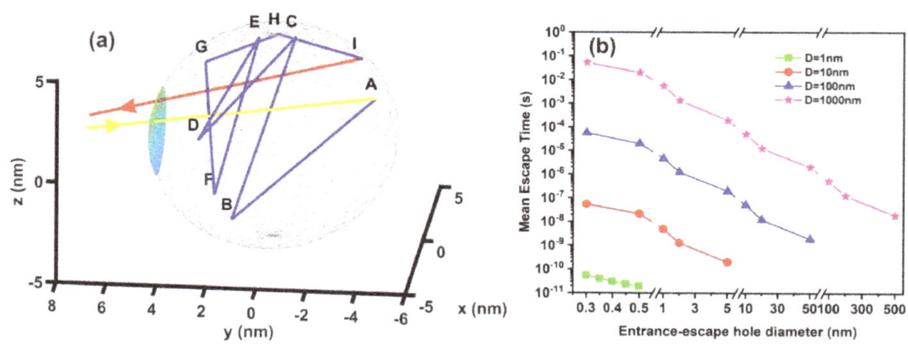

Figure 10. Non-interactive random walk of one water molecule in a nanocavity. (**a**) The water molecule enters the cavity (yellow arrow) and then it collides with the inside walls of the spherical cavity (10 nm) several times (A–I points and blue lines) before escaping from the entrance-escape hole (3 nm, red line); (**b**) mean escape time for 10^3 different random walk runs in 1 nm (green), 10 nm (red), 10^2 nm (blue), and 10^3 nm (magenta) spherical cavities for different entrance-escape hole diameters (0.3 nm–500 nm). The y-axis represents a logarithmic time scale.

4. Discussion

4.1. 157 nm Molecular Photodissociation of PAM Polymeric Chains

Initially, surface and fractal analytical methods were used to typify surface cavitations crafted by 157 nm laser photons on PAM surfaces. Diverging texture morphologies of 2 μm × 2 μm PAM areas irradiated with 157 nm with a different number of laser pulses (photon fluence) are shown in Figure 2. Major conformational changes of photon processed PAM surfaces are evident through a diversity of fractal dimensionalities and surface parameters. Specifically, irradiated areas exhibit either a uniform or heterogeneous surface structural networks, according to the laser fluence (Figure 2b–e). Different size nano/microstructures including "hills and lakes" and fractal dimensionality diversity, nano aggregations (1–10^3 nm) and cavitations are shreds of evidence of significant photochemical topological matrix alterations (Figure 2c–e). Similar structures were previously observed on PAM hydrogel surfaces by cross-link concentration variations [78].

The energy of 157 nm laser photons is used to excite a molecular site in the polymeric chain from an electronic ground state (A) vibrational level to an excited electronic state (B) vibrational level, Figure 11a. The excitation is followed by a rapid internal transition to a dissociative (repulsive) state (Γ), and the parent molecule is disintegrated fast to a number of smaller size photo-fragments, Figure 11b,c. Consequently, surface irradiation with 157 nm laser photons modifies the morphology of the PAM matrix by creating defective molecular sites (DE) and micro/nano cavitations, Figure 11c. The volatile compounds, such as carbon-hydrogen monomers, ions, or larger polymer fragments, are moving away from the matrix at high velocities [37,41,42]. Carbon cluster (CL) formation (Figure 11d) also appears on the surface from re-deposited photo-dissociated products on the matrix (Figure 11e) and the photo-dissociated cycle profoundly modifies the chemical and the morphological features of the exposed polymeric surface. Because each 157 nm laser photon destroys via photo-dissociation one chemical bond of the polymeric matrix, Figure 11, it is reasonable to accept diverging cavitations and local nano-matrix volume diversities [37] in agreement with surface and fractal analysis results.

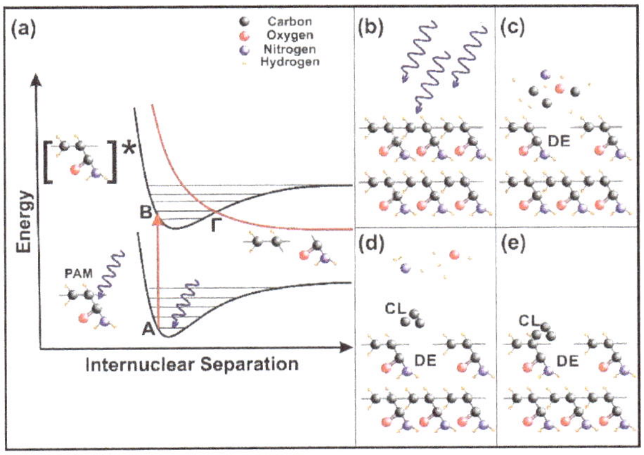

Figure 11. Nanocavitation by 157 nm laser photodissociation of polyacrylamide (PAM) matrixes: (**a**) Molecular photodissociation at 157 nm. Vertical arrows indicate photon transitions between two vibrational levels of the ground (A) and an excited electronic state (B). A transition from the excited to a repulsive electronic state (red curve) through an avoided crossing via a vibration state at the point (Γ) is very fast (< 1 ps) and breaks a molecular bond in the polymeric chain; (**b**) PAM surface irradiation with 157 nm photons; (**c**) (DE): a bond break is followed by molecular decomposition and nanocavitation; (**d**) (CL): possible recombination of carbon dissociative products, and cluster formation in the gas phase; (**e**) carbon cluster deposition on the polymeric matrix and possible structure of carbon nanocavitation.

4.2. Trapping of Water Molecules in Nanocavities

Water molecular confinement is a complex issue having great importance in life sciences [8,83,84]. Because the nature of the H-bond undergoes a diverging number of structural conformations at surface boundaries, and also, inside tiny spaces, the water confinement is hindered with the long-range fluctuations of both the water networks [85] and single molecules, Figure 12a. The dynamics and the time scale of interactions in confined spaces are notably diverging with the spatial scale length and the local geometries. For example, terahertz spectroscopy of water molecules in gemstone nanocavities identify quantum water molecular tunnelling through a six-well potential caused by the interaction of the water molecule with the cavity walls [86]. The length and directionality of H-bonds are highly susceptible to the type of confining surfaces and the degree of confinement [87]. In addition, atomistic molecular dynamics simulations of dipolar fluids confined to spherical nanocavities of radii ranging from 1 to 4 nm reveal a surprisingly small Kirkwood correlation factor in water, but not so in a dipolar fluid because of ultrafast relaxation of the total dipole moment time correlation function of water [6]. The static dielectric constant of confined water exhibits a strong dependence on the size too, with a remarkably low value even at 3 nm and a slow convergence to the bulk value because of surface-induced long-range orientation correlations [6]. The trapped water experiences peculiar thermodynamic properties and under confinement unexpectedly shows high pressures (GPa) [88]. Because the mean escape time is independent of the number of molecules, inside the cavity, Figure 12, the average mechanical pressure exerted on the walls of the cavity is independent of the number of molecules. Therefore, the molecular state inside the cavity deviates from an equilibrium thermodynamic state because the escape time in "equilibrium thermodynamic" cavities should be pressure dependent. In addition, the extensive thermodynamic properties of confined molecules in tiny spaces might be disproportional to the volume of the system, and instead, they could be higher-order functions of size and shape [89–92].

It is also known that for tiny empty spaces, equal or below the atomic dimensions, stressing fields are emerging from electromagnetic vacuum fluctuations. The repulsive Casimir stress $\sigma_c(R,t)$ within a conductive spherical cavity of radius R at time t was calculated to be $\frac{0.09\hbar c}{8\pi R^4(t)}$ [93]. For balancing the Casimir stress with the atmospheric pressure, a 10 nm spherical cavity has the proper size, if the equation of ideal gases is used. On the other hand, for a spherical cavity in thermal equilibrium with the matrix that bears a small hole on its surface connecting the inside with the outside volume space of the cavity, in the case of a pressure balance outside and inside the cavity, four molecules are confined in the cavity, if the equation of ideal gases is used as a first approximation. However, for an average molecular thermal energy of $E_{kin} \sim kT$ and a spherical volume V of 5.34×10^{-25} m^3, the volume stress exerted on the walls of the cavity from the collisions of a molecule with the walls of a cavity should be of the order of $\sim \frac{kT}{V} = 7.9 \times 10^4$ Pa, a value that almost matches the atmospheric pressure outside the cavity. By increasing the number of molecules inside a small cavity, the volume stress should be increased proportionally to the number of molecules because of mechanical collisions with the cavity walls. Consequently, extremely high pressures should be developed inside small cavities, in agreement with [88]. In addition, for small size cavities, there is rather an entropic than an energy barrier that balances the flow kinetics of molecules in and out the cavity [94,95]. Previous studies indicated that in the case of elastic collisions in the cavity, the molecular dynamics depends on the number of molecules inside the cavity and is either frictionless (inertial dynamics), moderately frictional (Langevin dynamics), or strongly frictional (Brownian dynamics) [96], where the noise term should be properly taken into account. For small entrance-escape holes, the number correlation function generally decays exponentially with time. The transition rate in the frictionless limit is given by a microcanonical ensemble. As the strength of the friction is increased, the rate of collisions approaches the diffusive limit without a Kramers turnover. In this work, random-walk calculations of non-interactive and interactive molecules in the cavity for 10^3 and 10^2 runs, point to variable escape times of water molecules from different size nanocavities (1–10^3 nm) and entrance-escape holes (0.3–5×10^2 nm), Figure 10 and Appendix A, Tables A1 and A2. For the same cavity size the mean escape time falls with large entrance-escape hole size, extended over a wide dynamic range of 10 orders of magnitude. The mean escape time for the interactive model is independent of the number of molecules inside small cavities and interestingly, the mean escape time fluctuates a great deal inside tiny cavities, Figure 12a–d and Tables A3–A6; suggesting that the system is in non-thermal equilibrium, a state that dominates the statistics and the dynamics of molecules inside small cavities.

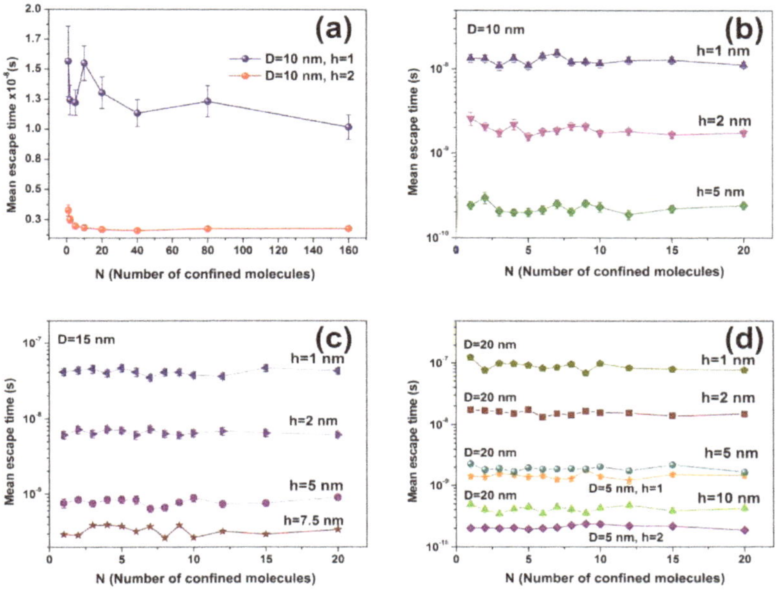

Figure 12. Mean escape time for a different number of non-interactive molecules calculated for 10^3 runs. (**a**) Fluctuations of mean molecular escape time are prominent for small size cavities, while remains constant for a large number of molecules; (**b**–**d**) mean escape time for 5, 10, 15 and 20 nm spherical cavities with different entrance-escape holes and number of molecules in the cavity. The mean escape time is independent of the number of molecules and is a function only of the cavity geometry (diameter and entrance-escape hole).

On the contrary, for larger size cavities the volume stress is diminished and the state of molecules inside the cavities approaches the thermodynamic limit, in agreement with [86]. he gradient of the mean escape time, the mean escape time distribution and the mean distance a molecule travels inside a cavity before its escape through the entrance-escape holes is diverging for very small size cavities (1 nm) cavities, Figures 12a, 13 and 14a.

The mean escape time and the mean travelling distance retain a constant ratio for large size cavities (10–10^3 nm), while the ratio deviates for small ones, suggesting again a non-equilibrium thermal state and large fluctuations inside small size cavities, Figure 14. Most interesting, the local fluctuations of mean molecular escape time are prominent for small size cavities and small number of molecules, while the mean molecular escape time remains steadier for a larger number of molecules, Figure 12a–d, in agreement with molecular dynamic results [6,85,89–92,96] and general nanothermodynamic considerations [31]. In addition, the mean escape time distribution of molecules for both the non-interactive and interactive models (1 and 150 molecules) inside different size small cavities reveals a rather non-thermal distribution and the absence of a thermal equilibrium state inside the cavities, Figures 13 and 15.

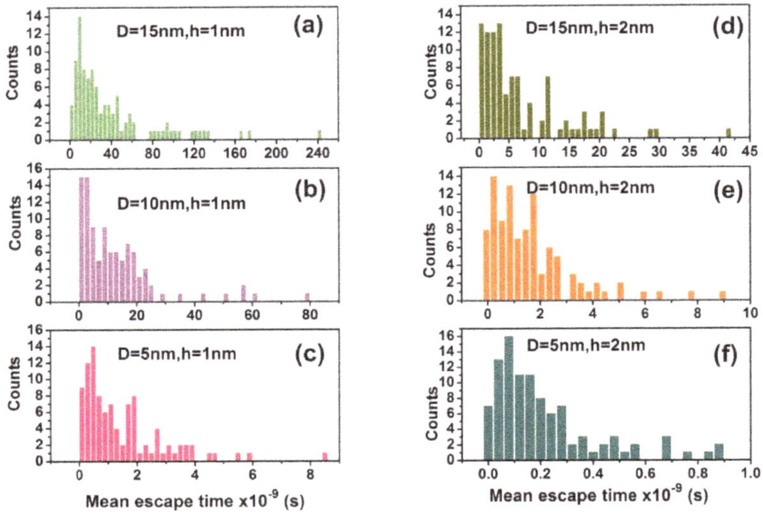

Figure 13. Distribution of mean escape times for 10^2 random walk runs of a water molecule (non-interactive model) for different cavity geometries (cavity size D and entrance-escape hole size h). (**a**) D = 15 nm, h = 1 nm; (**b**) D = 10 nm, h = 1 nm; (**c**) D = 5 nm, h = 1 nm; (**d**) D = 15 nm, h = 2 nm; (**e**) D = 10 nm, h = 2 nm (**f**) D = 5 nm, h = 1 nm. Distributions are non-normal and besides that skewness and long tails indicate non-equilibrium processes inside the cavities.

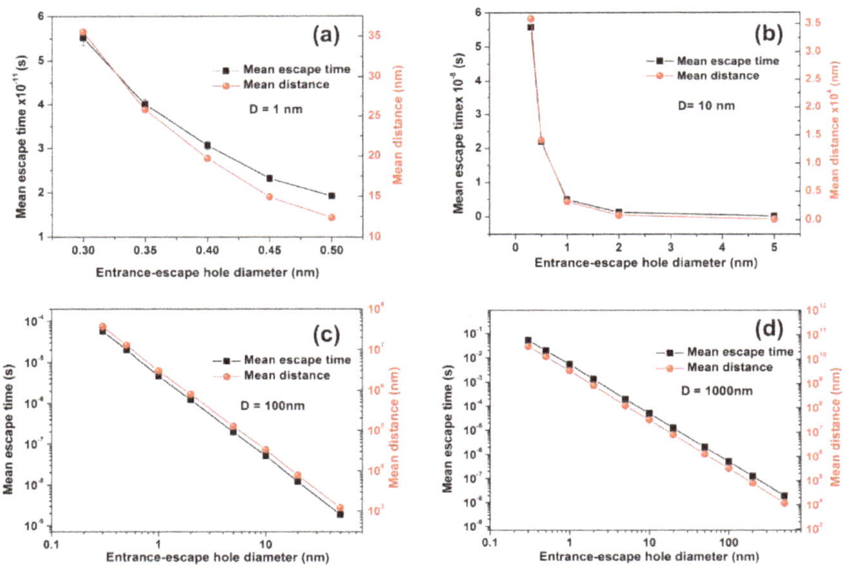

Figure 14. Mean escape time and mean travelling distance of a molecule within cavities of different geometries. (**a**,**b**) Gradient of the mean escape time and the mean distance that a molecule travels in the cavity before it escapes with different entrance-escape holes is diverging for very small size cavities (1, 10 nm); (**c**,**d**) mean escape time and the travelling distance gradients are constant for large size cavities (10^2, 10^3 nm), suggesting a non-thermal equilibrium state and large fluctuations for small size cavities.

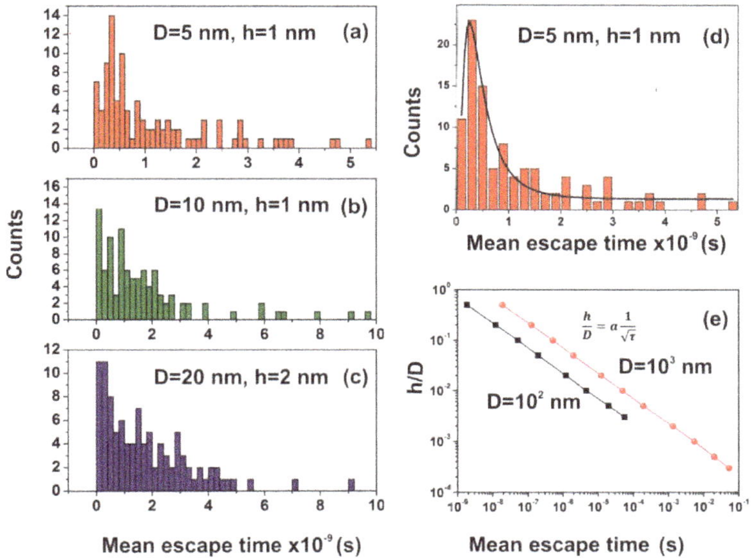

Figure 15. (**a–c**) Mean escape time distribution of 150 interactive molecules for different cavity and entrance-escape hole size for 10^2 runs; (**d**) best-fitting of Figure 15a is for a log-normal distribution; (**e**) mean escape time vs. the cavity ratio $\frac{h}{D}$. The time differentiation of molecular movements inside and outside cavities is provided by the dependence of the ratio $\frac{h}{D}$ on the waiting time τ, which for D=10^2 and 10^3 nm cavities goes as a power law of the waiting time with exponent −0.5.

The mean escape time of water molecules in the cavity is diverging from the mean collision time (~70 ns) and the thermal de Broglie time outside the cavity by many orders of magnitude, according to the geometry of the cavity. Therefore, the "molecular time" inside the cavities "flows differently" than the physical time of the events on the PAM surface. This spatial "time differentiation" across a thin layer in the PAM surface is responsible for the excess entropic potential, arising from a state of ordered arrangements between nanocavities and the water molecular ensemble of fixed molecular length near the PAM surface after 157 nm irradiation. A further sign of time differentiation of molecular movements inside and outside cavities is provided by the dependence of the ratio $\frac{h}{D}$ on the waiting time. It goes as a power law of the waiting time with exponent −0.5, Figure 15. Finally, the configured number of microstates from confinement properly counts for the experimental surface entropy deviation during the trapping of water molecules (vide infra).

4.3. Stress-Strain Response in Polymeric Matrixes-A Relation between Physics and Mechanics

During the interaction of a system with a thermal bath, the exchange of energy appears in the form of heat or work. The first law of thermodynamics states that the infinitesimal change of the heat absorbed by a system per unit volume Q is equal to the increase of the differential of its internal energy change U minus the infinitesimal change of the work done on the system W:

$$\delta Q = dU - \delta W \tag{2}$$

The departure from a state of equilibrium will be governed by the second law of thermodynamics and the direction of entropy change. Any thermodynamic system is described by its extensive and intensive variables, U, S, σ_{ij}, e_{ij}, T, where U is the internal energy, S is the entropy, σ_{ij} and ε_{ij} stand for second-rank stress and strain tensors acting on the volume element dV and T is the temperature of the system. Following Cauchy's theory of stress, the equilibrium of elements requires the balance of forces acting on a volume element, [97]

$$t_i^n = \frac{d\vec{F}_i}{dA} = \sigma_{ji} n_j \tag{3}$$

t_i^n is the ith component of a traction force $d\vec{F}$ along the *i*-axis, along a unit vector $\vec{n}$ perpendicular on an infinitesimal surface area $d\vec{A}$, σ_{ij} is the $(i,\ j)$ component of the stress tensor, and n_j is the j_{th} component of the $\vec{n}$ vector that is perpendicular to the surface area $d\vec{A}$.

The force component F_i acting on the volume element dV and bounded by the surface area A is given by the equation

$$F_i = \int_A t_i^n \, dA + \int_V X_i dV \tag{4}$$

where X_i are the body forces (e.g., the weight of the volume element dV). From Equation (4) and the Gauss theorem, the surface integral of the components of the traction forces is transformed into a volume integral

$$F_i = \int_V \left(\frac{\partial t_i^n}{\partial x_i} + X_i \right) dV = \int_V \left(\frac{\partial \sigma_{ji}}{\partial x_j} + X_i \right) dV \tag{5}$$

For the infinitesimal theory of elasticity the strain tensor ε_{ij} is reduced to a linear form

$$\varepsilon_{ij} = \left(\frac{\partial u_i}{\partial x_j} + \frac{\partial u_j}{\partial x_i} \right) \tag{6}$$

where the displacement gradient of the volume element dV along one axis $\frac{\partial u_i}{\partial x_j}$ is a second-order tensor, and the derivative of u_i is a second rank tensor

$$\frac{\partial u_i}{\partial x_j} = \frac{1}{2}\left(\frac{\partial u_i}{\partial x_j} + \frac{\partial u_j}{\partial x_i} \right) + \frac{1}{2}\left(\frac{\partial u_i}{\partial x_j} - \frac{\partial u_j}{\partial x_i} \right) = \frac{1}{2}\varepsilon_{ij} - \frac{1}{2}\omega_{ij} \tag{7}$$

where $\omega_{ij} = \frac{\partial u_i}{\partial x_j} - \frac{\partial u_j}{\partial x_i}$ is the rotational skew-symmetric tensor.

The infinitesimal displacement du_i along the *j* direction and for $\omega_{ij} = 0$ is

$$du_i = \frac{\partial u_i}{\partial x_j} dx_j = \frac{1}{2}\varepsilon_{ij} dx_j \tag{8}$$

Using Gauss' theorem and Equations (4) and (8) the total mechanical work is done on the system by the traction and the body forces

$$W = \int_A (t_i^n \, du_i) dA + \int_V (X_i du_i) dV = \int_A \sigma_{ji} n_j du_i dA + \int_V X_i du_i dV \tag{9}$$

For an isothermal and isobaric process during sorption, the infinitesimal mechanical work δW along the direction n_i outwards the boundaries of a surface A enclosing the volume $dV = dx_i dA$ is equal with

$$\delta W = (\frac{1}{2}\sigma_{ij} n_j \varepsilon_{ij} + X_i du_i) dV \tag{10}$$

Neglecting the body forces X_i , the mechanical work is

$$\delta W = \frac{1}{2}\sigma_{ij} n_j \varepsilon_{ij} dV \tag{11}$$

A superposition of the three normal stress components uniformly distributed over the volume dV is used to calculate the strain in a given direction, say the z-axis:

$$\varepsilon_z = \frac{1}{E(n)}(\sigma_x + \sigma_y + \sigma_z) \tag{12}$$

where $E(n)$ is the modulus of elasticity in tension or Young's modulus.

$$\varepsilon_x = -v\frac{\sigma_z}{(n)}, \ \varepsilon_y = -v\frac{\sigma_z}{(n)} \tag{13}$$

in which v is a constant called Poisson's ratio, equal to ~ 0.3. Using Equations (12) and (13) we obtain the strain equations along the principal axes

$$\varepsilon_x = \frac{1}{E(n)}\left[\sigma_x - v\left(\sigma_y + \sigma_z\right)\right] \tag{14}$$

$$\varepsilon_y = \frac{1}{E(n)}\left[\sigma_y - v(\sigma_z + \sigma_x)\right]$$

$$\varepsilon_z = \frac{1}{E(n)}\left[\sigma_z - v\left(\sigma_x + \sigma_y\right)\right]$$

For relatively thick and isotropic layers and for small linear and homogeneous elastic deformations along the axes, the normal stresses are equal and Equation (14) reads as

$$\varepsilon_z = \frac{1-2v}{E(n)}\sigma_z \tag{15}$$

Because a contraction of a volume element in the z-direction in an elastic medium is accompanied by lateral extensions

$$\varepsilon_z = -\varepsilon_x = -\varepsilon_y \tag{16}$$

and using Equation (16) in Equation (11) the mechanical work along the principal axes is

$$\delta W = \frac{1}{2}\sigma_z\varepsilon_z dV \tag{17}$$

From the first and second law of thermodynamics, the mechanical work W done on a system is

$$W = T\,S - U + \mu_i N_i \ + \Psi A \tag{18}$$

and the infinitesimal mechanical work per unit volume before and after sorption is

$$\frac{(n)}{2(1-2v)}\varepsilon_z^2 dV = -\delta U + T\delta S + \mu_i\delta N_i + \delta(\Psi A) \tag{19}$$

where δU, δS, and δN_i stand for the variations of the internal energy, entropy and the number of bind water molecules on active polymeric sites prior and after sorption, e_z is the strain of the volume element along the z-axis from the confinement of water molecules, μ_i is the chemical potential of δN_i absorbed particle on the polymeric matrix. The term $\delta(\Psi(n)) = \delta[\gamma(n) + (n)_s \iint n_k dA_k]$ is the algebraic sum of the surface energy $\delta(\gamma(n))$ plus the elastic energy strain $(n)_s$ of the nanocavities per unit area, from surface irradiation with some n laser pulses at 157 nm. The last term is zero under isothermal and isobaric sorption, $\delta(\Psi(n)) = 0$, [97]. The term δN_i is relatively negligible because of a small number of absorbed water molecules. Finally, the strain of a volume dV along the z-axis before and after water confinement is given by the equation

$$\varepsilon_z = \left(\frac{2(1-2\nu)(T\delta S - dU)}{(n)dV}\right)^{1/2} \tag{20}$$

4.4. Internal Energy Variation during Molecular Water Confinement

Besides confinement, molecular water molecules are attached to polymeric sites via electric dipole interactions. The internal energy variation is the outcome of the photon-escalating number of active dipole binding sites spawn via VUV matrix irradiation, Figure 11.

For surface irradiation with n laser pulses, the internal energy variation δU_b is given by the relation [13,98]

$$\delta U_b = -\lambda l N_b(n) < \Phi > = -\lambda l N_b(n) \frac{5d_{xy}^4}{64\pi^4 \varepsilon_0^2 \varepsilon_1^2 k_B T r^6} \tag{21}$$

$N_b(n)$ is the number of water molecules attached on the active sites, λ is the probability that a water molecule will overcome an energy threshold barrier and bind in a photon-activated dipole binding site and l is the average number of adsorbed water molecules on each photon-activated dipole binding site. d_{xy}= er is the x, y component of the electric dipole moment between a water molecule and a photo-activated dipole binding site. $e = 1.6 \times 10^{-19}$ C is the electron charge and $r \sim 0.1$ nm is a mean separating distance between a water molecule and a photon-activated dipole binding site, ε_0 is the vacuum permittivity equal to 8.85×10^{-12} Fm^{-1}, $\varepsilon_1 \sim 80$ is the relative electric permittivity of the polymer-water system, $k_B = 1.38 \times 10^{-23}$ J K^{-1} is Boltzmann's constant and T = 300 K is the absolute temperature. Because the energy of each laser pulse at 157 nm is 28 mJ, the number of photons carried in one laser pulse is $n = 2.26 \times 10^{16}$ photons/laser pulse, and this number equals to the number of photon-activated dipole binding sites. Each VUV photon at 157 nm dissociates one molecular bond and creates one active site on the polymeric matrix, Figure 11. For a 1.12×10^{-4} m^2 cross-sectional area of the 157 nm laser beam and 426 nm layer thickness, it is found that 4.73×10^{26} photon-activated dipole binding sites are generated within 1 m^3 per laser pulse. For a cross-section area of the WLRS beam of 4.90×10^{-8} m^2 and 426 nm matrix thickness, the volume dV of the polymeric matrix occupied by the white light beam is 4.09×10^{-14} m^3 and thus the total number of active binding sites per laser pulse within the volume occupied by the white beam is $N_b = 2.31 \times 10^{13}$. From Equation (21) $< \Phi > \approx 1.51 \times 10^{-23}$ J for $\lambda l = 0.05$ (vide infra) and finally $\delta U_d = 1.43 \times 10^{-11}$ J.

4.5. Entropic Energy Variation during Molecular Confinement

Photon-induced nanocavitations are also responsible for surface entropic variation at the boundary between inside and outside nanocavity areas. The entropic variation at the interphase has its origin from time differentiation between the inside and outside areas of nanocavities. Actually, the mean collision time (~70 ns) of water molecules outside the nanocavities within the matrix or near the surface is specified by the laws of ideal gases. On the contrary, the mean escape time of water molecules inside the nanocavities is specified by the hole geometry and the interplay between entrance-escape hole size with cavity diameter. The waiting times follow an inverse power law behavior because thermal equilibrium does not apply in tiny spaces, Figures 12–16 and Tables A1 and A2. In addition, VUV laser irradiation locally ablates the polymeric material, crafting photon-guided "hill-lake" morphologies. The total number of lakes (cavities) vs. the surface area follows a power-law behavior. In this dependency, the number of laser pulses is present through a pre-factor term, Figure 4b. A schematic layout of this modified interphase between photon processed PAM surface and water vapor domain is illustrated in Figure 16.

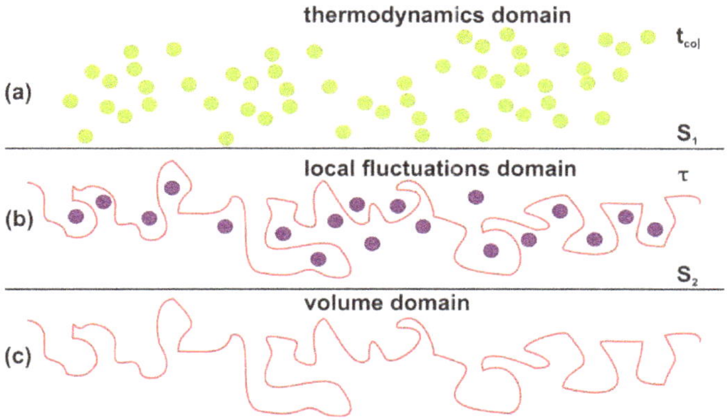

Figure 16. Schematic layout of the interphase between the photon processed polyacrylamide PAM surface and the water vapor domain. (**a**) Thermal equilibrium domain. Reference time and space scales are determined by the mean collision time t_{col} between the water and air molecules and the entropy of the ideal gases and the mean collision distance. The entropy S_1 is given by the Sackur–Tetrode equation for the ideal gases [99]; (**b**) local fluctuations domain. Nanocavitations on the surface with confined molecules. The time scale is determined by the mean escape time τ of water molecules. The entropy S_2 in this domain is determined by the number of microstates $\Omega(N_b(n), N_c(n), E_\alpha)$, which specify a state of ordered arrangements between nanocavities in one hand and molecular water ensembles of fixed molecular length near the surface on the other; (**c**) volume matrix domain.

Random movements in such complex landscapes could be modeled in the frame of continuous time random walk [100,101] by also taking into account the fractal properties of the modified polymeric material [102]. We leave this challenging task for future work where both analytical and extensive numerical calculations combined with experimental results will be presented. Because different water molecules enter and escape the nanocavitations, the number of different microstates $\Omega(N_b(n), N_c(n), E_\alpha)$ per unit time is specified by the frequency of water molecules confined in the nanocavities. The rate of visits is regulated by the mean escape time of water molecules. n and $N_c(n)$ is the number of laser pulses and nanocavities, respectively, N_a is the number of water molecules outside the nanocavities with energy E_α and $N_c(n)$ is the number of nanocavities. The number of microstates is equal to the number of indistinguishable permutations $\{N_a(n),\ N_b\ (n) + N_c(n)\}$ between the number of water molecules N_a and the number of nanocavities $N_c(n)$ and the photon-induced dipole binding sites $N_b(n)$

$$\begin{aligned}\Omega(_b(n),\ N_c(n),\ _\alpha) &= \{N_a(n),\ N_b\ (n) + N_c(n)\} \\ &= \frac{N_a!}{(N_b\ (n)+N_c(n))!(N_a-(N_b\ (n)+N_c(n))!}\ \text{for}\ \ N_b\ (n) + N_c(n) < N_a\end{aligned} \tag{22}$$

$$\begin{aligned}\Omega(_b(n),\ N_c(n),\ _\alpha) &= \{N_b\ (n) + N_c(n), N_a(n)\} \\ &= \frac{(N_b\ (n)+N_c(n))!}{N_a!((N_b\ (n)+N_c(n))-N_a)!}\ \ \text{for}\ N_b\ (n) + N_c(n) > N_a\end{aligned} \tag{23}$$

To arrive in Equations (22) and (23) it is considered that only one water molecule per unit time is either trapped to a specific nanocavity or attached a photon-induced polar binding site. An escalating number of nanocavities is building up in the matrix after each laser pulse, and the ratio of the sum of the number of dipole binding sites and nanocavities to the number of water molecules near the surface is a function of the number of laser pulses

$$x(n) = \frac{N_b\ (n) + N_c(n)}{N_a} \tag{24}$$

From Equations (22)–(24), the entropy from the confinement and the attachment of water molecules in nanocavities and photon-induced polar adhesion binding sites is [13,26]

$$\delta S = k_B ln\Omega(_b(n),\ N_c(n),\ _\alpha) = k_B ln\{N_a(n),\ N_b\ (n) + N_c(n)\},\ \ x(n) < 1 \tag{25}$$

$$\delta S = k_B ln\Omega(_b(n),\ N_c(n),\ _\alpha) = k_B ln\{N_b\ (n) + N_c(n), N_a(n)\},\ \ x(n) > 1 \tag{26}$$

Using Equation (22), Equation (25) read as

$$\delta S = k_B\{\ ln(N_a!) - ln((N_b\ (n) + N_c(n)))! - ln(N_a - (N_b\ (n) + N_c(n))! \tag{27}$$

By using Stirling's formula

$$lnN! = NlnN - N \tag{28}$$

Equation (27) transforms to

$$\begin{aligned} \delta S = k_B\{N_a ln(N_a) - N_a - (N_b\ (n) +\ N_c(n))\ \ln(N_b\ (n) +\ N_c(n)) + (N_b\ (n) +\ N_c(n)) \\ -(N_a - (N_b\ (n) +\ N_c(n) \ln(N_a - (N_b\ (n) +\ N_c(n) + (N_a - (N_b\ (n) +\ N_c(n)))\} \\ = k_B\{N_a ln(N_a) - (N_b\ (n) +\ N_c(n))\ ln(N_b\ (n) +\ N_c(n)) \\ -(N_a - (N_b\ (n) +\ N_c(n))\ \ln(N_a - (N_b\ (n) +\ N_c(n))\} \end{aligned} \tag{29}$$

Using Equation (24), Equation (29) becomes

$$\delta S = k_B(N_b(n) + N_c(n))\left\{\ln\left(\frac{1 - x(n)}{x(n)}\right) - \frac{1}{x(n)}\ln(1 - x(n))\right\},\ x(n) < 1 \tag{30}$$

Similarly from Equations (23) and (24)

$$\delta S = k_B(N_b(n) + N_c(n))\left\{\ln\left(\frac{x(n)}{x(n) - 1}\right) + \frac{1}{x(n)}\ln(x(n) - 1)\right\},\qquad x(n) > 1 \tag{31}$$

In the case of a constant attachment of water molecules in the photon-induced binding sites, Equations (30) and (31) are modified accordingly

$$\delta S = k_B N_c(n)\left\{\ln\left(\frac{1 - y(n)}{y(n)}\right) - \frac{1}{y(n)}\ln(1 - y(n))\right\},\ y(n) < 1 \tag{32}$$

$$\delta S = k_B N_b(n)\beta(n)\left\{\ln\left(\frac{y(n)}{y(n) - 1}\right) + \frac{1}{y(n)}\ln(y(n) - 1)\right\},\qquad y(n) > 1 \tag{33}$$

where

$$y(n) = \frac{N_c(n)}{N_a},\ N_c(n) = \beta(n)N_b(n) \tag{34}$$

In the case where some nanocavities are not visited by the water molecules, then $y(n) > 1$. This condition is fulfilled under the current experimental configuration, Figure 14. For $\beta(n) \sim 0.2$, $N_b = 2.31 \times 10^{13}$, $y(n) = 2$, the entropic energy at 300 K is $k_B T\delta S = 1.31 \times 10^{-8}$ J, which is almost three orders of magnitude larger than δU_d. Equations (32), (33) properly reflect the extensive variable character of the entropy as it should be.

4.6. Surface Strain from the Confinement of Water Molecules

Using Equations (20) and (32)–(34) the surface strain following 157 nm laser irradiation takes the form

$$\varepsilon_z = \left(\frac{N_b(n)}{E(n)dV}\right)^{\frac{1}{2}}[-\lambda l < \Phi > + k_B T\beta(n)[\ \ln\left(\frac{1 - y(n)}{y(n)}\right) - \frac{1}{y(n)}\ln(1 - y(n))]]^{\frac{1}{2}}\ \ y(n) < 1 \tag{35}$$

$$\varepsilon_z = \left(\frac{N_b(n)}{E(n)dV}\right)^{\frac{1}{2}}\left[-\lambda l < \Phi > + k_B T\beta(n)\left[ln\frac{y(n)}{y(n)-1} + \frac{1}{y(n)}\ln(y(n)-1)\right]\right]^{\frac{1}{2}} \quad y(n) > 1 \tag{36}$$

Equations (35) and (36) shape the main result. The equations relate the surface strain ε_z and Young's modulus $E(n)$ with the number of nanocavities, the photon-induced dipole binding sites in the matrix, and the water vapor molecules near nanocavities. For the current experimental configuration $y(n) > 1$. From Equation (36), the strain at 400 laser pulses is ~ 0.1 in agreement with the experimental results of Figure 17.

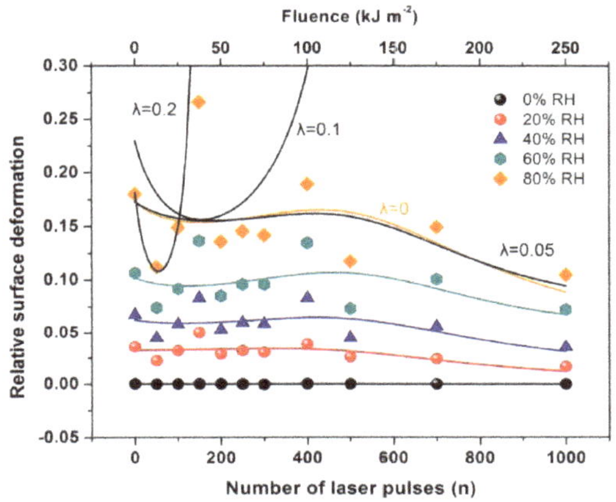

Figure 17. Relative surface deformation (strain) of the 426 nm polyacrylamide (PAM) layers measured with white light reflectance spectroscopy (WLRS) at different 157 nm irradiating conditions of the PAM matrix and relative humidity (RH). The solid lines at different RH represent the best fit of Equation (36) to the experimental data. The black lines at 80 % RH are the fittings for different λl values of 0, 0.05, 0.1 and 0.2. The best fit (orange line) respectively, is for $0 \leq \lambda l < 0.05$ suggesting a small contribution to the relative surface deformation from electric dipole attachment of water molecules to active binding sites in the PAM matrix and a substantial contribution from the confinement of water molecules in nanocavities, Equation (36).

By fitting Equation (36) to the experimental data of Figure 17, the functional dependence of $y(n)$ on the number of photons n is determined at different relative humidity (RH) values. Because $y(n)$ is proportional to the number of dipole binding sites and the number of nanocavities $N_c\ (n)$, $y(n)$ is a measure of the surface carbonization. By using a linear functional for both $y(n)$ and $E(n)$, the best fit of Equation (36) to the experimental data of Figure 17 for relative humidity 80% is for $\beta(n) = 0.2$ and $0 \leq \lambda l < 0.05$. The above fitting values suggest a small and large contribution from the electric dipole interactions and the entropic variation in surface strain, respectively. From Equation (36), the surface strain is proportional to the square root of the number of nanocavities and the concentration of the water molecules (RH) and inversely proportional to the square root of Young's modulus of the surface, in agreement with the experimental results of Figure 17. Finally, the entropic jump, probed by WLRS, trails the confinement of water molecules in nanocavities, while the deep physical root of surface entropy variation originates from the different "time flow and scales" and the validity and invalidity of thermal equilibrium outside and inside the nanocavities, respectively, Figures 15–17.

The experimental approach permits to monitor water confinement on surfaces, including biological ones.

5. Conclusions

Water molecules confined inside laser photon crafted nanocavities on PAM polymeric matrixes are in a state of non-thermal equilibrium. The mean escape time of water molecules from the nanocavities diverges from the mean collision time of water molecules outside the nanocavities (ideal gas state). The time differentiation inside and outside the nanocavities reveals an additional state of ordered arrangements between nanocavities and molecular water ensembles of fixed molecular length near the surface. The configured number of microstates correctly counts for the experimental surface entropy deviation during molecular water confinement.

Author Contributions: Conceptualization, V.G., E.S. and A.-C.C.; methodology, E.S., V.G. and A.-C.C.; software, V.G.; validation, V.G., E.S., A.-C.C., Z.K., E.B., D.C., G.A.K. and M.C.; data curation, V.G., E.S., Z.K. and M.C.; visualization, V.G., A.-C.C. and E.S.; formal analysis, V.G., E.B., E.S. and A.-C.C.; resources, E.S., V.G., Z.K., A.C.-C. and M.C.; investigation, V.G.,E.S., Z.K. and M.C.; writing-original draft preparation, V.G. and E.S.; writing-review and editing, V.G., E.S., E.B., D.C. and A.-C.C.; supervision, E.S.; project administration, E.S.; funding acquisition, A.-C.C. and E.S. All authors have read and agreed to the published version of the manuscript.

Funding: The work was funded under the frame of the projects "ELI – LASERLAB Europe Synergy, HiPER and IPERION-CH.gr" (MIS 5002735) which is implemented under the Action "Reinforcement of the Research and Innovation Infrastructure", funded by the Operational Programme "Competitiveness, Entrepreneurship and Innovation" (NSRF 2014–2020) and co-financed by Greece and the European Union (European Regional Development Fund and "Advanced Materials and Devices" (MIS 5002409) which is implemented under the "Action for the Strategic Development on the Research and Technological Sector", funded by the Operational Programme "Competitiveness, Entrepreneurship and Innovation" (NSRF 2014–2020) and co-financed by Greece and the European Union (European Regional Development Fund).

Conflicts of Interest: The authors declare no conflict of interest.

Appendix A

Table A1. Mean escape time (τ) of a given molecule, errors and total traveling distance of the confined molecule in a spherical cavity (cavity diameter D of 1 nm and 10 nm) for different entrance–escape hole diameters (h) calculated by the diffusion and the non-interactive random walk models.

h (nm)	Cavity Diameter D = 1 nm					Cavity Diameter D = 10 nm				
	Diffusion	Random Walk				Diffusion	Random Walk			
	τ (s)	τ (s)	Error (s)	Distance (nm)	Error (nm)	τ(s)	τ (s)	Error (s)	Distance (nm)	Error (nm)
0.3	3.98×10^{-14}	5.51×10^{-11}	1.7×10^{-12}	35.5	1.1	3.98×10^{-11}	5.57×10^{-8}	1.7×10^{-9}	35,931	1133
0.5	2.39×10^{-14}	1.91×10^{-11}	5×10^{-13}	12.4	0.3	2.39×10^{-11}	2.19×10^{-8}	7×10^{-10}	14,151	465
1						1.19×10^{-11}	4.93×10^{-9}	1.6×10^{-10}	3182	107
2						5.98×10^{-12}	1.28×10^{-9}	4×10^{-11}	822	26
5						2.39×10^{-12}	1.90×10^{-10}	5×10^{-12}	128	4

Table A2. Mean escape time (τ) of a given molecule, errors and total traveling distance of the confined molecule in a spherical cavity (cavity diameter 10^2 nm and 10^3 nm) for different entrance–escape hole diameters (h) calculated by the diffusion and the non-interactive random walk models.

h (nm)	Cavity Diameter D = 10^2 nm					Cavity Diameter D = 10^3 nm				
	Diffusion	Random Walk				Diffusion	Random Walk			
	τ (s)	τ (s)	Error (s)	Distance (nm)	Error (nm)	τ (s)	τ (s)	Error (s)	Distance (nm)	Error (nm)
0.3	3.98×10^{-8}	5.71×10^{-5}	1.8×10^{-6}	3.68×10^{7}	1.2×10^{6}	3.98×10^{-5}	5.3×10^{-2}	1.7×10^{-3}	3.42×10^{10}	1.1×10^{9}
0.5	2.39×10^{-8}	2.00×10^{-5}	6×10^{-7}	1.29×10^{7}	4.1×10^{5}	2.39×10^{-5}	2.01×10^{-2}	6.3×10^{-4}	1.3×10^{10}	4.1×10^{8}
1	1.19×10^{-8}	4.63×10^{-6}	1.4×10^{-7}	2.99×10^{6}	9×10^{4}	1.19×10^{-5}	5.43×10^{-3}	1.6×10^{-4}	3.5×10^{9}	1.1×10^{8}
2	5.98×10^{-9}	1.22×10^{-6}	4×10^{-8}	7.89×10^{5}	2.5×10^{4}	5.98×10^{-6}	1.33×10^{-3}	4.1×10^{-5}	8.56×10^{8}	2.7×10^{7}
5	2.39×10^{-9}	1.95×10^{-7}	6×10^{-9}	1.26×10^{5}	4×10^{3}	2.39×10^{-6}	1.92×10^{-4}	6.7×10^{-6}	1.24×10^{8}	4.3×10^{6}
10	1.19×10^{-9}	5.18×10^{-8}	1.6×10^{-9}	3.34×10^{4}	1060	1.19×10^{-6}	4.99×10^{-5}	1.6×10^{-6}	3.22×10^{7}	1.1×10^{6}
20	5.97×10^{-10}	1.21×10^{-8}	3×10^{-10}	7821	244	5.97×10^{-7}	1.23×10^{-5}	4.1×10^{-7}	7.93×10^{7}	2.6×10^{5}
50	2.39×10^{-10}	1.87×10^{-9}	5×10^{-11}	1204	35	2.39×10^{-7}	1.99×10^{-6}	6.4×10^{-8}	1.28×10^{6}	4.1×10^{4}
100						1.19×10^{-7}	5.07×10^{-7}	1.5×10^{-8}	3.27×10^{5}	9.8×10^{3}
200						5.97×10^{-8}	1.25×10^{-7}	3.7×10^{-9}	8.05×10^{4}	2.4×10^{3}
500						2.39×10^{-8}	1.88×10^{-8}	5.6×10^{-10}	1.21×10^{4}	3.6×10^{2}

Table A3. Mean escape time (τ) and associated errors of different number molecules confined in a 5-nm spherical cavity for different entrance–escape hole diameters (h) calculated by the interactive random walk models for 10^2 runs.

	Cavity Diameter = 5 nm			
N (Number of Molecules)	**h = 1 nm**		**h = 2 nm**	
	τ (s)	**Error (s)**	**τ (s)**	**Error (s)**
1	1.41×10^{-9}	1.4×10^{-10}	2.02×10^{-10}	2.0×10^{-11}
2	1.38×10^{-9}	1.4×10^{-10}	2.04×10^{-10}	2.2×10^{-11}
3	1.57×10^{-9}	1.6×10^{-10}	2.00×10^{-10}	2.1×10^{-11}
4	1.50×10^{-9}	1.4×10^{-10}	2.05×10^{-10}	1.9×10^{-11}
5	1.39×10^{-9}	1.4×10^{-10}	1.95×10^{-10}	1.8×10^{-11}
6	1.46×10^{-9}	1.3×10^{-10}	2.01×10^{-10}	2.2×10^{-11}
7	1.27×10^{-9}	1.2×10^{-10}	2.06×10^{-10}	2.1×10^{-11}
8	1.30×10^{-9}	1.4×10^{-10}	2.25×10^{-10}	2.4×10^{-11}
9	1.82×10^{-9}	1.8×10^{-10}	2.37×10^{-10}	2.1×10^{-11}
10	1.41×10^{-9}	1.4×10^{-10}	2.35×10^{-10}	2.4×10^{-11}
12	1.23×10^{-9}	1.5×10^{-10}	2.20×10^{-10}	2.1×10^{-11}
15	1.53×10^{-9}	1.5×10^{-10}	2.18×10^{-10}	2.0×10^{-11}
20	1.47×10^{-9}	1.3×10^{-10}	1.87×10^{-10}	1.9×10^{-11}

Table A4. Mean escape time (τ) and associated errors of different number molecules confined in a 10-nm spherical cavity for different entrance–escape hole diameters (h) calculated by the interactive random walk models for 10^2 runs.

	Cavity Diameter = 10 nm					
N (Number of Molecules)	**h = 1 nm**		**h = 2 nm**		**h = 5 nm**	
	τ (s)	**Error (s)**	**τ (s)**	**Error (s)**	**τ (s)**	**Error (s)**
1	1.35×10^{-8}	1.5×10^{-9}	2.59×10^{-9}	4.6×10^{-10}	2.42×10^{-10}	2.7×10^{-11}
2	1.33×10^{-8}	1.3×10^{-9}	2.07×10^{-9}	2.3×10^{-10}	2.97×10^{-10}	4.7×10^{-11}
3	1.09×10^{-8}	1.3×10^{-9}	1.74×10^{-9}	1.7×10^{-10}	2.06×10^{-10}	2.5×10^{-11}
4	1.35×10^{-8}	1.5×10^{-9}	2.21×10^{-9}	3.2×10^{-10}	2.00×10^{-10}	1.8×10^{-11}
5	1.10×10^{-8}	1.1×10^{-9}	1.57×10^{-9}	1.7×10^{-10}	2.00×10^{-10}	2.4×10^{-11}
6	1.43×10^{-8}	1.4×10^{-9}	1.78×10^{-9}	1.7×10^{-10}	2.14×10^{-10}	2.5×10^{-11}
7	1.55×10^{-8}	1.6×10^{-9}	1.85×10^{-9}	1.9×10^{-10}	2.51×10^{-10}	3.1×10^{-11}
8	1.21×10^{-8}	1.1×10^{-9}	2.10×10^{-9}	2.2×10^{-10}	2.03×10^{-10}	2.1×10^{-11}
9	1.23×10^{-8}	1.1×10^{-9}	2.07×10^{-9}	2.0×10^{-10}	2.54×10^{-10}	2.4×10^{-11}
10	1.17×10^{-8}	1.3×10^{-9}	1.73×10^{-9}	1.5×10^{-10}	2.31×10^{-10}	3.1×10^{-11}
12	1.28×10^{-8}	1.2×10^{-9}	1.82×10^{-9}	1.7×10^{-10}	1.88×10^{-10}	2.4×10^{-11}
15	1.29×10^{-8}	1.2×10^{-9}	1.66×10^{-9}	1.6×10^{-10}	2.21×10^{-10}	2.3×10^{-11}
20	1.13×10^{-8}	1.0×10^{-9}	1.74×10^{-9}	1.5×10^{-10}	2.40×10^{-10}	2.5×10^{-11}

Table A5. Mean escape time (τ) and associated errors of different number molecules confined in a 15-nm spherical cavity for different entrance–escape hole diameters (h) calculated by the interactive random walk models for 10^2 runs.

N (Number of Molecules)	Cavity Diameter = 15 nm							
	h = 1 nm		h = 2 nm		H = 5 nm		h = 7.5 nm	
	τ	Error (s)	τ (s)	Error (s)	τ (s)	Error (s)	τ (s)	Error (s)
1	1.41×10^{-9}	1.4×10^{-10}	6.17×10^{-9}	7.0×10^{-10}	7.61×10^{-10}	1.1×10^{-10}	2.95×10^{-10}	3.6×10^{-11}
2	1.38×10^{-9}	1.4×10^{-10}	7.26×10^{-9}	8.3×10^{-10}	8.49×10^{-10}	8.1×10^{-11}	2.88×10^{-10}	2.7×10^{-11}
3	1.57×10^{-9}	1.6×10^{-10}	6.39×10^{-9}	7.5×10^{-10}	7.52×10^{-10}	7.7×10^{-11}	3.88×10^{-10}	5.9×10^{-11}
4	1.50×10^{-9}	1.4×10^{-10}	7.29×10^{-9}	6.8×10^{-10}	8.47×10^{-10}	8.8×10^{-11}	3.93×10^{-10}	5.2×10^{-11}
5	1.39×10^{-9}	1.4×10^{-10}	7.09×10^{-9}	6.8×10^{-10}	8.47×10^{-10}	9.2×10^{-11}	3.77×10^{-10}	4.3×10^{-11}
6	1.46×10^{-9}	1.3×10^{-10}	6.13×10^{-9}	7.7×10^{-10}	8.37×10^{-10}	1.1×10^{-10}	3.21×10^{-10}	4.2×10^{-11}
7	1.27×10^{-9}	1.2×10^{-10}	7.26×10^{-9}	7.4×10^{-10}	6.41×10^{-10}	7.0×10^{-11}	3.72×10^{-10}	4.3×10^{-11}
8	1.30×10^{-9}	1.4×10^{-10}	6.37×10^{-9}	6.8×10^{-10}	6.60×10^{-10}	6.8×10^{-11}	2.65×10^{-10}	3.3×10^{-11}
9	1.82×10^{-9}	1.8×10^{-10}	6.11×10^{-9}	6.1×10^{-10}	7.77×10^{-10}	7.6×10^{-11}	3.88×10^{-10}	4.1×10^{-11}
10	1.41×10^{-9}	1.4×10^{-10}	6.49×10^{-9}	6.9×10^{-10}	8.88×10^{-10}	1.0×10^{-10}	2.67×10^{-10}	3.3×10^{-11}
12	1.23×10^{-9}	1.5×10^{-10}	6.90×10^{-9}	7.8×10^{-10}	7.38×10^{-10}	8.1×10^{-11}	3.21×10^{-10}	4.0×10^{-11}
15	1.53×10^{-9}	1.5×10^{-10}	6.52×10^{-9}	6.9×10^{-10}	7.58×10^{-10}	7.7×10^{-11}	2.96×10^{-10}	3.2×10^{-11}
20	1.47×10^{-9}	1.3×10^{-10}	6.19×10^{-9}	5.3×10^{-10}	9.07×10^{-10}	8.9×10^{-11}	3.39×10^{-10}	4.5×10^{-11}

Table A6. Mean escape time (τ) and associated errors of different number molecules confined in a 20-nm spherical cavity for different entrance–escape hole diameters (h) calculated by the interactive random walk models for 10^2 runs.

N (Number of Molecules)	Cavity Diameter = 20 nm							
	h = 1 nm		h = 2 nm		h = 5 nm		h = 10 nm	
	τ (s)	Error (s)	τ (s)	Error (s)	τ (s)	Error (s)	τ (s)	Error (s)
1	1.26×10^{-7}	1.6×10^{-8}	1.74×10^{-8}	2.0×10^{-9}	2.27×10^{-9}	3.85×10^{-10}	4.89×10^{-10}	7.03×10^{-11}
2	7.59×10^{-8}	7.5×10^{-9}	1.69×10^{-8}	1.8×10^{-9}	1.82×10^{-9}	2.24×10^{-10}	4.07×10^{-10}	4.78×10^{-11}
3	9.88×10^{-8}	9.5×10^{-9}	1.63×10^{-8}	1.7×10^{-9}	1.91×10^{-9}	2.04×10^{-10}	3.46×10^{-10}	4.19×10^{-11}
4	9.80×10^{-8}	8.5×10^{-9}	1.51×10^{-8}	1.7×10^{-9}	1.70×10^{-9}	1.85×10^{-10}	4.17×10^{-10}	5.17×10^{-11}
5	9.22×10^{-8}	9.6×10^{-9}	1.75×10^{-8}	1.8×10^{-9}	1.96×10^{-9}	2.24×10^{-10}	4.54×10^{-10}	4.66×10^{-11}
6	8.20×10^{-8}	7.7×10^{-9}	1.32×10^{-8}	1.3×10^{-9}	1.87×10^{-9}	2.11×10^{-10}	3.51×10^{-10}	4.63×10^{-11}
7	8.51×10^{-8}	8.3×10^{-9}	1.51×10^{-8}	1.2×10^{-9}	1.87×10^{-9}	1.87×10^{-10}	4.48×10^{-10}	5.82×10^{-11}
8	9.64×10^{-8}	8.5×10^{-9}	1.43×10^{-8}	1.5×10^{-9}	1.90×10^{-9}	2.05×10^{-10}	4.10×10^{-10}	4.13×10^{-11}
9	6.85×10^{-8}	5.9×10^{-9}	1.66×10^{-8}	1.6×10^{-9}	1.88×10^{-9}	2.15×10^{-10}	3.56×10^{-10}	3.69×10^{-11}
10	9.94×10^{-8}	9.6×10^{-9}	1.56×10^{-8}	1.5×10^{-9}	2.05×10^{-9}	2.18×10^{-10}	4.32×10^{-10}	4.07×10^{-11}
12	8.36×10^{-8}	9.8×10^{-9}	1.55×10^{-8}	1.5×10^{-9}	1.77×10^{-9}	1.81×10^{-10}	4.77×10^{-10}	5.87×10^{-11}
15	8.01×10^{-8}	7.7×10^{-9}	1.40×10^{-8}	1.4×10^{-9}	2.23×10^{-9}	2.01×10^{-10}	3.88×10^{-10}	4.80×10^{-11}
20	7.68×10^{-8}	7.8×10^{-9}	1.49×10^{-8}	1.6×10^{-9}	1.69×10^{-9}	1.89×10^{-10}	4.26×10^{-10}	4.87×10^{-11}

References

1. Levinger, N.E. Water in confinement. *Science* **2002**, *298*, 1722–1723. [CrossRef] [PubMed]
2. Martin, D.R.; Forsmo, J.E.; Matyushov, D.V. Complex Dynamics of Water in Protein Confinement. *J. Phys. Chem.* **2018**, *122*, 3418–3425. [CrossRef]
3. Rasaiah, J.C.; Garde, S.; Hummer, G. Water in nonpolar confinment: From Natotubes to Proteins and Beyond. *Annu. Rev. Phys. Chem.* **2008**, *59*, 713–740. [CrossRef] [PubMed]
4. Hummer, G.; Garde, S. Cavity expulsion and weak dewetting of hydrophobic soluts in water. *Phys. Rev. Lett.* **1998**, *80*, 4193–4196. [CrossRef]
5. Muñoz-Santiburcio, D.; Marx, D. Nanoconfinement in Slit Pores Enhances Water Self-Dissociation. *Phys. Rev. Lett.* **2017**, *119*, 056002. [CrossRef] [PubMed]
6. Mondal, S.; Acharya, S.; Bagchi, B. Altered polar character of nanoconfined liquid water. *Phys. Rev. Res.* **2019**, *1*, 033145. [CrossRef]
7. Gladovic, M.; Bren, U.; Urbic, T. Thermodynamic properties of water in confined environments: A Monte Carlo study. *Mol. Phys.* **2018**, *116*, 1133–1144. [CrossRef]
8. Ball, P. Water is an active matrix of life for cell and molecular biology. *Proc. Natl. Acad. Sci. USA* **2017**, *114*, 13327–13335. [CrossRef]
9. Moscatelli, A. The power of one atom. *Nat. Nanotech.* **2016**. [CrossRef]

10. Lucia, U.A. Link between Nano- and Classical Thermodynamics: Dissipation Analysis (The Entropy Generation Approach in Nano-Thermodynamics). *Entropy* **2015**, *17*, 1309–1328. [CrossRef]
11. Ernst, J.A.; Clubb, R.T.; Zhou, H.X.; Gronenborn, A.M.; Clore, G.M. Demonstration of positionally disordered water within a protein hydrophobic cavity by NMR. *Science* **1995**, *267*, 1813–1817. [CrossRef] [PubMed]
12. Chaban, V.V.; Prezhdo, V.V.; Prezhdo, O.V. Confinement by Carbon Nanotubes Drastically Alters the Boiling and Critical Behavior of Water Droplets. *ACS Nano* **2012**, *6*, 2766–2773. [CrossRef] [PubMed]
13. Cefalas, A.C.; Sarantopoulou, E.; Kollia, Z.; Kitsara, M.; Raptis, I.; Bakalis, E. Entropic nanothermodynamic potential from molecular trapping within photon induced nano-voids in photon processed PDMS layers. *Soft Matter* **2012**, *8*, 5561–5574. [CrossRef]
14. Hill, T.L. A Different Approach to Nanothermodynamics. *Nano Lett.* **2001**, *1*, 273–275. [CrossRef]
15. Giebultowicz, T. Nanothermodynamics: Breathing life into an old model. *Nature* **2000**, *408*, 299–301. [CrossRef]
16. Li, Z.H.; Truhlar, D.G. Nanothermodynamics of metal nanoparticles. *Chem. Sci.* **2014**, *5*, 2605–2624. [CrossRef]
17. Wang, C.X.; Yang, G.W. Thermodynamics of metastable phase nucleation at the nanoscale. *Mater. Sci. Eng. R.* **2005**, *49*, 157–202. [CrossRef]
18. Hill, T.L. Extension of Nanothermodynamics to Include a One-Dimensional Surface Excess. *Nano Lett.* **2001**, *1*, 159–160. [CrossRef]
19. Schnell, S.K.; Liu, X.; Simon, J.M.; Bardow, A.; Bedeaux, D.; Vlugt, T.J.H.; Kjelstrup, S. Calculating Thermodynamic Properties from Fluctuations at Small Scales. *J. Phys. Chem.* **2011**, *115*, 10911–10918. [CrossRef]
20. Gupta, S.; Zhang, Q.; Emrick, T.; Balazs, A.C.; Russell, T.P. Entropy-driven segregation of nanoparticles to cracks in multilayered composite polymer structures. *Nat. Mater.* **2006**, *5*, 229–233. [CrossRef]
21. Ruggeri, F.; Krishnan, M. Entropic Trapping of a Singly Charged Molecule in Solution. *Nano Lett.* **2018**, *18*, 3773–3779. [CrossRef] [PubMed]
22. Guisbiers, G. Advances in thermodynamic modelling of nanoparticles. *Adv. PhysX* **2019**, *4*. [CrossRef]
23. Dai, X.; Hou, C.; Xu, Z.; Yang, Y.; Zhu, G.; Chen, P.; Huang, Z.; Yan, L.T. Entropic Effects in Polymer Nanocomposites. *Entropy* **2019**, *21*, 186. [CrossRef]
24. Garcia-Morales, G.; Cervera, J.; Manzanares, J.A. Nanothermodynamcis. In *Handbook of Nanophysics: Principles and Methods*, 1st ed.; Sattler, K.D., Ed.; CRC Press Taylor & Francis Group: Boca Raton Florida, FL, USA, 2011. [CrossRef]
25. Gavriil, V.; Chatzichristidi, M.; Kollia, Z.; Cefalas, A.C.; Spyropoulos-Antonakakis, N.; Semashko, V.V.; Sarantopoulou, E. Photons Probe Entropic Potential Variation during Molecular Confinement in Nanocavities. *Entropy* **2018**, *20*, 545. [CrossRef]
26. Huber, P. Soft matter in hard confinement: Phase transition thermodynamics, structure, texture, diffusion and flow in nanoporous media. *J. Phys. Condens. Matter* **2015**, *27*, 103102. [CrossRef]
27. Galteland, O.; Bedeaux, D.; Kjelstrup, S. Pressures Inside a Nano-Porous Medium. The Case of a Single Phase Fluid. *Front. Phys.* **2019**, *7*, 1–10. [CrossRef]
28. Stefi, A.L.; Sarantopoulou, E.; Kollia, Z.; Spyropoulos-Antonakakis, N.; Bourkoula, A.; Petrou, P.S.; Kakabakos, S.; Soras, G.; Trohopoulos, P.N.; Nizamutdinov, A.S.; et al. Nanothermodynamics Mediates Drug Delivery. In *GeNeDis 2014 Neurodegeneration*; Vlamos, P., Alexiou, A., Eds.; Springer: Cham, Switzerland, 2015; Volume 822, pp. 213–220. [CrossRef]
29. Cefalas, A.C.; Kollia, Z.; Spyropoulos-Antonakakis, N.; Gavriil, V.; Christofilos, D.; Kourouklis, G.; Shemashko, V.V.; Pavlov, V.; Sarantopoulou, E. Surface profile gradient in amorphous Ta_2O_5 semi conductive layers regulates nanoscale electric current stability. *Appl. Surf. Sci.* **2017**, *396*, 1000–1019. [CrossRef]
30. Tranchida, D.; Sperotto, E.; Chateauminois, A.; Schönherr, H. Entropic Effects on the Mechanical Behavior of Dry Polymer Brushes During Nanoindentation by Atomic Force Microscopy. *Macromolecules* **2011**, *44*, 368–374. [CrossRef]
31. Nicolis, S.J. *Dynamics of Hierarchical Systems: An. Evolutionary Approach*, 1st ed.; Springer: Berlin/Heidelberg, Germany, 1986; ISBN 978-3-642-69694-7.
32. Nash, T. The Role of Entropy in Molecular Self-Assembly. *J. Nanomed. Res.* **2017**, *5*, 00126. [CrossRef]

33. Cefalas, A.C.; Kobe, S.; Dražic, G.; Sarantopoulou, E.; Kollia, Z.; Stražišar, J.; Medend, A. Nanocrystallization of $CaCO_3$ at solid/liquid interfaces in magnetic field: A quantum approach. *Appl. Surf. Sci.* **2008**, *254*, 6715–6724. [CrossRef]
34. Grzybowski, B.A.; Wilmer, C.E.; Kim, J.; Browne, K.P.; Bishop, K.J.M. Self-assembly: From crystals to cells. *Soft Matter* **2009**, *5*, 1110–1128. [CrossRef]
35. Dauenhauer, P.J.; Abdelrahman, O.A. A Universal Descriptor for the Entropy of Adsorbed Molecules in Confined Spaces. *ACS Cent. Sci.* **2018**, *4*, 1235–1243. [CrossRef]
36. Campbell, C.T.; Sellers, J.R.V. The Entropies of Adsorbed Molecules. *J. Am. Chem. Soc.* **2012**, *134*, 18109–18115. [CrossRef] [PubMed]
37. Sarantopoulou, E.; Petrou, P.S.; Kollia, Z.; Palles, D.; Spyropoulos-Antonakakis, N.; Kakabakos, S.; Cefalas, A.C. Protein immobilization and detection on laser processed polystyrene surfaces. *J. Appl. Phys.* **2011**, *110*, 064309-17. [CrossRef]
38. Nguyen, A.K.; Narayan, R.J. Two-photon polymerization for biological applications. *Mater. Today* **2017**, *20*, 314–322. [CrossRef]
39. Zheng, L.; Kurselis, K.; El-Tamer, A.; Hinze, U.; Reinhardt, C.; Overmeyer, L.; Chichkov, B. Nanofabrication of High-Resolution Periodic Structures with a Gap Size Below 100 nm by Two-Photon Polymerization. *Nanoscale Res. Lett.* **2019**, *14*, 134. [CrossRef]
40. Yang, T.H.; Kao, C.R.; Shigetou, A. Organic-Inorganic Solid-State Hybridization with High-Strength and Anti-Hydrolysis Interface. *Sci. Rep.* **2019**, *9*, 504. [CrossRef]
41. Sarantopoulou, E.; Kollia, Z.; Cefalas, A.C.; Siokou, A.E.; Argitis, P.; Bellas, V.; Kobe, S. Surface modification of polyhedral oligomeric silsesquioxane block copolymer films by 157 nm laser light. *J. Appl. Phys.* **2009**, *105*, 114305. [CrossRef]
42. Sarantopoulou, E.; Kollia, Z.; Cefalas, A.C.; Manoli, K.; Sanopoulou, M.; Goustouridis, D.; Chatzandroulis, S.; Raptis, I. Surface nano/micro functionalization of PMMA thin films by 157 nm irradiation for sensing applications. *Appl. Surf. Sci.* **2008**, *254*, 1710–1719. [CrossRef]
43. Douvas, A.M.; Petrou, P.S.; Kakabakos, S.E.; Misiakos, K.; Argitis, P.; Sarantopoulou, E.; Kollia, Z.; Cefalas, A.C. 157-nm Laser ablation of polymeric layers for fabrication of biomolecule microarrays. *Anal. Bioanal. Chem.* **2005**, *381*, 1027–1032. [CrossRef]
44. Wu, C.T.; Soliman, A.I.A.; Utsunomiya, T.; Ichii, T.; Sugimura, H. Formation of submicron-sized silica patterns on flexible polymer substrates based on vacuum ultraviolet photo-oxidation. *RSC Adv.* **2019**, *9*, 32313–32322. [CrossRef]
45. Kim, Y.J.; Sugimura, H. Effects of irradiation atmosphere on vacuum ultraviolet-induced surface modification of cyclo-olefin polymer substrates. *Appl. Phys. Express* **2019**, *12*, 101002. [CrossRef]
46. Kirchner, R.; Chidambaram, N.; Schift, H. Benchmarking surface selective vacuum ultraviolet and thermal postprocessing of thermoplastics for ultrasmooth 3-D-printed micro-optics. *Opt. Eng.* **2018**, *57*, 041403. [CrossRef]
47. Kumar, K.V.; Srinivas, G.; Wood, B.; Ramisetty, K.A.; Stewart, A.A.; Howard, C.A.; Brett, D.J.L.; Rodriguez-Reinoso, F. Characterization of adsorption site energies and heterogeneous surfaces of porous materials. *J. Mater. Chem.* **2019**, *7*, 10104–10137. [CrossRef]
48. Lorzing, G.R.; Gosselin, E.J.; Trump, B.A.; York, A.H.P.; Sturluson, A.; Casey, R.; Glenn, P.A.; Brown, C.M.; Simon, C.M.; Bloch, E. Understanding Gas Storage in Cuboctahedral Porous Coordination Cages. *J. Am. Chem. Soc.* **2019**, *141*, 12128–12138. [CrossRef]
49. Sadasivuni, K.K.; Cabibihan, J.J.; Deshmukh, K.; Goutham, S.; Abubasha, M.K.; Gogoi, J.P.; Klemenoks, I.; Sakale, G.; Sekhar, B.S.; Sreekanth, P.S.R.; et al. A review on porous polymer composite materials for multifunctional electronic applications. *Polymer Plast. Tech. Eng.* **2019**, *58*, 1253–1294. [CrossRef]
50. Liu, X.; Xu, Y.; Jiang, D. Conjugated Microporous Polymers as Molecular Sensing Devices: Microporous Architecture Enables Rapid Response and Enhances Sensitivity in Fluorescence-On and Fluorescence-off Sensing. *J. Am. Chem. Soc.* **2012**, *134*, 8738–8741. [CrossRef]
51. Reguera, D.; Luque, A.; Burada, P.S.; Schmid, G.; Rubi, J.M.; Hänggi, P. Entropic splitter for particle separation. *Phys. Rev. Lett.* **2012**, *108*, 020604-5. [CrossRef]
52. Zwanzig, R. Diffusion past an entropy barrier. *J. Phys. Chem.* **1992**, *96*, 3926–3930. [CrossRef]
53. Pullumbi, P.; Brandani, F.; Brandani, S. Gas separation by adsorption: Technological drivers and opportunities for improvement. *Curr. Opin. Chem. Eng.* **2019**, *24*, 131–142. [CrossRef]

54. Chena, X.Y.; Kaliaguinea, S.; Rodriguea, D. Polymer hollow fiber membranes for gas separation: A comparison between three commercial resins. *AIP Conf. Proc.* **2019**, *2139*, 070003. [CrossRef]
55. Szałaj, U.; Świderska-Środa, A.; Chodara, A.; Gierlotka, S.; Łojkowski, W. Nanoparticle Size Effect on Water Vapour Adsorption by Hydroxyapatite. *Nanomaterials* **2019**, *9*, 1005. [CrossRef] [PubMed]
56. Zainuddin, T.; Chirila, T.V.; Barnard, Z.; Watson, G.S.; Toh, C.; Blakey, I.; Whittaker, A.K.; Hill, D.J.T. F_2 excimer laser (157nm) radiation modification and surface ablation of PHEMA hydrogels and the effects on bioactivity: Surface attachment and proliferation of human corneal epithelial cells. *Rad. Phys. Chem.* **2011**, *80*, 219–229. [CrossRef]
57. Shi, Q.; Liu, H.; Tang, D.; Li, Y.; Li, X.J.; Xu, F. Bioactuators based on stimulus-responsive hydrogels and their emerging biomedical applications. *NPG Asia Mater.* **2019**, *11*, 64. [CrossRef]
58. Spyropoulos-Antonakakis, N.; Sarantopoulou, E.; Trohopoulos, P.N.; Stefi, A.L.; Kollia, Z.; Gavriil, V.E.; Bourkoula, A.; Petrou, P.S.; Kakabakos, S.; Semashko, V.V.; et al. Selective aggregation of PAMAM dendrimernanocarriers and PAMAM/ZnPc nanodrugs on human atheromatous carotid tissues: A photodynamic therapy for atherosclerosis. *Nanoscale Res. Lett.* **2015**, *10*, 210. [CrossRef] [PubMed]
59. Muzzio, M.; Li, J.; Yin, Z.; Delahunty, I.; Xie, J.; Sun, S. Monodisperse nanoparticles for catalysis and nanomedicine. *Nanoscale* **2019**, *11*, 18946–18967. [CrossRef]
60. Sanzari, I.; Callisti, M.; Grazia, A.; Evans, D.J.; Polcar, T.; Prodromakis, T. Parylene topographic micropattern as a template for patterning PDMS and Polyacrylamide hydrogel. *Sci. Rep.* **2017**, *7*, 5764. [CrossRef]
61. Xiong, B.; Loss, R.D.; Shields, D.; Pawlik, T.; Hochreiter, R.; Zydney, A.L.; Kumar, M. Polyacrylamide degradation and its implications in environmental systems. *NPJ Clean Water* **2018**, *1*, s41545-18. [CrossRef]
62. Kundu, B.; Kundu, S.C. Silk sericin/polyacrylamide in situ forming hydrogels for dermal reconstruction. *Biomaterials* **2012**, *33*, 7456–7467. [CrossRef]
63. Ahad, I.U.; Bartnik, A.; Fiedorowicz, H.; Kostecki, J.; Korczyc, B.; Ciach, T.; Brabazon, D. Surface modification of polymers for biocompatibility via exposure to extreme ultraviolet radiation. *J. Biomed. Mater. Res.* **2014**, *102*, 3298–3310. [CrossRef]
64. Sarantopoulou, E.; Kollia, Z.; Cefalas, A.C.; Douvas, A.M.; Chatzichristidi, M.; Argitis, P.; Kobe, S. Nano-scale spatial control over surface morphology of biocompatible fluoropolymers at 157 nm. *Mater. Sci. Eng.* **2007**, *27*, 1191–1196. [CrossRef]
65. Tran, K.T.M.; Nguyen, T.D. Lithography-based methods to manufacture biomaterials at small scales. *J. Sci. Adv. Mater. Dev.* **2017**, *2*, 1–14. [CrossRef]
66. Nzenguet, A.M.; Aqlil, M.; Essamlali, Y.; Amadine, O.; Snik, A.; Larzek, M.; Zahouily, M. Novel bionanocomposite films based on graphene oxide filled starch/polyacrylamide polymer blend: Structural, mechanical and water barrier properties. *J. Polym. Res.* **2018**, *25*, 86. [CrossRef]
67. Smithson, R.L.W.; Evans, D.F.; Monfils, J.D.; Guire, P.E. Ultrathin film characterization: Photoreactive polyacrylamide. *Colloids Surf. Biointerfaces* **1993**, *1*, 349–355. [CrossRef]
68. Vijayalakshmi, S.P.; Senapati, D.; Madras, G. Pulsed laser degradation of polyethylene oxide and polyacrylamide in aqueous solution. *Polym. Degrad. Stab.* **2005**, *87*, 521–526. [CrossRef]
69. Rong, J.; Zhu, K.; Chen, M. Study on purification technology of polyacrylamide wastewater by non-thermal plasma. *Plasma Sci. Technol.* **2019**, *21*, 054008. [CrossRef]
70. Song, W.; Zhang, Y.; Yu, J.; Gao, Y.; Naitoc, T.; Oinumac, G.; Inanagac, Y.; Yang, M. Rapid removal of polyacrylamide from wastewater by plasma in the gas–liquid interface. *J. Environ. Sci.* **2019**, *83*, 1–7. [CrossRef]
71. Caulfield, M.J.; Hao, X.; Qiao, G.G.; Solomon, D.H. Degradation of polyacrylamides. Part, I. Linear polyacrylamide. *Polymer* **2003**, *44*, 1331–1337. [CrossRef]
72. Timoshenko, S. *Theory of Elasticity*, 1st ed.; McCraw-Hill Book Company, Inc.: New York, NY, USA; London, UK, 1934; p. 7.
73. Nečas, D.; Klapetek, P. Gwyddion: An open-source software for SPM data analysis. *Cent. Eur. J. Phys.* **2012**, *10*, 181–188. [CrossRef]
74. Zahn, W.; Zösch, A. Characterization of thin film surfaces by fractal geometry. *Fresenius J. Anal. Chem.* **1997**, *358*, 119–121. [CrossRef]
75. Raptis, I.; Kovač, J.; Chatzichristidi, M.; Sarantopoulou, E.; Kollia, Z.; Kobe, S.; Cefalas, A.C. Enhancement of sensing properties of thin poly(methyl methacrylate) films by VUV modification. *J. Laser Micro Nanoeng.* **2007**, *2*, 200–205. [CrossRef]

76. Friedrich, J. Accelerated Plasma-Aging of Polymers. In *The Plasma Chemistry of Polymer Surfaces: Advance Techniques for Surface Design*, 1st ed.; Wiley-VCH Verlag GmbH & Co. KGaA: Weinheim, Germany, 2012; pp. 239–248. [CrossRef]
77. Park, C.S.; Jung, E.Y.; Jang, H.J.; Bae, G.T.; Shin, B.J.; Tae, H.S. Synthesis and Properties of Plasma-Polymerized Methyl Methacrylate via the Atmospheric Pressure Plasma Polymerization Technique. *Polymers* **2019**, *11*, 396. [CrossRef]
78. Denisin, A.K.; Pruitt, B.L. Tuning the Range of Polyacrylamide Gel Stiffness for Mechanobiology Applications. *ACS Appl. Mater. Interfaces* **2016**, *8*, 21893–21902. [CrossRef] [PubMed]
79. Ranganathan, N. *Materials Characterization Modern Methods and Applications*, 1st ed.; CRC Press, Taylor & Francis Group: New York, NY, USA, 2015; ISBN 978-9814613064.
80. Constantinides, G.; Kalcioglu, Z.I.; McFarland, M.; Smith, J.F.; Van Vliet, K.J. Probing mechanical properties of fully hydrated gels and biological tissues. *J. Biomech.* **2008**, *41*, 3285–3289. [CrossRef] [PubMed]
81. An, M.; Demir, B.; Wan, X.; Meng, H.; Yang, N.; Walsh, T.R. Predictions of Thermo-Mechanical Properties of Cross-Linked Polyacrylamide Hydrogels Using Molecular Simulations. *Adv. Theory Simul.* **2019**, *2*, 1800153. [CrossRef]
82. Grigoriev, I.V.; Makhnovskii, Y.A.; Berezhkovskii, A.M.; Zitserman, V.Y. Kinetics of escape through a small hole. *J. Chem. Phys.* **2002**, *116*, 9574–9577. [CrossRef]
83. Wohlfromm, T.; Vogel, M. On the coupling of protein and water dynamics in confinement: Spatially resolved molecular dynamics simulation studies. *J. Chem. Phys.* **2019**, *150*, 245101. [CrossRef]
84. Jonchhe, S.; Pandey, S.; Emura, T.; Hidaka, K.; Hossain, M.A.; Shrestha, P.; Sugiyama, H.; Endo, M.; Mao, H. Decreased water activity in nanoconfinement contributes to the folding of G-quadruplex and i-motif structures. *Proc. Natl. Acad. Sci. USA* **2018**, *115*, 9539–9544. [CrossRef]
85. Biswas, R.; Bagchi, B. Anomalous water dynamics at surfaces and interfaces: Synergistic effects of confinement and surface interactions. *J. Phys. Condens. Matter* **2018**, *30*, 013001. [CrossRef]
86. Gorshunov, B.P.; Zhukova, E.S.; Torgashev, V.I.; Lebedev, V.V.; Shakurov, G.S.; Kremer, R.K.; Pestrjakov, E.V.; Thomas, G.; Fursenko, D.A.; Dressel, M. Quantum Behavior of Water Molecules Confined to Nanocavities in Gemstones. *J. Phys. Chem. Lett.* **2013**, *4*, 2015–2020. [CrossRef]
87. Oh, M.I.; Gupta, M.; Oh, C.I.; Weaver, D.F. Understanding the effect of nanoconfinement on the structure of water hydrogen bond networks. *Phys. Chem. Chem. Phys.* **2019**, *21*, 26237–26250. [CrossRef] [PubMed]
88. Yang, L.; Guo, Y.; Diao, D. Structure and dynamics of water confined in a graphene nanochannel under gigapascal high pressure: Dependence of friction on pressure and confinement. *Phys. Chem. Chem. Phys.* **2017**, *19*, 14048. [CrossRef] [PubMed]
89. Strøm, B.A.; Simon, J.M.; Schnell, S.K.; Kjelstrup, S.; He, J.; Bedeaux, D. Size and shape effects on the thermodynamic properties of nanoscale volumes of water. *Phys. Chem. Chem. Phys.* **2017**, *19*, 9016. [CrossRef] [PubMed]
90. Chakraborty, S.; Kumar, H.; Dasgupta, C.; Maiti, P.K. Confined Water: Structure, Dynamics, and Thermodynamics. *Acc. Chem. Res.* **2017**, *50*, 2139–2146. [CrossRef] [PubMed]
91. Leoni, F.; Franzese, G. Effects of confinement between attractive and repulsive walls on the thermodynamics of an anomalous fluid. *Phys. Rev.* **2016**, *94*, 062604. [CrossRef]
92. Firat, C.; Sisman, A.; Ozturk, Z.F. Thermodynamics of gases in nano cavities. *Energy* **2010**, *35*, 814–819. [CrossRef]
93. Boyer, T.H. Quantum electromagnetic zero point energy of a conducting spherical cell and the Casimir model for a charged particle. *Phys. Rev.* **1968**, *174*, 1764. [CrossRef]
94. Cheng, K.L.; Sheng, Y.J.; Tsao, H.K. Brownian escape and force-driven transport through entropic barriers: Particle size effect. *J. Chem. Phys.* **2008**, *129*, 184901. [CrossRef]
95. Pascal, T.A.; Goddard, W.A.; Junga, Y. Entropy and the driving force for the filling of carbon nanotubes with water. *Proc. Natl. Acad. Sci. USA* **2011**, *108*, 11794–11798. [CrossRef]
96. Zhou, H.X.; Zwanzig, R. A rate process with an entropy barrier. *J. Chem. Phys.* **1991**, *94*, 6147–6152. [CrossRef]
97. Hetnarski, R.B.; Eslami, M.R. *Thermal Stresses-Advanced Theory and Applications*, 1st ed.; Springer: New York, NY, USA, 2009; ISBN 978-3-030-10436-8.
98. Grosman, A.; Ortega, C. Influence of elastic deformation of porous materials in adsorption-desorption process: A thermodynamic approach. *Phys. Rev.* **2008**, *78*, 085433. [CrossRef]

99. Pathria, R.K.; Beale, P.B. *Statistical Mechanics*, 3rd ed.; Academic Press: Burlington, MA, USA, 2011; ISBN 978-0-12-382188-1.
100. Montroll, E.W.; Weiss, G.H. Random Walks on Lattices II. *J. Math. Phys.* **1965**, *6*, 167. [CrossRef]
101. Montroll, E.W.; Scher, H. Random Walks on Lattices. IV. Continuous-Time walks and influence of Absorbing Boundaries. *J. Stat. Phys.* **1973**, *9*, 101–135. [CrossRef]
102. Blumen, A.; Klafter, J.; White, B.S.; Zumofen, G. Continuous-Time Random Walks on fractals. *Phys. Rev. Lett.* **1984**, *53*, 1301. [CrossRef]

nanomaterials

MDPI

Article

Quantum Characteristics of a Nanomechanical Resonator Coupled to a Superconducting LC Resonator in Quantum Computing Systems

Jeong Ryeol Choi * and Sanghyun Ju

Department of Physics, Kyonggi University, Yeongtong-gu, Suwon, Gyeonggi-do 16227, Korea; shju@kyonggi.ac.kr

* Correspondence: jrchoi@kyonggi.ac.kr or choiardor@hanmail.net; Tel.: +82-31-249-1320

Received: 26 November 2018; Accepted: 21 December 2018; Published: 24 December 2018

Abstract: The mechanical and quantum properties of a nanomechanical resonator can be improved by connecting it to a superconducting resonator in a way that the resonator exhibits new phenomena that are possibly available to novel quantum technologies. The quantum characteristics of a nanomechanical resonator coupled to a superconducting resonator have been investigated on the basis of rigorous quantum solutions of the combined system. The solutions of the Schrödinger equation for the coupled system have been derived using the unitary transformation approach. The analytic formula of the wave functions has been obtained by applying the adiabatic condition for time evolution of the coupling parameter. The behavior of the quantum wave functions has been analyzed for several different values of parameters. The probability densities depicted in the plane of the two resonator coordinates are distorted and rotated due to the coupling between the resonators. In addition, we have shown that there are squeezing effects in the wave packet along one of the two resonator coordinates or along both the two depending on the magnitude of several parameters, such as mass, inductance, and angular frequencies.

Keywords: nanomechanical resonator; superconducting resonator; wave function; unitary transformation; Hamiltonian; probability density; adiabatic condition; quantum solution

1. Introduction

A rapidly developing field in nano-based science and technology is optomechanics which deals with the interaction of light with a mechanical motion [1]. Especially, optomechanics combined with nanomechanical resonators can be practically applied to a broad scientific domain such as quantum information processing [2], biological sensing [3], wave detections [4], measurements of mechanical displacement [5], and quantum metrologies [6]. Such optomechanics developed in the quantum regime with low phonon occupation states will play a major role in future-oriented quantum technologies with nanodevices.

The investigation of nanodevices regarding their application in quantum information science, including quantum computing, is a promising research topic. In particular, research into nanomechanical resonators in which the parameters are dependent on time is quite necessary for the advancement of the quantum information technology [7–11]. Now, it is possible to design quantum computing devices with a reliable architecture for multi-qubit operations in the GHz-frequency range by coupling mechanical resonators to Josephson phase qubits [12]. Onchip-integrated hybrid systems, i.e., mechanical resonators combined with phase qubits which are composed of Josephson-junction superconducting circuits, can be used as a compact quantum information storage with a high quality factor [13]. Recent advances in nanotechnology in the past decade enabled the fabrication of nanomechanical resonators, of which quality factors are high in the desired frequency ranges.

By connecting nanomechanical resonators to the superconducting resonators, it is possible to enhance the properties of nanomechanical resonators. Thanks to this, many relevant experiments have been developed and renewed, such as, for example, preparing quantum ground state in resonators [10], a frequency up/down conversion [11], and maintaining longer coherence times [14]. In particular, coherent feedback control of the nanoresonators can be used for cooling them to the zero-point temperature. The cooling of a mechanical system to the ground-state has already been achieved in several laboratories [15,16].

As the size of the resonator reaches below a critical value that is the limiting one from a quantum mechanical point of view, there emerge distinguishing quantum features that are totally absent in the classical world [5]. Quantum behaviors of a nanomechanical resonator were observed through its sideband laser-cooling over the quantum ground state [17]. A deeper understanding of quantum characteristics of nanomechanical resonators, where the device-size is within the Heisenberg uncertainty principle limit, is crucial for utilizing them in quantum computing systems. We investigate quantum properties of a nanomechanical resonator coupled to a superconducting resonator. This is important for providing theoretical knowledge as background information for manipulating quantum computing processes. Furthermore, this research may help to achieve robust quantum computations on the basis of the restriction of the decoherence and noise.

2. Materials and Methods

Cooling of nanomechanical resonators can be achieved by the methods of sideband cooling or measurement-based feedback cooling [18]. The former is classified as coherent feedback controlling for a nanomechanical system, that can be carried out by coupling it to an auxiliary one such as an optical cavity and a superconducting oscillator, whereas the systems in the letter case are controlled by means of the feedback technique with the data which are continuously measured from homodyne detection. In this work, we consider sideband cooling using a high-frequency superconducting LC oscillator as an auxiliary mode in order to control the motion of the nanomechanical resonator. For more details of the nanomechanical-resonator design and the related mechanics, refer to Refs. [10,11].

To see quantum features of the system, we first need to derive relevant quantum solutions in a rigorous way. We introduce the Hamiltonian for describing the quantum dynamics of the nanomechanical resonator coupled to the superconducting resonator. The Hamiltonian involves a coupling term which is associated with the coupling of the two resonators. Due to not only such a coupling term but the time-dependence of the parameters as well, it may be not an easy task to solve the Schrödinger equation on the basis of the conventional separation of variables method. For this reason, we will derive quantum solutions by making use of another method which is the unitary transformation method [7,8,19].

It may be convenient that we unfold quantum theory after mathematically transforming the system described by a time-dependent Hamiltonian into a simple one that can be easily managed. By introducing a unitary operator, we will transform the original system to a system composed of two decoupled harmonic oscillators of which quantum solutions are well known. By inversely transforming the quantum solutions associated to the transformed system, it is possible to obtain the complete quantum solutions in the (original) system. This is the main strategy used in this work for deriving quantum solutions of the system whose Hamiltonian is a fairly complicated form. We will analyze quantum properties of the system in detail on the basis of the quantum solutions evaluated in such a way.

3. Results and Discussion

3.1. Hamiltonian and the Unitary Transformation

Superconducting systems can be used to improve the properties of other quantum systems by connecting them to the target systems [20]. From there, we can explore new phenomena which could possibly bring about the deveopment of novel quantum technologies for quantum information processing. For instance, a sideband cooling of a nanoresonator is possible through its modulation via coupling it to a superconducting LC oscillator. In this case, we can regard the nanoresonator as the target device and the superconducting LC resonator as an auxiliary one.

The Hamiltonian that describes the nanomechanical resonator combined with the superconducting LC resonator is given by [10,11]

$$\hat{H} = \hbar\omega(\hat{a}^\dagger\hat{a} + 1/2) + \hbar\Omega(\hat{b}^\dagger\hat{b} + 1/2) + \hbar\lambda(t)(\hat{a} + \hat{a}^\dagger)(\hat{b} + \hat{b}^\dagger), \tag{1}$$

where $\hat{a}$ and $\hat{b}$ are annihilation operators in the nanomechanical resonator and the superconducting resonator, respectively, and ω is the frequency of the nanoresonator, while Ω is the frequency of the superconducting resonator. Here, we assume that the parameter $\lambda(t)$ is a slowly varying function so that we can apply the adiabatic theorem. Although we are interested in the quantum description of the nanomechanical resonator coupled to the superconducting resonator with Equation (1), the Hamiltonian in Equation (1) can also be used for other purposes such as the gauge field theory of two superconducting resonators which are tunably coupled to each other [21] and quantum simulation of bosonic modes utilizing superconducting circuits [22].

Because the operators $\hat{a}$ and $\hat{b}$ are represented in the form

$$\hat{a} = \sqrt{\frac{m\omega}{2\hbar}}\hat{x} + \frac{i\hat{p}_x}{\sqrt{2\hbar m\omega}}, \tag{2}$$

$$\hat{b} = \sqrt{\frac{L\Omega}{2\hbar}}\hat{q} + \frac{i\hat{p}_q}{\sqrt{2\hbar L\Omega}}, \tag{3}$$

with $\hat{p}_x = -i\hbar\partial/\partial x$ and $\hat{p}_q = -i\hbar\partial/\partial q$, Equation (1) can be rewritten as

$$\hat{H} = \frac{\hat{p}_x^2}{2m} + \frac{1}{2}m\omega^2\hat{x}^2 + \frac{\hat{p}_q^2}{2L} + \frac{1}{2}L\Omega^2\hat{q}^2 + 2\lambda(t)\sqrt{mL\omega\Omega}\hat{x}\hat{q}. \tag{4}$$

Due to the coupling term (the last term) in the above equation, it may be somewhat difficult to treat this system from a quantum mechanical point of view. In order to overcome this, we will mathematically decouple the two sub-systems by means of the unitary transformation method [7,8,19]. For this purpose, we introduce a unitary operator of the form

$$\begin{aligned}\hat{U} = {} & \exp\left[\frac{i}{4\hbar}(\hat{p}_x\hat{x} + \hat{x}\hat{p}_x)\ln\left(\sqrt{m/L}\right)\right] \\ & \times \exp\left[\frac{i}{4\hbar}(\hat{p}_q\hat{q} + \hat{q}\hat{p}_q)\ln\left(\sqrt{L/m}\right)\right] \\ & \times \exp\left[-\frac{i\theta(t)}{\hbar}(\hat{p}_x\hat{q} - \hat{p}_q\hat{x})\right],\end{aligned} \tag{5}$$

where

$$\theta(t) = \frac{1}{2}\tan^{-1}\left(\frac{4\lambda(t)\sqrt{\omega\Omega}}{\Omega^2 - \omega^2}\right). \tag{6}$$

By transforming $\hat{H}$ in Equation (4) using this operator:

$$\hat{H}' = \hat{U}^{-1}\hat{H}\hat{U} - i\hbar\hat{U}^{-1}\frac{\partial\hat{U}}{\partial t}, \tag{7}$$

we have the Hamiltonian in the transformed system as

$$\hat{H}' = \frac{\hat{p}_x^2 + \hat{p}_q^2}{2\sqrt{mL}} + \frac{1}{2}\sqrt{mL}\mathrm{w}_1^2(t)\hat{x}^2 + \frac{1}{2}\sqrt{mL}\mathrm{w}_2^2(t)\hat{q}^2 + \dot{\lambda}(t)\beta(t)(\hat{p}_x\hat{q} - \hat{p}_q\hat{x}), \tag{8}$$

where

$$\mathrm{w}_1^2(t) = \omega^2\cos^2\theta(t) + \Omega^2\sin^2\theta(t) - 4\lambda(t)\sqrt{\omega\Omega}\cos\theta(t)\sin\theta(t), \tag{9}$$

$$\mathrm{w}_2^2(t) = \omega^2\sin^2\theta(t) + \Omega^2\cos^2\theta(t) + 4\lambda(t)\sqrt{\omega\Omega}\cos\theta(t)\sin\theta(t), \tag{10}$$

$$\beta(t) = -\frac{2\sqrt{\omega\Omega}}{\Omega^2 - \omega^2 + 16\omega\Omega\lambda^2(t)/(\Omega^2 - \omega^2)}. \tag{11}$$

According to the adiabatic condition which is that $\lambda(t)$ is a sufficiently slowly time-varying function, we can neglect the last term in Equation (8), leading to

$$\hat{H}' \simeq \frac{\hat{p}_x^2 + \hat{p}_q^2}{2\mu} + \frac{1}{2}\mu\mathrm{w}_1^2(t)\hat{x}^2 + \frac{1}{2}\mu\mathrm{w}_2^2(t)\hat{q}^2, \tag{12}$$

where $\mu = \sqrt{mL}$. Thus, the two sub-systems are decoupled as can be confirmed from the above Hamiltonian. The quantum treatment of the system relying on the transformed Hamiltonian, Equation (12), may be much simpler than that relying on the original Hamiltonian given in Equation (4). Two independent classical equations of motion that correspond to $\hat{H}'$ are given by

$$\frac{d^2x}{dt^2} + \mathrm{w}_1^2(t)x = 0, \tag{13}$$

$$\frac{d^2q}{dt^2} + \mathrm{w}_2^2(t)q = 0. \tag{14}$$

In the next section, we will derive the quantum wave solutions associated to the Hamiltonian given in Equation (12). By using the unitary relation between the wave functions in the transformed system and those in the original system, the quantum wave solutions in the original system will be obtained and analyzed.

3.2. Quantum Wave Solutions

From the knowledge of the formulae of quantum wave functions in the transformed system, we can obtain the wave functions in the original system because the two systems are connected by a unitary operator. If we denote the wave functions in the transformed system associated with Equations (13) and (14) as $\psi_n'(x,t)$ and $\tilde{\psi}_l'(q,t)$, respectively, they can be divided into kernel and phase parts such that

$$\psi_n'(x,t) = \phi_n'(x,t)\exp[i\alpha_n(t)], \tag{15}$$

$$\tilde{\psi}_l'(q,t) = \tilde{\phi}_l'(q,t)\exp[i\tilde{\alpha}_l(t)], \tag{16}$$

where $\alpha_n(t)$ and $\tilde{\alpha}_l(t)$ are time-dependent phases.

Let us write the corresponding Schrödinger equation as

$$i\hbar\frac{\partial\Psi'_{n,l}(x,q,t)}{\partial t} = \hat{H}'\Psi'_{n,l}(x,q,t), \tag{17}$$

where $\Psi'_{n,l}(x,q,t)$ are wave functions in the transformed system, that are of the form

$$\Psi'_{n,l}(x,q,t) = \psi_n'(x,t)\tilde{\psi}_l'(q,t). \tag{18}$$

Because the transformed system is composed of the two decoupled harmonic oscillators, we can easily identify the corresponding quantum solutions, Equation (18). Then, the wave functions $\Psi_{n,l}(x,q,t)$ in the original system are obtained from such solutions through the use of the unitary relation $\Psi_{n,l}(x,q,t) = \hat{U}\Psi'_{n,l}(x,q,t)$. According to this, the wave functions in the original system are represented as

$$\Psi_{n,l}(x,q,t) = \psi_n(x,q,t)\tilde{\psi}_l(x,q,t), \tag{19}$$

where

$$\psi_n(x,q,t) = \phi_n(x,q,t)\exp[i\alpha_n(t)], \tag{20}$$
$$\tilde{\psi}_l(x,q,t) = \tilde{\phi}_l(x,q,t)\exp[i\tilde{\alpha}_l(t)], \tag{21}$$

while

$$\phi_n(x,q,t) = \hat{U}\phi'_n(x,t), \tag{22}$$
$$\tilde{\phi}_l(x,q,t) = \hat{U}\tilde{\phi}'_l(q,t). \tag{23}$$

Let us now further see for the case that the coupling parameter is a positive real constant, $\lambda(t) = \lambda_0$. In this case, w_i ($i = 1,2$) and θ become constants. As a consequence, the corresponding quantum solutions are easily identified to be

$$\alpha_n(t) = -(n+1/2)\mathrm{w}_1 t, \tag{24}$$
$$\tilde{\alpha}_l(t) = -(l+1/2)\mathrm{w}_2 t, \tag{25}$$
$$\phi'_n(x) = \left(\frac{\sqrt{k_1/\pi}}{2^n n!}\right)^{1/2} H_n\left(\sqrt{k_1}x\right)\exp\left[-k_1x^2/2\right], \tag{26}$$
$$\tilde{\phi}'_l(q) = \left(\frac{\sqrt{k_2/\pi}}{2^l l!}\right)^{1/2} H_l\left(\sqrt{k_2}q\right)\exp\left[-k_2q^2/2\right], \tag{27}$$

where $k_i = \mu\mathrm{w}_i/\hbar$, and $H_{n(l)}$ are Hermite polynomials. The probability density, $|\Psi'_{n,l}(x,q,t)|^2$, in the transformed system is illustrated in Figure 1 with the choice of $(n, l) = (3, 5)$ under the limit that the coupling parameter is constant. This density is not deformed because the Hamiltonian of the transformed system does not involve the coupling term.

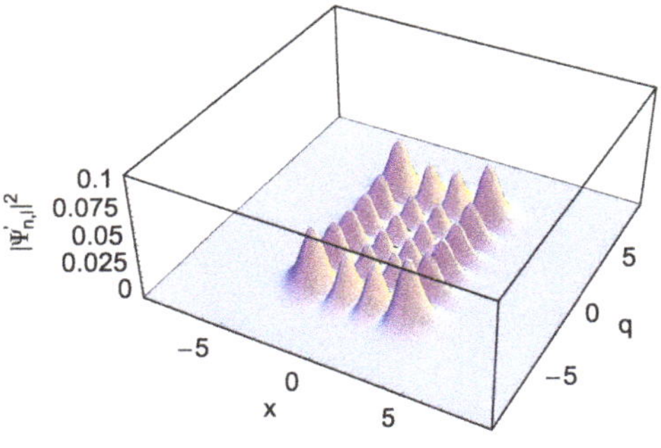

Figure 1. The probability density, $|\Psi'_{n,l}(x,q,t)|^2$, in the transformed system with the choice of $(m, L) = (1.0, 1.0)$ under the limit $\lambda(t) = \lambda_0$. This is associated to the wave functions, Equation (18) with Equations (15), (16), and (24)–(27). We used $(n, l) = (3, 5)$, $(\omega, \Omega) = (0.5, 0.49)$, $\hbar = 1$, and $\lambda_0 = 0.1$.

According to the relations, Equations (22) and (23), the eigenfunctions in the original system can be derived to be

$$\phi_n(x,q,t) = \left(\frac{\sqrt{k_1/\pi}}{2^n n!}\right)^{1/2} H_n\left(\sqrt{k_1}Q_1\right)\exp\left[-k_1Q_1^2/2\right], \tag{28}$$

$$\tilde{\phi}_l(x,q,t) = \left(\frac{\sqrt{k_2/\pi}}{2^l l!}\right)^{1/2} H_l\left(\sqrt{k_2}Q_2\right)\exp\left[-k_2Q_2^2/2\right], \tag{29}$$

where

$$Q_1 = \left(\frac{m}{L}\right)^{1/4}\cos\theta(t)x - \left(\frac{L}{m}\right)^{1/4}\sin\theta(t)q, \tag{30}$$

$$Q_2 = \left(\frac{m}{L}\right)^{1/4}\sin\theta(t)x + \left(\frac{L}{m}\right)^{1/4}\cos\theta(t)q. \tag{31}$$

Notice that the wave packets in the original system in the x-p coordinate have more or less been deformed depending on the values of parameters such as m and L, and rotated in proportion to θ from those in the transformed system. If θ is positive, the direction of rotation in the x-p plane is clockwise. The probability densities, $|\Psi_{n,l}(x,q,t)|^2$, are illustrated in Figures 2–4. By comparing these figures with that in the transformed system represented in Figure 1, we can confirm the effects of the coupling on the behavior of the wave functions. Figure 2 is for several different values of ω with the choice of $(m, L) = (1.0, 0.5)$ under the condition $\omega > \Omega$. The rotation angle θ can be evaluated from Equation (6) and is given by -0.76 rad for Figure 2A, -0.18 rad for Figure 2B, and -0.05 rad for Figure 2C.

We see from Figure 2 that the uncertainty of x is relatively small than that of q, which means that the wave packet is squeezed along the x coordinate. Such a squeezing effect becomes large as the difference $\omega - \Omega$ increases. On the other hand, Figure 3, which is for the case that $\omega - \Omega$ is negative, shows the squeeze of waves along the q-coordinate.

For the case of Figure 4 which is depicted under $(m, L) = (0.5, 2.0)$, the wave packets exhibit q-squeezing. However, a weak x-squeezing for the waves also takes place as ω increases (see Figure 4C).

For the case that $\lambda(t)$ is not a constant, it is necessary to use the quantum theory of time-dependent harmonic oscillators [9,23,24] in order to manage the Schrödinger equation, Equation (17) with Equation (12). According to that theory, the phases and the eigenfunctions in the transformed system are given in terms of time functions as [24]

$$\alpha_n(t) = -(n+1/2)\gamma_1(t), \tag{32}$$

$$\tilde{\alpha}_l(t) = -(l+1/2)\gamma_2(t), \tag{33}$$

$$\phi_n'(x,t) = \left(\frac{\sqrt{\kappa_1(t)/\pi}}{2^n n!}\right)^{1/2} H_n\left(\sqrt{\kappa_1(t)}x\right)\exp\left[-\frac{\kappa_1(t)}{2}\left(1-\frac{i\dot{s}_1(t)}{\dot{\gamma}_1(t)s_1(t)}\right)x^2\right], \tag{34}$$

$$\tilde{\phi}_l'(q,t) = \left(\frac{\sqrt{\kappa_2(t)/\pi}}{2^l l!}\right)^{1/2} H_l\left(\sqrt{\kappa_2(t)}q\right)\exp\left[-\frac{\kappa_2(t)}{2}\left(1-\frac{i\dot{s}_2(t)}{\dot{\gamma}_2(t)s_2(t)}\right)q^2\right], \tag{35}$$

where $\kappa_i = \mu\dot{\gamma}_i/\hbar$, while the time functions $s_i(t)$ and $\gamma_i(t)$ satisfy

$$\ddot{s}_i + \mathrm{w}_i^2(t)s_i - C_i^2/s_i^3 = 0, \tag{36}$$

$$\dot{\gamma}_i = C_i/s_i^2, \tag{37}$$

where C_i are arbitrary real constants.

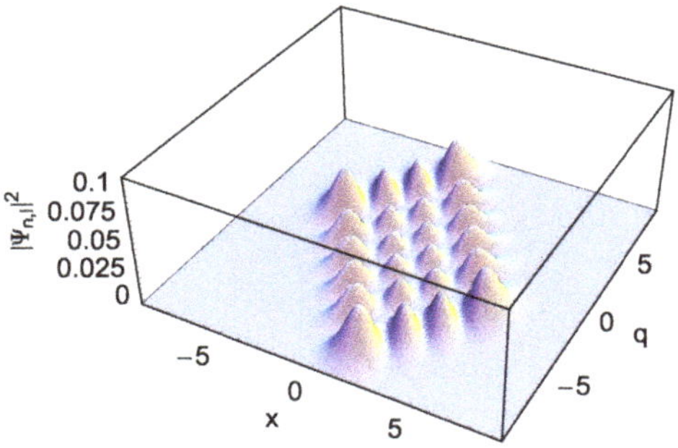

A

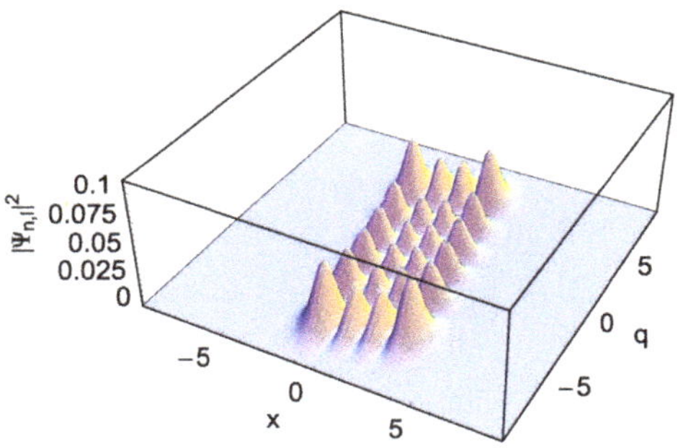

B

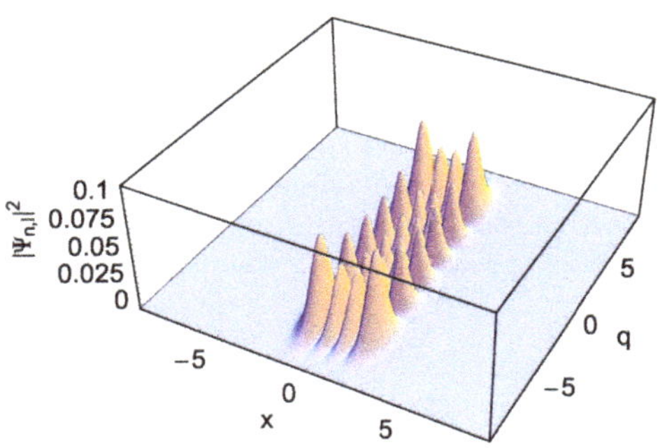

C

Figure 2. The probability density $|\Psi_{n,l}(x,q,t)|^2$ in the original system for several different values of ω with the choice of $(m, L) = (1.0, 0.5)$ under the limit $\lambda(t) = \lambda_0$. This is associated to the wave functions given in Equation (19) with Equations (20), (21), (24), (25), (28), and (29). The values of (ω, Ω) are (0.5, 0.49) for (**A**), (1.0, 0.49) for (**B**), and (2.0, 0.49) for (**C**). We used $(n, l) = (3, 5)$, $\hbar = 1$, and $\lambda_0 = 0.1$.

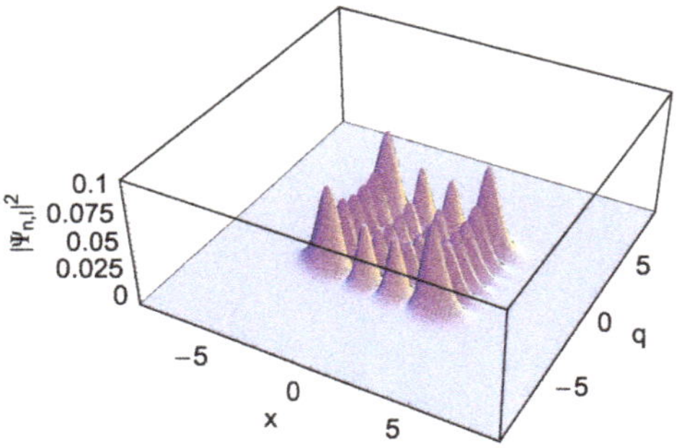

Figure 3. This figure is the same as Figure 2C, but for a different choice of angular frequencies. The values of (ω, Ω) are (0.49, 2.0).

From a minor evaluation through Equations (22) and (23), we have the wave solutions in the original system:

$$\phi_n(x,q,t) = \left(\frac{\sqrt{\kappa_1(t)/\pi}}{2^n n!}\right)^{1/2} H_n\left(\sqrt{\kappa_1(t)}Q_1\right) \exp\left[-\frac{\kappa_1(t)}{2}\left(1-\frac{i\dot{s}_1(t)}{\dot{\gamma}_1(t)s_1(t)}\right)Q_1^2\right], \quad (38)$$

$$\tilde{\phi}_l(x,q,t) = \left(\frac{\sqrt{\kappa_2(t)/\pi}}{2^l l!}\right)^{1/2} H_l\left(\sqrt{\kappa_2(t)}Q_2\right) \exp\left[-\frac{\kappa_2(t)}{2}\left(1-\frac{i\dot{s}_2(t)}{\dot{\gamma}_2(t)s_2(t)}\right)Q_2^2\right], \quad (39)$$

where Q_i are given by Equations (30) and (31), but in terms of $\lambda(t)$ which is not a constant.

The overall wave function is given by

$$\Psi(x,q,t) = \sum_{n=0}^{\infty}\sum_{l=0}^{\infty} c_{n,l}\Psi_{n,l}(x,q,t), \quad (40)$$

where $c_{n,l}$ are complex numbers that yield $\sum_{n=0}^{\infty}\sum_{l=0}^{\infty}|c_{n,l}|^2 = 1$. Thus, we have obtained the full wave function of the system which is given by Equation (40) with Equations (19)–(21), (32), (33), (38), and (39). This wave function is crucial in unfolding quantum theory of the system and can be used to investigate diverse quantum characteristics that the system exhibits. We can use the wave function in estimating various quantum variables, such as energy eigenvalues, expectation values and fluctuations of the canonical variables, uncertainty products, Wigner distribution functions, and so on.

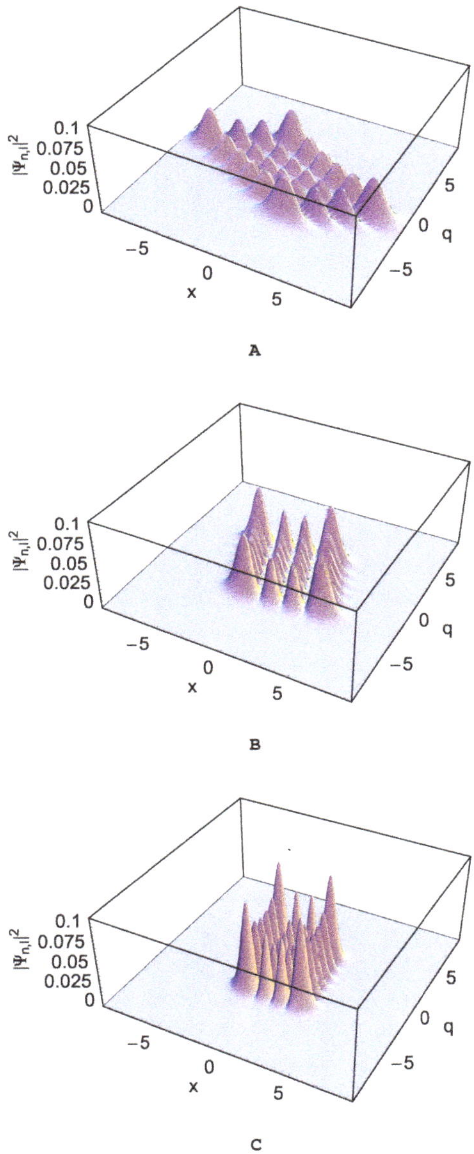

Figure 4. The probability density $|\Psi_{n,l}(x,q,t)|^2$ in the original system for several different values of ω with the choice of $(m, L) = (0.5, 2.0)$ under the limit $\lambda(t) = \lambda_0$. This is associated to the wave functions given in Equation (19) with Equations (20), (21), (24), (25), (28), and (29). The values of (ω, Ω) are (0.5, 0.49) for (**A**), (1.0, 0.49) for (**B**), and (2.0, 0.49) for (**C**). We used $(n, l) = (3, 5)$, $\hbar = 1$, and $\lambda_0 = 0.1$.

4. Conclusions

The quantum properties of the nanomechanical resonator coupled to a superconducting resonator via a small time-varying coupling constant have been investigated. We have used an adiabatic condition under the assumption that the time-variation of the coupling parameter is sufficiently slow. Due to the presence of the coupling term in the Hamiltonian, we cannot develop quantum theory for x and q coordinates independently in the original system. Hence, by means of the unitary transformation approach, we have decoupled x and q coordinates from each other in the expression of the Hamiltonian.

Through the transformation performed using a unitary operator, the Hamiltonian became a manageable one. Not only did the coupling term no longer appear in the transformed Hamiltonian, but the transformed system has also become much more simplified compared to the original one. More precisely speaking, the transformed system is composed of two independent harmonic oscillators with time-dependent angular frequencies. We accordingly have easily identified quantum solutions in the transformed system. By inverse transformation of those quantum solutions, we have obtained complete quantum solutions in the original system.

The exact wave functions derived here can be used for evaluating various quantum quantities of the coupled resonators, such as expectation values and fluctuations of the canonical variables, uncertainties, and energy eigenvalues, which are necessary for identifying the quantum characteristics of the system. It is also possible to derive Wigner distribution function of the system, which serves in describing signal processing, from the wave functions given in the text. Wigner distribution function can be used not only in estimating quantum corrections from classical statistical mechanics but also in demonstrating nonclassicalities of the system through quantum probability distribution. Terraneo et al. used Wigner distribution functions for the purpose of developing an efficient way for extracting information from the wave functions of quantum algorithms associated with quantum computation [25].

The probability densities which are the absolute square of the wave functions were illustrated in detail in the limit that λ is a constant. The wave packets in the x-q plane are rotated and deformed compared to those in the transformed system. The rotation direction is counterclockwise when $\omega - \Omega$ is positive. Due to the coupling of the two sub-systems, the wave packets of the system underwent x-squeezing, q-squeezing, or both depending on the values of m, L, and the frequency difference $\omega - \Omega$. For the situation of x-squeezing (q-squeezing), the wave amplitude of the nanomechanical (the superconducting) resonator is small and, as a consequence, the nanomechanical (the superconducting) resonator has less energy. The energy of the nanomechanical resonator flows to the superconducting resonator, and vice versa [10,18]. From this process, the superconducting resonator extracts information about the nanomechanical resonator in order to control the nanoresonator.

The techniques for sideband cooling, frequency conversion, and state-swapping through coherent control of the resonator are important in quantum information systems, especially, in the realization of quantum computers. For the purpose of developing such techniques, the use of nanomechanical resonators is more effective than the use of nanophotonic resonators [11]. It is also noticeable that quantum data bus in quantum computers could be designed using nanomechanical resonators through their controllable coupling with qubits [2,26].

Our analysis regarding the wave functions of the system is useful for understanding the quantum behavior of the combined system. If we consider that theoretical analysis of a quantum system starts from the wave functions rigorously derived from the Schrödinger equation, the wave functions developed in this work are the basic tool for elucidating quantum properties of the system. Exact analyses of the quantum behaviors of nanoresonators based on such clarified quantum properties are necessary for realizing efficient quantum information processing.

The advantage of information processing and physical simulations using quantum computing devices rather than classical computers is that they make it possible to solve time-consuming problems within a polynomial time, such as factorizing large numbers which is classically very difficult to solve. The theoretical research for demonstrating the characteristics of quantum devices and their operations are important stepping stones for the development of quantum information science. While the system we treated here is a completely solvable one using our method, we also expect that quantum characteristics of more complicated namomecanical systems, including nanoresonators described by higher-order nonlinearity such as Duffing nonlinearity [27] and/or newly designed nanophotonic resonators [28,29], would be analyzed in the near future using the same method. Nanomechanical resonators are used for the detection of quantum states, spins, thermal fluctuations, etc., whereas nanophotonic resonators as passive optical components are mainly used for detecting light–matter interactions.

The coupled system investigated in this work can be used as a basic component of quantum computing systems. Quantum computation as a next generation technology should be developed together with other quantum information science such as quantum communication and quantum cryptography. Moreover recent development of neural computation through neural network with multi-core optical fibers [30,31] enhances the performance of the technology in quantum information science.

As a next task, the investigation of geometric phases that appear in the wave functions when parameters vary over time may be a good research topic which can be fulfilled on the basis of the quantum theory developed in the present work. Geometric phase can be applied not only to fundamental physics [23] but also to various next generation quantum technologies, such as quantum computing [32], interferometric imaging of microstructures [33], and beam steering in virtual/augmented reality displays [34].

Author Contributions: J.R.C. proposed the idea of the research. J.R.C. and S.J. co-wrote the manuscript. The graphics in the text have been prepared by J.R.C. All authors read and approved the final manuscript.

Funding: This research was supported by the Basic Science Research Program of the year 2018 through the National Research Foundation of Korea (NRF) funded by the Ministry of Education (Grant No.: NRF-2016R1D1A1A09919503).

Conflicts of Interest: The authors declare no conflict of interest.

References

1. Favero, I.; Marquardt, F. Focus on optomechanics. *New J. Phys.* **2014**, *16*, 085006. [CrossRef]
2. Rabl, P.; Kolkowitz, S.J.; Koppens, F.H.L.; Harris, J.G.E.; Zoller, P.; Lukin, M.D. A quantum spin transducer based on nanoelectromechanical resonator arrays. *Nat. Phys.* **2010**, *6*, 602–608. [CrossRef]
3. Tetard, L.; Passian, A.; Venmar, K.T.; Lynch, R.M.; Voy, B.H.; Shekhawat, G.; Dravid, V.P.; Thundat, T. Imaging nanoparticles in cells by nanomechanical holography. *Nat. Nanotechnol.* **2008**, *3*, 501–505. [CrossRef] [PubMed]
4. Sawadsky, A.; Kaufer, H.; Nia, R.M.; Tarabrin, S.P.; Khalili, F.Y.; Hammerer, K.; Schnabel, R. Observation of generalized optomechanical coupling and cooling on cavity resonance. *Phys. Rev. Lett.* **2015**, *114*, 043601. [CrossRef] [PubMed]
5. LaHaye, M.D.; Buu, O.; Camarota, B.; Schwab, K.C. Approaching the quantum limit of a nanomechanical resonator. *Science* **2004**, *304*, 74–77. [CrossRef] [PubMed]
6. Woolley, M.J.; Milburn, G.J.; Caves, C.M. Nonlinear quantum metrology using coupled nanomechanical resonators. *New J. Phys.* **2008**, *10*, 125018. [CrossRef]
7. Liao, J.-Q.; Law, C.K.; Kuang, L.-M.; Nori, F. Enhancement of mechanical effects of single photons in modulated two-mode optomechanics. *Phys. Rev. A* **2015**, *92*, 013822. [CrossRef]
8. Wang, D.-Y.; Bai, C.-H.; Liu, S.; Zhang, S.; Wang, H.-F. Optomechanical cooling beyond the quantum backaction limit with frequency modulation. *Phys. Rev. A* **2018**, *98*, 023816. [CrossRef]
9. Choi, J.R. Hamiltonian dynamics and adiabatic invariants for time-dependent superconducting qubit-oscillators and resonators in quantum computing systems. *Adv. Math. Phys.* **2015**, *2015*, 120573. [CrossRef]
10. Tian, L. Ground state cooling of a nanomechanical resonator via parametric linear coupling. *Phys. Rev. B* **2009**, *79*, 193407. [CrossRef]
11. Jacobs, K.; Nurdin, H.I.; Strauch, F.W.; James, M. Frequency conversion: Side-band cooling, state-swapping, and coherent control of mechanical resonators. *arXiv* **2010**, arXiv:1003.2653v3.
12. Cleland, A.N.; Geller, M.R. Mechanical quantum resonators. *AIP Conf. Proc.* **2005**, *786*, 396–400.
13. Pechal, M.; Arrangoiz-Arriola, P.; Safavi-Naeini, A.H. Superconducting circuit quantum computing with nanomechanical resonators as storage. *Quantum Sci. Technol.* **2018**, *4*, 015006. [CrossRef]

14. Armour, A.D.; Blencowe, M.P. Probing the quantum coherence of a nanomechanical resonator using a superconducting qubit: I. Echo scheme. *New J. Phys.* **2008**, *10*, 095004. [CrossRef]
15. O'Connell, A.D.; Hofheinz, M.; Ansmann, M.; Bialczak, R.C.; Lenander, M.; Lucero, E.; Neeley, M.; Sank, D.; Wang, H.; Weides, M.; et al. Quantum ground state and single-phonon control of a mechanical resonator. *Nature* **2010**, *464*, 697–703. [CrossRef] [PubMed]
16. Teufel, J.D.; Donner, T.; Li, D.; Harlow, J.W.; Allman, M.S.; Cicak, K.; Sirois, A.J.; Whittaker, J.D.; Lehnert, K.W.; Simmonds, R.W. Sideband cooling of micromechanical motion to the quantum ground state. *Nature* **2011**, *475*, 359–363. [CrossRef] [PubMed]
17. Safavi-Naeini, A.H.; Chan, J.; Hill, J.T.; Alegre, T.P.M.; Krause, A.; Painter, O. Observation of quantum motion of a nanomechanical resonator. *Phys. Rev. Lett.* **2012**, *108*, 033602. [CrossRef]
18. Jacobs, K.; Nurdin, H.I.; Strauch, F.W.; James, M. Comparing resolved-sideband cooling and measurement-based feedback cooling on an equal footing: Analytical results in the regime of ground-state cooling. *Phys. Rev. A* **2015**, *91*, 043812. [CrossRef]
19. Choi, J.R. Exact solution of a quantized LC circuit coupled to a power source. *Phys. Scr.* **2006**, *73*, 587–595. [CrossRef]
20. Xiang, Z.-L.; Ashhab, S.; You, J.Q.; Nori, F. Hybrid quantum circuits: Superconducting circuits interacting with other quantum systems. *Rev. Mod. Phys.* **2013**, *85*, 623–653. [CrossRef]
21. Peropadre, B.; Zueco, D.; Wulschner, F.; Deppe, F.; Marx, A.; Gross, R.; García-Ripoll, J.J. Tunable coupling engineering between superconducting resonators: From sidebands to effective gauge fields. *Phys. Rev. B* **2013**, *87*, 134504. [CrossRef]
22. Fedortchenko, S.; Felicetti, S.; Marković, D.; Jezouin, S.; Keller, A.; Coudreau, T.; Huard, B.; Milman, P. Quantum simulation of ultrastrongly coupled bosonic modes using superconducting circuits. *Phys. Rev. A* **2017**, *95*, 042313. [CrossRef]
23. Lakehal, H.; Maamache, M.; Choi, J.R. Novel quantum description for nonadiabatic evolution of light wave propagation in time-dependent linear media. *Sci. Rep.* **2016**, *6*, 19860. [CrossRef] [PubMed]
24. Khandekar, D.C.; Lawande, S.V. Exact propagator for a time-dependent harmonic oscillator with and without a singular perturbation. *J. Math. Phys.* **1975**, *16*, 384–388. [CrossRef]
25. Terraneo, M.; Georgeot, B.; Shepelyansky, D.L. Quantum computation and analysis of Wigner and Husimi functions: Toward a quantum image treatment. *Phys. Rev. E* **2005**, *71*, 066215. [CrossRef] [PubMed]
26. Xue, F.; Wang, Y.D.; Sun, C.P.; Okamoto, H.; Yamaguchi, H.; Semba, K. Controllable coupling between flux qubit and nanomechanical resonator by magnetic field. *New J. Phys.* **2007**, *9*, 35. [CrossRef]
27. Lü, X.-Y.; Liao, J.-Q.; Tian, L.; Nori, F. Steady-state mechanical squeezing in an optomechanical system via Duffing nonlinearity. *Phys. Rev. A* **2015**, *91*, 013834. [CrossRef]
28. Malka, D.; Cohen, M.; Turkiewicz, J.; Zalevsky, Z. Optical micro-multi-racetrack resonator filter based on SOI waveguides. *Photonics Nanostruct. Fundam. Appl.* **2015**, *16*, 16–23. [CrossRef]
29. McGehee, W.R.; Michels, T.; Aksyuk, V.; McClelland, J.J. Two-dimensional imaging and modification of nanophotonic resonator modes using a focused ion beam. *Optica* **2017**, *4*, 1444–1450. [CrossRef]
30. Cohen, E.; Malka, D.; Shemer, A.; Shahmoon, A.; Zalevsky, Z.; London, M. Neural networks within multi-core optic fibers. *Sci. Rep.* **2016**, *6*, 29080. [CrossRef]
31. Shabairou, N.; Cohen, E.; Wagner, O.; Malka, D.; Zalevsky, Z. Color image identification and reconstruction using artificial neural networks on multimode fiber images: Towards an all-optical design. *Opt. Lett.* **2018**, *43*, 5603–5606. [CrossRef] [PubMed]
32. Liu, T.; Cao, X.-Z.; Su, Q.-P.; Xiong, S.-J.; Yang, C.-P. Multi-target-qubit unconventional geometric phase gate in a multi-cavity system. *Sci. Rep.* **2016**, *6*, 21562. [CrossRef] [PubMed]
33. Tavrov, A.V.; Miyamoto, Y.; Kawabata, T.; Takeda, M.; Andreev, V.A. Interferometric microimaging based on geometrical spin-redirection phase. *Opt. Lett.* **2000**, *25*, 460–462. [CrossRef] [PubMed]
34. Lee, Y.-H.; Tan, G.; Zhan, T.; Weng, Y.; Liu, G.; Gou, F.; Peng, F.; Tabiryan, N.V.; Gauza, S.; Wu, S.-T. Recent progress in Pancharatnam-Berry phase optical elements and the applications for virtual/augmented realities. *Opt. Data Process. Storage* **2017**, *3*, 79–88. [CrossRef]

nanomaterials

MDPI

Article

On-Chip Real-Time Chemical Sensors Based on Water-Immersion-Objective Pumped Whispering-Gallery-Mode Microdisk Laser

Qijing Lu [1], Xiaogang Chen [1], Liang Fu [1], Shusen Xie [1,*] and Xiang Wu [2,*]

1 Key Laboratory of Optoelectronic Science and Technology for Medicine of Ministry of Education, Provincial Key Laboratory for Photonics Technology, Institute of Laser and Optoelectronics Technology, Fujian Normal University, Fuzhou 350007, China; qjlu@fjnu.edu.cn (Q.L.); xgchen01@139.com (X.C.); fuliang1995@126.com (L.F.)

2 Department of Optical Science and Engineering, Fudan University, Shanghai 200433, China

* Correspondence: ssxie@fjnu.edu.cn (S.X.); wuxiang@fudan.edu.cn (X.W.); Tel.: +86-1386-061-6010 (S.X.); +86-1381-609-4633 (X.W.)

Received: 21 February 2019; Accepted: 19 March 2019; Published: 24 March 2019

Abstract: Optical whispering-gallery-mode (WGM) microresonator-based sensors with high sensitivity and low detection limit down to single unlabeled biomolecules show high potential for disease diagnosis and clinical application. However, most WGM microresonator-based sensors, which are packed in a microfluidic cell, are a "closed" sensing configuration that prevents changing and sensing the surrounding liquid refractive index (RI) of the microresonator immediately. Here, we present an "open" sensing configuration in which the WGM microdisk laser is directly covered by a water droplet and pumped by a water-immersion-objective (WIO). This allows monitoring the chemical reaction progress in the water droplet by tracking the laser wavelength. A proof-of-concept demonstration of chemical sensor is performed by observing the process of salt dissolution in water and diffusion of two droplets with different RI. This WIO pumped sensing configuration provides a path towards an on-chip chemical sensor for studying chemical reaction kinetics in real time.

Keywords: whispering-gallery-mode (WGM); sensor; water-immersion-objective (WIO); microdisk; laser

1. Introduction

Light whispering-gallery-modes (WGMs) in microresonators, which are the analogy of acoustic waves travelling in a closed concave surface such as St. Paul's Cathedral, support ultrasmall mode volumes and ultrahigh quality (Q) factors. WGM microresonators with high local field intensities make them an excellent platform for enhancing the interaction between light and matter, based on which many applications have been demonstrated over the past two decades, such as low-threshold lasers [1,2], opto-mechanics [3,4], integrated optical devices [5,6], non-linear optics processes [7–9], and optical sensors [10–12]. Particularly, the WGM microresonator with strong evanescent waves has been proven as a versatile label-free biochemical sensor [13–17] with high sensitivity and low detection limit down to single molecules and nanoparticles by monitoring the shift [18], split [16,19–21], or broadening [22] of the resonance spectra. Various types of WGM microresonators including microsphere [18], microtoroid [16], microbubble [23,24], microdisk/ring [25], and microcapillaries [26,27] have been used in biosensing experiments; they can be catalogued into passive resonators and active resonators. In passive resonator-based sensor configuration, the resonator is driven by a waveguide, such as a relatively fragile tapered fiber or prism. The relative position between them must be adjusted precisely by a nano-positioning platform that makes the sensing system costly and cumbersome. While in active resonator sensor configuration, the fluorescence or laser spectrum is collected by free-space optics that makes the sensing system relatively practical.

In most WGM microresonator (both for passive and active) based biochemical sensor configurations, the microcavity is immersed in water and sealed in a microfluidic cell. This sensing configuration is a closed system. To change the refractive index (RI) of liquid around the microresonator, a micro syringe pump is used to pump the biochemical sample under test through the microfluidic channel. Thus, this sensing configuration cannot change the RI of liquid around the microresonator and sensing it immediately because sensing samples must be crossed through the microfluidic channel. Here, we report on the demonstration of a biochemical sensor based on a water-immersion-objective (WIO) pumped on-chip microdisk laser. This WIO pumped configuration is an open sensing system. The microdisk is directly immersed in the water environment that is what the WIO requires. By monitoring the laser wavelength, any feeble disturbance of the RI of the water can be detected in real time. Thus, the WIO pumped configuration provides an excellent platform for monitoring the RI change of the water that is induced by chemical reactions. In this paper, we demonstrate that the salt dissolution in water process and diffusion process of two droplets with different RI can be tracked by the lasing mode wavelength in real time. We believe that this work opens up a novel on-chip biochemical sensor based on a WIO pumped configuration, which provides potential accesses to study chemical reaction kinetics.

2. Laser Operation

2.1. Optical System Setup

The optical system setup is shown in Figure 1a, which is a typical optically pumped microcavity laser setup, except that a water-immersion objective (WIO) instead of conventional air objective is used. The pump laser is focused on the active microdisk resonator, which is directly immersed in water, through this WIO (LUMPlanFLN 40× W, Olympus, Tokyo, Japan). In this study, an active SU-8 microdisk resonator doped with Rhodamine B (RhB) is used for the experiment. The microdisk laser array (Figure 1b) is fabricated on silica-on-silicon chip and the detailed fabrication process can be found in previous studies [28,29]. The diameter and thickness of the microdisk are controlled to 40 μm and 1 μm, respectively. To avoid thermal wavelength drift induced by the fluctuation of the lab temperature, the sample chip is mounted on a home-made TEC (TZET Co. Ltd., Tianjin, China). The TEC is connected to a temperature controller (TED4015, Thorlabs Inc., Newton, NJ, USA) with a controlling precision of 2 mK. The pump light used here is an ns-pulsed OPO laser with a repetition rate of 10 Hz. The wavelength of pump laser is tuned to 532 nm for matching the absorption band of RhB. The laser emitted from the active microdisk is collected through free-space and recorded by the monochromator (HORIBA iHR550, Kyoto, Japan) equipped with a cooled CCD detector. A 1200 mm^{-1} grating is installed in the monochromator.

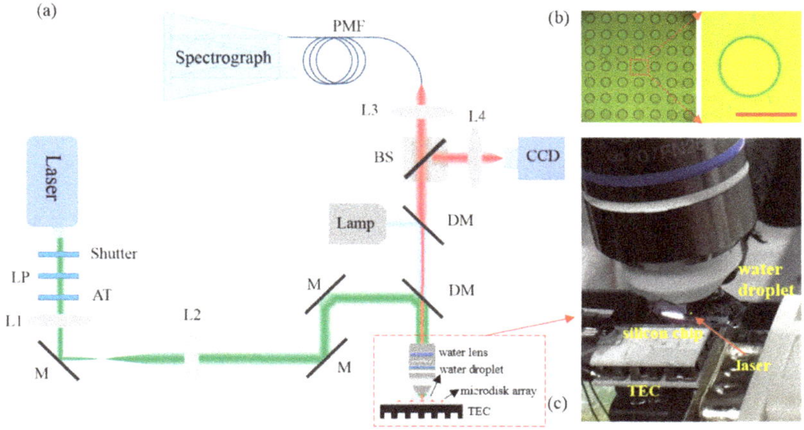

Figure 1. Optical system setup. (**a**) Schematic of lasing chemical sensor pumped by water-immersion-objective (WIO). LP: linear polarizer; AT: attenuator; L: lens; M: mirror; DM: dichroic mirror; BS: beam splitter; PMF: polarization-maintaining fiber. (**b**) Optical images of microdisk array and single microdisk, the scale bar is 40 μm. (**c**) Close-up image of the microdisk laser-based chemical sensor under WIO. The silicon chip is mounted on a home-made TEC which is connected to a temperature controller (Thorlabs, TED4015).

2.2. Threshold Measurement and Spectral Identification

The threshold of the laser emission is firstly measured by tuning the power of pump light and the light emission form the active microdisk is recorded correspondingly. Both the air objective (LMPlanFLN 20×, Olympus, Tokyo, Japan) pumped configuration and WIO pumped configuration are performed for comparison; the result is shown in Figure 2. Under both configurations, there are obvious kinks (Figure 2a,b), which indicate that the lasing thresholds are about 5.2 $\mu J/mm^2$ and 13.9 $\mu J/mm^2$, respectively. The linewidth of the laser peak above the threshold is about 0.08 nm. The corresponding Q factor is about 7800. However, this is not the real Q factor which is limited by the grating of the monochromator. The laser threshold pumped by WIO is higher than that pumped by the air objective. This is attributed to that the RI contrast is lower, which increases radiation loss in the WIO pumped configuration.

To verify this, the eigenmodes of a microdisk surrounded by air and water are calculated by Finite Element Method (FEM) with COMSOL Multiphysics 3.5a [30]. The perfectly matched layers (PMLs) are introduced to accurately calculate the radiation-loss-related Q_{rad}. The Q_{rad} of the fundamental radial mode (radial quantum number $p = 1$) is 1.5×10^8 when the microdisk is exposed in air, while it drops to 2.4×10^7 when the microdisk is immersed in water.

The Q_{rad} and field distribution of different radial modes are calculated and plotted in Figure 3. The Q_{rad} decreases exponentially with the increment of p. Thus, we can conclude that the laser peaks shown in Figure 2c,d are the fundamental modes with adjacent azimuthal mode numbers. The free spectral range (FSR) is slightly larger in the WIO pumped configuration, because of the RI increment of surrounding materials of the microdisk. Little discrepancy between the calculated FSR and measured FSR may result from the inaccuracy of the microdisk diameter during the fabrication process.

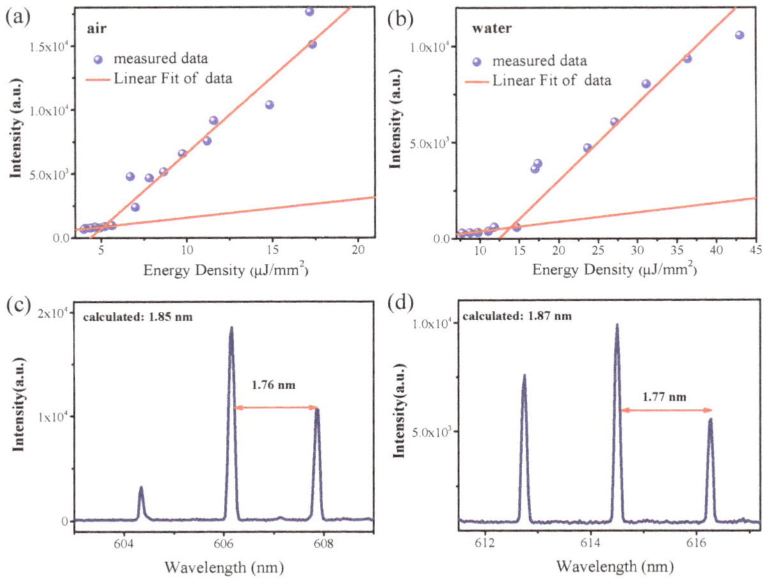

Figure 2. Comparisons of the output laser intensity as a function of the energy density (light-light curve) around the single mode lasing wavelength pumped by (**a**) air objective and (**b**) WIO, respectively. Typical laser spectrum pumped by air objective (**c**) and WIO (**d**), respectively. Calculated free spectral range (FSR) and measured FSR are both denoted.

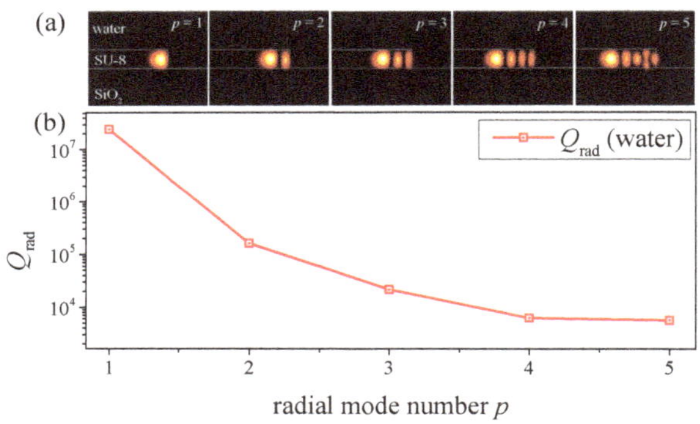

Figure 3. (**a**) The field distributions for different radial mode numbers. (**b**) Calculated radiation-loss-related Q_{rad} for different radial mode numbers.

3. Basic Element Sensing

3.1. Bulk Refractive Index Sensitivity (BRIS)

Under the WIO pumped configuration, the microdisk is directly immersed in a water environment, so the lasing spectra will be quite sensitive to the RI of the water. The BRIS of the microdisk laser is tested by directly changing the RI of the water droplet between the WIO and the silicon chip. Different RI solutions are obtained by dissolving different masses of NaCl into deionized water. Five NaCl solutions with concentrations of 1, 2.05, 3, 4.03, and 4.94 mol/liter are prepared for testing as shown in Figure 4a. To change in advance each droplet between the WIO and silicon chip, the sample is washed by the deionized water at least twice to eliminate the error of the droplet RI.

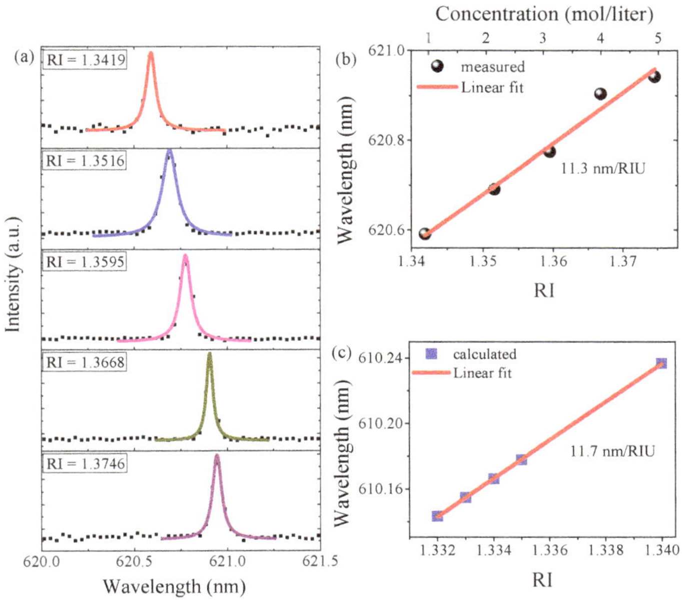

Figure 4. (**a**) Lasing spectrum shift by slightly increasing the refractive index (RI) of the water droplet. Insets show the RI of the NaCl solutions. (**b**) Measured wavelengths of the lasing modes as a function of the NaCl concentration (top x-axis) and RI of the water droplet (bottom x-axis); the measured data is linear fitted. (**c**) The calculated RI sensitivity.

The lasing spectra are recorded for each droplet, respectively, which are shown in Figure 4a. The central lasing wavelength is obtained by fitting the lasing peaks with Lorentz function. The lasing wavelength shift is plotted as a function of RI (Figure 4b). By linear fitting the measured wavelength data, the slope, i.e., the BRIS is 11.3 nm/RIU.

The wavelengths of the eigenmodes are calculated by changing the RI of the surrounding environment of the microdisk in the simulation model, and are also plotted over RI in Figure 4c. As shown, the calculated BRIS is 11.7 nm/RIU. The measured BRIS matches very well with this calculated value. The relatively high BRIS guarantees the performance of microdisk laser as a chemical sensor.

3.2. Thermal Sensing

SU-8 material exhibits high thermal optics (TO) coefficient of -3.5×10^{-4} K^{-1} [31], which is higher than that of materials that are commonly used to fabricate WGM resonators, such as silica (1×10^{-5} K^{-1}) [32] and silicon (1.8×10^{-4} K^{-1}) [33]. Combining with on-chip integration, ease of laser probing and readout based on free-space optics against passive WGM thermal sensors using relatively fragile tapered microfiber [34,35], the fabricated all-polymer microdisk laser provides a good platform for ultrasensitive thermal sensing. Thus, the thermal sensing of the SU-8 microdisk laser is performed.

The laser emission spectra are recorded in real time when changing the TEC temperature by a step of 1 K. At each temperature, the spectra are recorded for about 2 mins. A home-made code is written to track the spectral positions of the laser peaks over time. The result of spectral shift over time with different temperatures is shown in Figure 5a, where the spectral shift (triangle dots) exhibits step change to the temperature. However, at each constant temperature, the lasing wavelength minor shifts linearly which can be seen from the enlarged view of spectral shift at 20 °C (inset in Figure 5a). This blue shift of laser wavelength is attributed to the decreased RI of the polymer-dye composites due to photobleaching [36–39]. As the spectral shift is linear over time [38], the spectral shift can be corrected by eliminating its slope (red fitting line in the inset of Figure 5a). The corrected spectral shift (circular dots) is almost flat over time at each constant temperature.

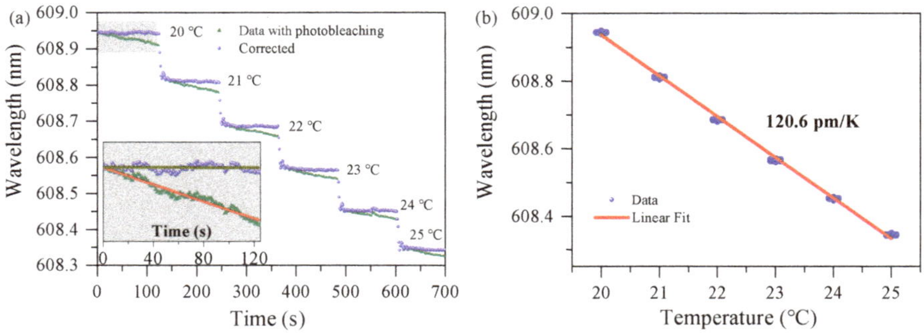

Figure 5. Temperature dependence of wavelengths of lasing mode. (**a**) Lasing emission wavelength shift with photobleaching and corrected data for photobleaching when changing the TEC temperature at a periodic interval. Inset shows the enlarged view at 20 °C. Brown and red line are the linear fitting curve. (**b**) Tracked laser wavelength with error bar vs temperature.

The average values of laser wavelength with error bar for each temperature is plotted in Figure 5b. The linear fit result indicates the thermal sensitivity is as high as 120.6 pm/K and has benefited from the high TO coefficient. Here the thermal expansion effect is ignored since its coefficient of ~ 10^{-6} K^{-1} is much lower than TO coefficient [40].

For bio-chemical sensing, the temperature fluctuation and photobleaching will cause the resonance wavelength drift and influence the quantification of the proposed analytes. To eliminate the effect of temperature fluctuation, the sensor chip is mounted on a TEC for stabilizing the temperature. The wavelength drift caused by the photobleaching can be corrected by eliminating its slope. Recently, S. F. Wondimu also reported a novel scheme that allows for simultaneous compensation of temperature drift and photobleaching by using microdisk laser arrays with on-chip references [39].

4. Chemical Sensor

As the microdisk is directly immersed in the water droplet and the microdisk sensor exhibits high sensitivities, so the lasing mode can quickly sense the local change (e.g., RI) of the water droplet in real time. This provides an excellent platform for studying some chemical reactions in water, e.g., dissolution and diffusion process, by monitoring the laser wavelength. Thus in the following study, a proof-of-principle chemical sensor is demonstrated based on a WIO pumped configuration by adding salt crystals or NaCl solution into the water droplet.

4.1. Salt Dissolution Process in Water

In this section, adding a certain quality of salt crystal into the water droplet between the WIO and silicon chip is performed. The water droplet with a fixed volume of 0.2 mL is firstly injected into the space between the WIO and silicon chip by a pipette. By tuning one microdisk just under the focus of pump light, the laser is generated above the threshold and the emission spectrum is automatically recorded with integration time of 1 s. Then salt crystal with a mass of 14 mg is cast into the edge of the droplet with a tweezer. In this process, the acquisition of emission spectra is ongoing for about 7 mins.

After data acquisition, one laser peak is monitored and its spectral wavelength shift is shown in Figure 6a (triangle dots). Slow linear change of wavelength induced by photobleaching is observed, after correction, and the spectral shift is denoted by circular dots. Before casting salt into the droplet, the lasing wavelength is quiet, which can be seen from Figure 6a (circular dots). After casting salt into the droplet, the lasing wavelength is red shifted dramatically. Then the lasing wavelength blue shifts slowly and finally approaches to equilibrium over 400 s. The whole process is explained below.

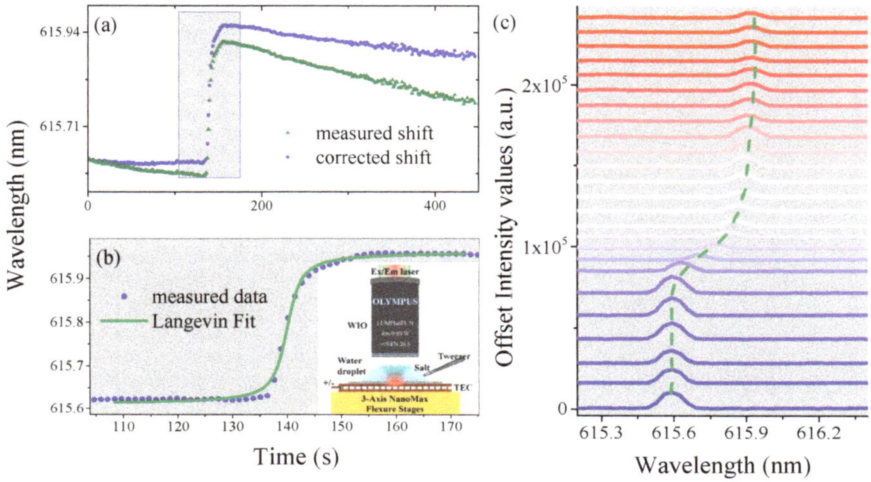

Figure 6. Monitoring the process of dissolution of salt in water using the lasing chemical sensor. (**a**) Measured lasing wavelength and corrected lasing wavelength shifts when adding the 14 mg salt in water under WIO. (**b**) Close-up view of corrected lasing wavelength and Langevin fitting. Inset shows the schematic diagram of operation process. (**c**) Dynamic change of lasing spectra in the process of dissolution of salt in water. The blue line (initial position) to red line (end position) is guided by the green dotted line.

Once the salt is cast into the droplet, the salt will dissolve and diffuse in the water, which induces the increment of RI of the droplet. As a result, the spectrum shifts to longer wavelength dramatically. Since the salt crystal is cast in one side of the droplet, there exists a gradient in concentration of NaCl solution from this side to the other side. Therefore, the RI in the surrounding volume of the lasing microdisk first increases and then decreases with the progress of diffusion. Consequently, the spectrum starts to shift slowly to shorter wavelength and finally no longer changes at the end of diffusion process (> 400 s), as shown in Figure 6a.

A close-up view of the dramatic change of laser wavelength is shown in Figure 6b. The trajectory of spectral shift is fitted well by Langevin function, which is written as: $L(t) = \coth(t) - 1/t$, where the "coth" is the hyperbolic cotangent, defined as $\coth(t) = (e^t + e^{-t})/(e^t - e^{-t})$. The dynamic change of the laser spectra is also plotted in Figure 6c (guided by the dotted line).

We found that the spectral does not shift as soon as the salt is cast into the droplet. From adding the salt into the droplet, the laser wavelength is almost stable for a certain time until the wavelength shifts obviously. Here, this time is named as dead time. In order to measure the dead time, the spectral data are recorded separately before and after adding the salt (8 mg) into the droplet, which is shown in Figure 7a. The time to add the salt into the droplet is set to 0 s. The dead time is about 6.5 s, as shown in Figure 7a (denoted as a yellow region). After the dead time, the laser wavelength drifts steeply. Dead time as a function of salt mass is also investigated and plotted in Figure 7b. As shown, the dead time decrease exponentially with the salt mass. This is attributed to the heavier salt possessing a larger surface area and the dissolution process is shorter.

Adding salt into the water actually contains two processes: dissolution of salt and diffusion of NaCl solution. The onset of the spectral shift means the RI of the surrounding volume of the lasing microdisk starts to change due to the diffusion of NaCl solution. The dead time is the sum of salt dissolution time and diffusion time of NaCl solution from one side of the droplet to the middle (where the lasing microdisk locates) of the droplet. Therefore, we perform next experiment by injecting NaCl solution directly into the water droplet to investigate the diffusion time.

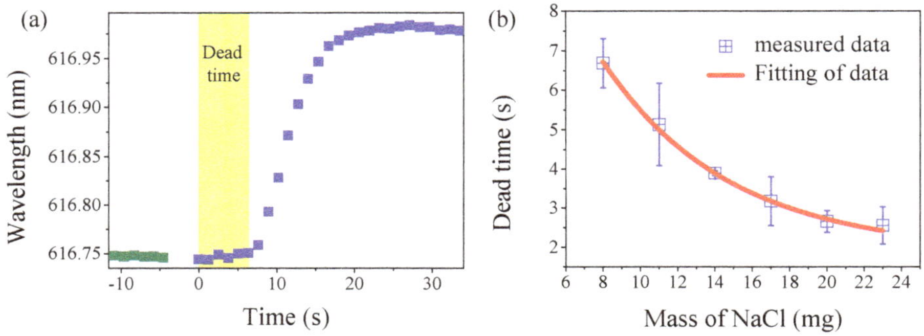

Figure 7. (**a**) Response of lasing wavelength recorded separately before (green dots) and after (blue dots) adding the salt (8 mg) into the water. The yellow region represents the dissolution process (dead time). (**b**) Dead time as function of the mass of the salt.

4.2. Diffusion Process of Two Different Solutions

In this section, the diffusion process of two different RI solutions is investigated. The water droplet with a fixed volume of 0.1 mL is firstly injected to the volume between the WIO and silicon chip. Then 0.1 mL NaCl solution with RI of 1.3516 is quickly injected into the water droplet by a pipette tip.

One laser peak was monitored and its spectral shift is shown in Figure 8. The spectral data are recorded separately before and after injecting NaCl solution into the droplet. The spectral response is different from that in adding salt into the droplet. As soon as the NaCl solution is injected into the droplet, the laser wavelength shifts immediately. This indicates the RI of the surrounding volume of the lasing microdisk changes immediately due to the diffusion of NaCl solution, and this diffusion process is very fast (less than 1 s as shown in Figure 8). Therefore, it can be confirmed that the dead time in Figure 7a is the reaction time of salt dissolution in water.

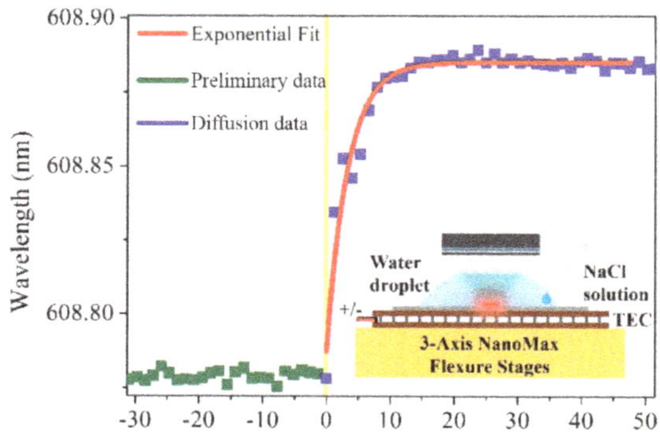

Figure 8. Monitoring the process of mixing NaCl solution with water using the lasing chemical sensor. The yellow line represents the diffusion process. Inset shows the schematic diagram of operation process.

5. Conclusion

We introduced and systematically studied on-chip versatile chemical sensors based on a WIO pumped WGM microlaser in which the microdisk is directly immersed in the water droplet. Laser properties including threshold and spectrum are firstly investigated and also compared with the microdisk pumped by WIO and air objective, respectively. Lasing mode identification is also performed by using FEM. Then, basic physical elements of RI and temperature sensing are tested, which shows that BRIS is as high as 11.3 nm/RIU and the thermal sensitivity can reach 120.6 pm/K. Finally, a proof-of-concept demonstration of the chemical sensor is carried out by locally changing the droplet RI with the help of typical chemical reactions. The tracking of the progression of dissolution and the diffusion process in real time are realized by monitoring the laser spectral shift.

We anticipate that our demonstration can be further extended to study various biochemical reactions, such as endothermic and exothermic reactions. The application of the WIO pumped configuration can be also broadened to multifunctional biological monitoring and treatment by directly immersing cells or tissues into the WIO droplet.

Author Contributions: Conceptualization, Q.L., X.C., X.W., and S.X.; methodology, Q.L., X.C., and L.F.; validation, X.C., L.F., and Q.L.; writing—original draft preparation, Q.L. and X.C.; writing—review and editing, X.W. and S.X.; funding acquisition, S.X. All authors discussed the results and commented on the manuscript.

Funding: This research was funded by the National Natural Science Foundation of China, grant number 61705039; National Key Basic Research Program of China (973 project), grant number 2015CB352006; China Postdoctoral Science Foundation, grant number 2017M610389; Fujian Provincial Program for Distinguished Young Scientists in University; Fujian Provincial Key Project of Natural Science Foundation for Young Scientists in University, grant number JZ160423; Program for Changjiang Scholars and Innovative Research Team in University, grant number IRT_15R10; Special Funds of the Central Government Guiding Local Science and Technology Development, grant number 2017L3009.

Conflicts of Interest: The authors declare no conflict of interest.

References

1. He, L.; Özdemir, Ş.K.; Yang, L. Whispering gallery microcavity lasers. *Laser Photonics Rev.* **2013**, *7*, 60–82. [CrossRef]
2. Jiang, X.F.; Zou, C.L.; Wang, L.; Gong, Q.; Xiao, Y.F. Whispering-gallery microcavities with unidirectional laser emission. *Laser Photonics Rev.* **2016**, *10*, 40–61. [CrossRef]
3. Kippenberg, T.J.; Vahala, K.J. Cavity opto-mechanics. *Opt. Express* **2007**, *15*, 17172–17205. [CrossRef] [PubMed]
4. Dong, C.; Fiore, V.; Kuzyk, M.C.; Wang, H. Optomechanical dark mode. *Science* **2012**, *338*, 1609–1613. [CrossRef] [PubMed]
5. Yang, K.Y.; Dong, Y.O.; Lee, S.H.; Yang, Q.F.; Yi, X.; Shen, B.; Wang, H.; Vahala, K. Bridging ultrahigh- Q devices and photonic circuits. *Nat. Photonics* **2018**, *12*, 297–302. [CrossRef]
6. Levy, J.S.; Gondarenko, A.; Foster, M.A.; Turner-Foster, A.C.; Gaeta, A.L.; Lipson, M. CMOS-compatible multiple-wavelength oscillator for on-chip optical interconnects. *Nat. Photonics* **2009**, *4*, 37. [CrossRef]
7. Armani, D.; Kippenberg, T.; Spillane, S.; Vahala, K. Ultra-high-Q toroid microcavity on a chip. *Nature* **2003**, *421*, 925–928. [CrossRef] [PubMed]
8. Jiang, X.; Shao, L.; Zhang, S.X.; Yi, X.; Wiersig, J.; Wang, L.; Gong, Q.; Lončar, M.; Yang, L.; Xiao, Y.F. Chaos-assisted broadband momentum transformation in optical microresonators. *Science* **2017**, *358*, 344–347. [CrossRef]
9. Lu, Q.; Liu, S.; Wu, X.; Liu, L.; Xu, L. Stimulated Brillouin laser and frequency comb generation in high-Q microbubble resonators. *Opt. Lett.* **2016**, *41*, 1736–1739. [CrossRef]
10. Foreman, M.R.; Swaim, J.D.; Vollmer, F. Whispering gallery mode sensors. *Adv. Opt. Photonics* **2015**, *7*, 168–240. [CrossRef]
11. Fan, X.; White, I.M.; Shopova, S.I.; Zhu, H.; Suter, J.D.; Sun, Y. Sensitive optical biosensors for unlabeled targets: A review. *Anal. Chim. Acta* **2008**, 6208–6226. [CrossRef] [PubMed]

12. Yang, J.; Guo, L.J. Optical sensors based on active microcavities. *IEEE J. Sel. Top. Quant. Electr.* **2006**, *12*, 143–147. [CrossRef]
13. Vollmer, F.; Braun, D.; Libchaber, A.; Khoshsima, M.; Teraoka, I.; Arnold, S. Protein detection by optical shift of a resonant microcavity. *Appl. Phys. Lett.* **2002**, *80*, 4057–4059. [CrossRef]
14. Vollmer, F.; Arnold, S. Whispering-gallery-mode biosensing: Label-free detection down to single molecules. *Nat. Methods* **2008**, *5*, 591–596. [CrossRef] [PubMed]
15. Lu, T.; Lee, H.; Chen, T.; Herchak, S.; Kim, J.-H.; Fraser, S.E.; Flagan, R.C.; Vahala, K. High sensitivity nanoparticle detection using optical microcavities. *Proc. Natl. Acad. Sci. USA* **2011**, *108*, 5976–5979. [CrossRef]
16. Zhu, J.; Ozdemir, S.K.; Xiao, Y.-F.; Li, L.; He, L.; Chen, D.-R.; Yang, L. On-chip single nanoparticle detection and sizing by mode splitting in an ultrahigh-Q microresonator. *Nat. Photonics* **2010**, *4*, 46–49. [CrossRef]
17. Baaske, M.D.; Foreman, M.R.; Vollmer, F. Single-molecule nucleic acid interactions monitored on a label-free microcavity biosensor platform. *Nat. Nanotechnol.* **2014**, *9*, 933–939. [CrossRef] [PubMed]
18. Dantham, V.R.; Holler, S.; Barbre, C.; Keng, D.; Kolchenko, V.; Arnold, S. Label-Free Detection of Single Protein Using a Nanoplasmonic-Photonic Hybrid Microcavity. *Nano Lett.* **2013**, *13*, 3347–3351. [CrossRef] [PubMed]
19. He, L.; Özdemir, Ş.K.; Zhu, J.; Kim, W.; Yang, L. Detecting single viruses and nanoparticles using whispering gallery microlasers. *Nat. Nanotechnol.* **2011**, *6*, 428–432. [CrossRef]
20. Özdemir, Ş.K.; Zhu, J.; Yang, X.; Peng, B.; Yilmaz, H.; He, L.; Monifi, F.; Huang, S.H.; Long, G.L.; Yang, L. Highly sensitive detection of nanoparticles with a self-referenced and self-heterodyned whispering-gallery Raman microlaser. *Proc. Natl. Acad. Sci. USA* **2014**, *111*, E3836–E3844. [CrossRef] [PubMed]
21. Li, B.-B.; Clements, W.R.; Yu, X.-C.; Shi, K.; Gong, Q.; Xiao, Y.-F. Single nanoparticle detection using split-mode microcavity Raman lasers. *Proc. Natl. Acad. Sci. USA* **2014**, *111*, 14657–14662. [CrossRef] [PubMed]
22. Shao, L.; Jiang, X.F.; Yu, X.C.; Li, B.B.; Clements, W.R.; Vollmer, F.; Wang, W.; Xiao, Y.F.; Gong, Q. Detection of single nanoparticles and lentiviruses using microcavity resonance broadening. *Adv. Mater.* **2013**, *25*, 5616–5620. [CrossRef]
23. Li, M.; Wu, X.; Liu, L.; Fan, X.; Xu, L. Self-referencing optofluidic ring resonator sensor for highly sensitive biomolecular detection. *Anal. Chem.* **2013**, *85*, 9328–9332. [CrossRef]
24. Ward, J.M.; Yang, Y.; Lei, F.; Yu, X.-C.; Xiao, Y.-F.; Chormaic, S.N. Nanoparticle sensing beyond evanescent field interaction with a quasi-droplet microcavity. *Optica* **2018**, *5*, 674–677. [CrossRef]
25. Liu, S.; Sun, W.; Wang, Y.; Yu, X.; Xu, K.; Huang, Y.; Xiao, S.; Song, Q. End-fire injection of light into high-Q silicon microdisks. *Optica* **2018**, *5*, 612–616. [CrossRef]
26. Li, H.; Fan, X. Characterization of sensing capability of optofluidic ring resonator biosensors. *Appl. Phys. Lett.* **2010**, *97*, 011105. [CrossRef]
27. Sun, Y.; Fan, X. Optical ring resonators for biochemical and chemical sensing. *Ana. Bioanal. Chem.* **2011**, *399*, 205–211. [CrossRef]
28. Wu, X.; Li, H.; Liu, L.; Xu, L. High Quality Direct Photo-Patterned Microdisk Lasers with Organic–Inorganic Hybrid Materials. *IEEE J. Quant. Electr.* **2007**, *44*, 75–80. [CrossRef]
29. Wang, H.; Liu, S.; Chen, L.; Shen, D.; Wu, X. Dual-wavelength single-frequency laser emission in asymmetric coupled microdisks. *Sci. Rep.* **2016**, *6*, 38053. [CrossRef]
30. Oxborrow, M. Traceable 2-D finite-element simulation of the whispering-gallery modes of axisymmetric electromagnetic resonators. *IEEE Trans. Microw. Theory* **2007**, *55*, 1209–1218. [CrossRef]
31. Wan, L.; Chandrahalim, H.; Chen, C.; Chen, Q.; Mei, T.; Oki, Y.; Nishimura, N.; Guo, L.J.; Fan, X. On-chip, high-sensitivity temperature sensors based on dye-doped solid-state polymer microring lasers. *Appl. Phys. Lett.* **2017**, *111*, 061109. [CrossRef]
32. Zhang, Y.; Liu, P.; Zhang, S.; Liu, W.; Chen, J.; Shi, Y. High sensitivity temperature sensor based on cascaded silicon photonic crystal nanobeam cavities. *Opt. Express* **2016**, *24*, 23037–23043. [CrossRef] [PubMed]
33. Guha, B.; Cardenas, J.; Lipson, M. Athermal silicon microring resonators with titanium oxide cladding. *Opt. Express* **2013**, *21*, 26557–26563. [CrossRef] [PubMed]
34. Kim, E.; Foreman, M.R.; Baaske, M.D.; Vollmer, F. Thermal characterisation of (bio)polymers with a temperature-stabilised whispering gallery mode microsensor. *Appl. Phys. Lett.* **2015**, *106*, 20701. [CrossRef]
35. Ward, J.M.; Yang, Y.; Chormaic, S.N. Highly sensitive temperature measurements with liquid-core microbubble resonators. *IEEE Photonic Technol. Lett.* **2013**, *25*, 2350–2353. [CrossRef]

36. Peng, G.D.; Xiong, Z.; Chu, P.L. Fluorescence decay and recovery in organic dye-doped polymer optical fibers. *J. Lightw. Technol.* **1998**, *16*, 2365–2372. [CrossRef]
37. Gupta, G.; Steier, W.H.; Liao, Y.; Luo, J.; Dalton, L.R.; Jen, A.K.Y. Modeling Photobleaching of Optical Chromophores: Light-Intensity Effects in Precise Trimming of Integrated Polymer Devices†. *J. Phys. Chem. C* **2008**, *113*, 8051–8060. [CrossRef]
38. Poon, J.K.; Huang, Y.; Paloczi, G.T.; Yariv, A.; Zhang, C.; Dalton, L.R. Wide-range tuning of polymer microring resonators by the photobleaching of CLD-1 chromophores. *Opt. Lett.* **2004**, *29*, 2584–2586. [CrossRef]
39. Wondimu, S.F.; Hippler, M.; Hussal, C.; Hofmann, A.; KrãMmer, S.; Lahann, J.; Kalt, H.; Freude, W.; Koos, C. Robust label-free biosensing using microdisk laser arrays with on-chip references. *Opt. Express* **2018**, *26*, 3161–3173. [CrossRef]
40. Liu, Z.; Liu, L.; Zhu, Z.; Zhang, Y.; Wei, Y.; Zhang, X.; Zhao, E.; Zhang, Y.; Yang, J.; Yuan, L. Whispering gallery mode temperature sensor of liquid microresonastor. *Opt. Lett.* **2016**, *41*, 4649–4652. [CrossRef]

nanomaterials

MDPI

Article

Spectral Modulation of Optofluidic Coupled-Microdisk Lasers in Aqueous Media

Zhihe Guo [1], Haotian Wang [2,*], Chenming Zhao [1], Lin Chen [1], Sheng Liu [1], Jinliang Hu [1], Yi Zhou [1] and Xiang Wu [1,*]

1 Key Laboratory of Micro and Nano Photonic Structures (Ministry of Education), Department of Optical Science and Engineering, Shanghai Engineering Research Center of Ultra Precision Optical Manufacturing, Fudan University, Shanghai 200433, China; 17110720004@fudan.edu.cn (Z.G.); 13110720009@fudan.edu.cn (C.Z.); 13110720001@fudan.edu.cn (L.C.); 14110720003@fudan.edu.cn (S.L.); 14110720001@fudan.edu.cn (J.H.); 18110720008@fudan.edu.cn (Y.Z.)

2 Jiangsu Key Laboratory of Advanced Laser Materials and Devices, School of Physics and Electronic Engineering, Jiangsu Normal University, Xuzhou 221116, China

* Correspondence: wanghaotian@jsnu.edu.cn (H.W.); wuxiang@fudan.edu.cn (X.W.)

Received: 2 September 2019; Accepted: 9 October 2019; Published: 11 October 2019

Abstract: We present the spectral modulation of an optofluidic microdisk device and investigate the mechanism and characteristics of the microdisk laser in aqueous media. The optofluidic microdisk device combines a solid-state dye-doped polymer microdisk with a microfluidic channel device, whose optical field can interact with the aqueous media. Interesting phenomena, such as mode splitting and single-mode lasing in the laser spectrum, can be observed in two coupled microdisks under the pump laser. We modulated the spectra by changing the gap of the two coupled microdisks, the refractive indices of the aqueous media, and the position of a pump light, namely, selective pumping schemes. This optofluidic microlaser provides a method to modulate the laser spectra precisely and flexibly, which will help to further understand spectral properties of coupled microcavity laser systems and develop potential applications in photobiology and photomedicine.

Keywords: microdisk laser; spectral modulation; optofluidic; coupled microdisk

1. Introduction

Microcavity lasers based on whispering-gallery mode (WGM), possessing small size and strong optical confinement, have been attracting great interest in fundamental research and for practical applications [1–4]. Compared to the laser wavelength, the size of the microcavity is usually large enough to cause multimode lasing, resulting in a noisy spectrum, which severely degrades the performance of microlasers in applications such as optical communication and ultrasensitive optical sensors [5,6]. Coupled microcavity configuration is a compact and effective method to purify and regulate the laser spectra of microcavities. For example, a single-mode laser can be realized, based on the Vernier effect, in two coupled cavities with slightly different sizes [6–11]. Non-Hermitian optical systems, which have emerged recently, for lasers are, in essence, open and non-conservative systems. This means the most common gains and losses of physical quantities can be distributed differently with artificial purposes in such coupled microcavities [12]. In particular, if the optical microcavity is tuned to the exceptional point (EP) [13], where the resonance frequencies of the coupled modes coalesce both in their real and imaginary parts and become degenerate, some unique phenomena appear, such as parity–time (PT) symmetry breaking [14–18], chiral modes [19], and enhanced perturbation at higher-order EPs [20,21]. Therefore, this principle provides an alternative approach to modulate the laser spectrum by controlling the gain and loss elements. By managing the gain and loss distribution for the coupled modes in microcavities, some studies have successfully achieved single-mode lasing in WGM microcavities [22–24].

One of the key points in such coupled microcavity systems is the flexible method for controlling the system parameters, for example, the coupling strength between the coupled laser modes. The coupling strength can be regulated by controlling the gap between different microcavities, which is predetermined for the desired modes in the fabrication process, or precisely controlled by the nano-positioning stages in real-time experiments [14,25]. However, these mechanisms all lack efficiency and flexibility, and are sensitive to external noise sources, such as acoustic waves. Moreover, the coupled cavities fabricated on chip have the fixed coupling gaps. They need a reliable and flexible means for controlling the coupling strengths on chip. Optofluidic devices can integrate some functional fluids into optical or photonics devices [26,27]. Such fluids usually have high transmittance and tunable optical properties, e.g., the refractive index [28,29]. Because the optofluidic devices have a good compatibility with aqueous media, they are widely used in biology and medical science [30–34]. Therefore, optofluidic devices have found great potential in on-chip microlaser systems for their highly integrated level and, more importantly, the microlaser cavity is given a liquid environment, which provides an additional means for tuning the microlaser.

In this study, solid-state dye-doped polymer microdisks are fabricated and integrated with a microfluidic channel, in order to investigate the mechanism and characteristics of the microdisk laser in aqueous media. Some phenomena, such as single-frequency laser and mode splitting in the coupled microcavities, were studied theoretically and experimentally in aqueous media for the first time. We observe that the resonance detuning of mode splitting is affected by changes in the gap between two coupled microdisks. Mode splitting and single-mode lasing in the laser spectrum can also be modulated by changing the refractive indices of the aqueous media and the position of the pump light. In addition, the aqueous media not only have a spectral modulation function but also a mode purification ability because they decrease the refractive index difference between the cavity and the background media, which eliminates some high-order WGM with larger reflection angles at the cavity surface. The results presented in this paper show that optofluidic microlasers possess potential for several promising applications, such as tunable single-mode lasers on chip and bio-photonic sensors.

2. Coupled-Mode Analysis

The coupled-mode theory could be applied to analyze such a coupled-microdisk laser system. We assume the two coupled modes in two respective cavities have amplitudes a and b. Considering the sinusoidally varying fields in the time domain, $e^{-i\omega_n t}$, a and b obey the coupled differential equation [20]:

$$i\frac{d}{dt}\begin{pmatrix} a \\ b \end{pmatrix} = \begin{pmatrix} ig_a & \kappa \\ \kappa & ig_b \end{pmatrix}\begin{pmatrix} a \\ b \end{pmatrix}, \tag{1}$$

where g_a and g_b represent the gain ($g > 0$) or loss ($g < 0$) for modes a and b, respectively. κ denotes the coupling strength between the two microcavities. It leads to an eigenvalue problem: $H_0\boldsymbol{V_n} = \omega_n\boldsymbol{V_n}$, where the eigenvalue ω_n and eigenvector $\boldsymbol{V_n}$ satisfy $\boldsymbol{V_n}\, e^{-i\omega_n t} = (a, b)^{\mathrm{T}}$, and the Hamiltonian H_0 of the system is

$$H_0 = \begin{pmatrix} ig_a & \kappa \\ \kappa & ig_b \end{pmatrix}, \tag{2}$$

To determine the eigenfrequency, we solve the equation $\det(H_0 - \omega_n I) = 0$ (where I is the identity matrix), which results in two eigenfrequencies:

$$\omega_{\pm} = \frac{i(g_a + g_b)}{2} \pm \sqrt{\kappa^2 - \frac{(g_a - g_b)^2}{4}}, \tag{3}$$

The difference between the two eigenfrequencies is

$$\Delta\omega \equiv \omega_{+} - \omega_{-} = 2\sqrt{\kappa^2 - (g_a - g_b)^2/4}, \tag{4}$$

where $\Delta\tilde{g} = (g_a - g_b)/2\kappa$ is the normalization gain difference between the two modes. If the two modes experience the same gain or loss, i.e., $\Delta\tilde{g} = 0$, the frequency difference equals 2κ, which is the well-known mode-splitting effect, as shown in Figure 1a. However, the two modes have different gains or losses in more general situations. In particular, such frequency difference changes into a purely imaginary number when $\Delta\tilde{g} > 1$. In this case, the difference between the two eigenfrequencies no longer shows two different resonance wavelengths in the spectrum, but instead, introduces two modes with different growth or decay rates, which often indicate different net gains in a laser system. Moreover, at the threshold where $\Delta\tilde{g} = 1$ two eigenfrequencies coalesce to a single one, with $\omega = i(g_a + g_b)/2$.

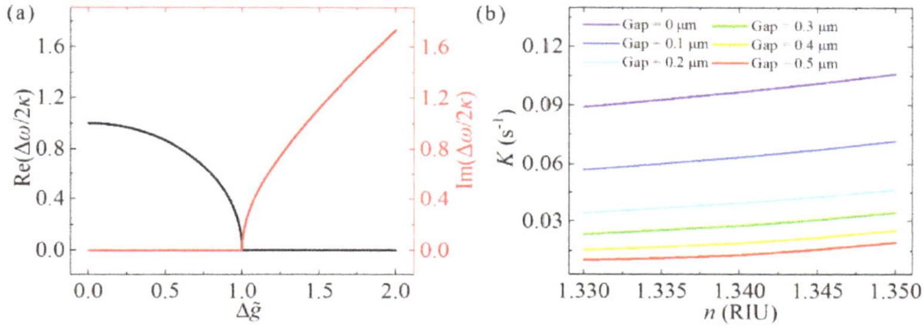

Figure 1. (**a**) Real and imaginary parts of $\Delta\omega/2\kappa$ as a function of $\Delta\tilde{g}$. The difference between the eigenfrequencies changes from purely real to purely imaginary at $\Delta\tilde{g}$= 0. (**b**) Coupling coefficient K as a function of refractive index under different gaps.

The coupling strength κ could be related to the coupling coefficient K; defined in the coupling between two parallel waveguides:

$$\kappa = \frac{KcL_{eff}}{n_{eff}2\pi R}, \tag{5}$$

where L_{eff} and n_{eff} are the effective interaction length and effective index of the cavity modes, respectively, and R is the radius of the microdisk. The coupling coefficient K; can be expressed by the overlap integral between the two coupling modes:

$$K = -\frac{i\omega}{4}\int_{-\infty}^{+\infty}\int_{-\infty}^{+\infty}({n_0}^2 - n^2)\boldsymbol{E_a}\bullet\boldsymbol{E_b}dxdy, \tag{6}$$

where $\boldsymbol{E_a}$ and $\boldsymbol{E_b}$ are the normalized field distributions in the two respective cavities, and n_0 and n are the refractive indices for the microcavities and the background medium, respectively. The coupling strength K in Equation (6) can be calculated numerically, and the relationships between the coupling strength K and refractive index n under different gaps are shown in Figure 1b. K grows with an increase in n, or with a decrease in the coupling gap. Because the increased n reduces the difference of the refractive indices between the microdisk and background medium, more of the optical field will leak into the coupling gap. This results in an improvement in the overlap between the optical fields in two microdisks, and accordingly increases the value of K. Similarly, a smaller coupling gap also improves the overlap of the optical fields and provides greater coupling strength.

3. Optofluidic Microdisk Lasers

3.1. Fabrication of the Optofluidic Microdisk Device and Experimental Setup

The experimental setup of the optofluidic microcavity laser is shown in Figure 2a. The optofluidic microdisk device consists of a microfluidic channel and solid-state dye-doped polymer microdisks as shown in Figure 2b,c. Rhodamine B dye (RhB, CAS: 81-88-9, J&K Scientific Ltd., Beijing, China)-doped SU-8 photoresist (SU-8 2002, MicroChem Corp. MA, USA) in a ratio of 1.5 mg:1 mL was used as gain media and mixed by a magnetic stirrer for 2 h. During stirring, attention should be paid to shading the photoresist, in order to avoid quality deterioration. A 2.4 μm active layer of the gain media was then deposited by a spin coater (Specialty Coating Systems Inc., Indianapolis, USA) onto a silicon wafer with a 2 μm thick thermal oxide layer. The geometric shapes of the single and coupled microdisks were pre-designed on the chromium mask, and the gap between each pair of coupled microdisks was increased from 0.2 to 1 μm, and 0.1 μm for each pair. Therefore, the solid-state dye-doped polymer microdisk arrays were fabricated via single-mask standard lithography (Karl Suss MJB 3 Mask Aligner, SUSS MicroTec SE, Garching, Germany). The geometric sizes of the microdisks were measured via scanning electron microscopy (SEM), as shown in Figure 2c and Figure S1 (details can be found in the Supplementary Materials).

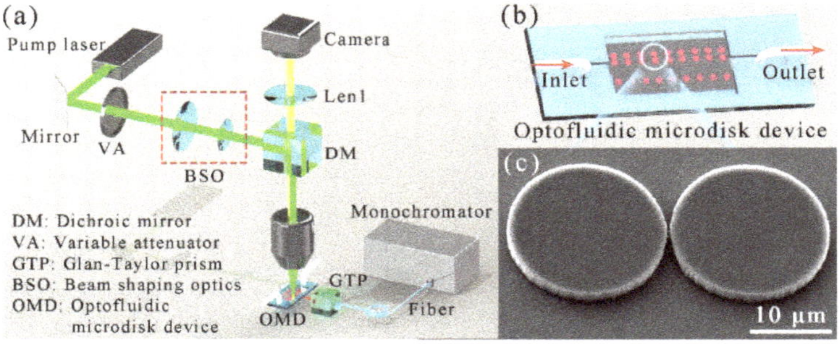

Figure 2. (**a**) Experimental setup schematic of the optofluidic microcavity laser. (**b**) 3-D schematic of the optofluidic microdisk device. (**c**) Scanning electron microscope (SEM) image of typical coupled microdisks. The scale bar is 10 μm.

The optofluidic microdisk device consists of a microfluidic channel device and solid-state dye-doped polymer microdisks (details can be found in Supplementary Materials). This microfluidic channel is a device that combines a clean glass slide and patterned polydimethylsiloxane (PDMS) leaf. The fabrication process of the optofluidic microdisk device is described in Figure S2 (in Supplementary Materials). The shape of the glass mold was spliced with glass sheets, and the convex structure is shown as in Figure S2a. After the PDMS leaf was formed from the glass mold, the inlet and outlet holes were punched by a needle. The solid-state dye-doped polymer microdisks were sandwiched between the microfluidic channel (PDMS leaf) and a glass slide, as shown in Figure S2d,e. Finally, an optofluidic microdisk device, consisting of a microfluidic channel and solid-state dye-doped polymer microdisks, was fabricated, as shown in Figure 2b and Figure S2e.

The laser spectra were measured using the experimental setup as shown in Figure 2a. The pump laser was a picosecond pulse laser (532 nm, PL2143A, EKSPLA, Vilnius, Lithuanian), and the pulse duration of the pump laser was 30 ps, with a repetition rate of 10 Hz. A monochromator (Acton SpectraPro 2750, Princeton Instruments, NJ, USA) was employed to collect the laser spectra. The spectra were measured by a monochromator with a 1200 grooves/mm grating and the spectral resolution was 0.01 nm. The size of the pump laser was controlled via tunable beam-shaping optics (BSO), and the pump laser energy was adjusted by a variable attenuator (VA). Firstly, the laser beam spot was coupled into the microscope system and further focused to nearly 50 μm under a 20× objective lens. Through the microscope with a camera, the pump position of the laser beam spot was able to be marked on a screen. The microcavity sample was then moved to the pump position using a 3-D stage. The output laser passed through a Glan–Taylor prism (GTP) and was collected by an optical multi-channel fiber bundle (core diameter 400 μm). The GTP was used as the polarization controller. The laser light was then coupled into the input slit (30 μm) of a monochromator (f = 0.75 m) and detected by an electron-multiplying charged coupled device (EMCCD, DV401A, Andor iDus, OX, UK). Finally, the laser spectra were recorded by a computer.

3.2. Modulation of the Microdisk Laser in Aqueous Media

A single microdisk was firstly investigated and the appearance of the microdisk sample was measured via SEM, as shown in Figure 3a. In this case, the diameter of the microdisk was set to 20.34 μm and the thickness was 2.4 μm, as measured using scanning electron microscopy (SEM) and a surface profiler, respectively. Compared to air cladding, the microdisk surrounded by water represents a much clearer spectrum. The water environment decreases the refractive index (RI) difference between the microdisk and the background media, where high-order WGMs with larger reflection angles at the cavity surface suffer from degradation of Q (quality) values and larger radiation losses. Therefore, they have a higher laser threshold and only fundamental-order WGMs with high Q values will exist in the laser spectra, which are shown in Figure 3b. The free spectral range (FSR) of this microdisk was 3.81 nm, agreeing well with the FSR calculated from $FSR = \lambda^2/(2\pi R n_{eff}) \approx 3.87$ nm, where $\lambda = 615$ nm, $n_{eff} = 1.53$, and $R = 10.17$ μm were the lasing wavelength, effective refractive index of the microdisk, and radius of the microdisk, respectively. The laser spectra were relatively clear in a water environment, which was conducive to sensing, single-frequency lasers, and other applications. Figure 3c,d show the simulation results of the electric field distribution of a single microdisk using the finite element method (FEM). To reduce consumption of computer memory, we divided the full 3-D model into two 2-D models, as shown in the top view and side view in Figure 3c,d, respectively. The refractive index of the microdisk in the simulation shown in Figure 3c was obtained by calculating the effective index of the three-layer planar waveguide of water-SU8-SiO_2. As shown in Figure 3d, the electric field distribution in the cross section of the microdisk was calculated using an axially symmetric model [35]. All the simulation regions for the two 2-D models were surrounded by the perfect matching layer (PML) to avoid reflecting from the boundary. The microdisk was immersed in water, and the polarization of the fundamental-order-radial modes was transverse electric (TE) polarization.

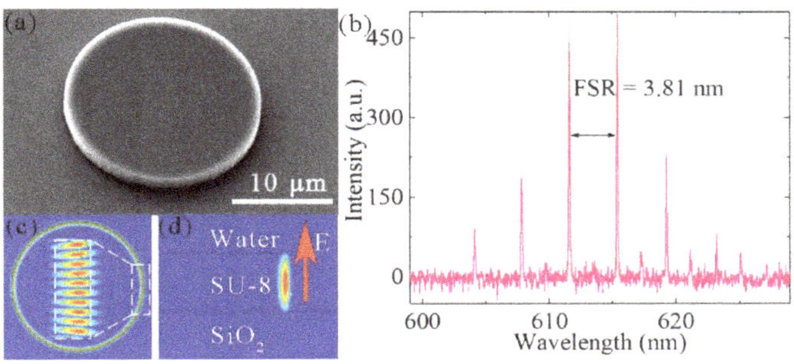

Figure 3. (**a**) SEM image of the microdisk. The scale bar is 10 μm. (**b**) Laser spectra of a microdisk dipped in water. The free spectral range (FSR) was 3.81 nm. (**c**) and (**d**) Field distributions of whispering-gallery modes (WGMs) in fundamental-order-radial mode with transverse electric (TE) polarization in the top view and side view. The direction indicated by the red arrow was the direction of electric field propagation. Finite element method (FEM) simulations were performed with the same parameters for the experimental data.

The long-term stability of the microdisk laser was determined by pumping the microdisk every 20 s over a period of 1800 s, with the laser spectra being collected simultaneously. As shown in Figure 4a,b, the laser wavelength maintained the same position during this period. The standard deviation of the curve in Figure 4b is 5.84 pm, which is much smaller than the resolution of the monochromator (0.01 nm). These results indicate that an optofluidic microdisk laser does not have a large wavelength shift over a long term and that the system stability is good enough for our next experiments.

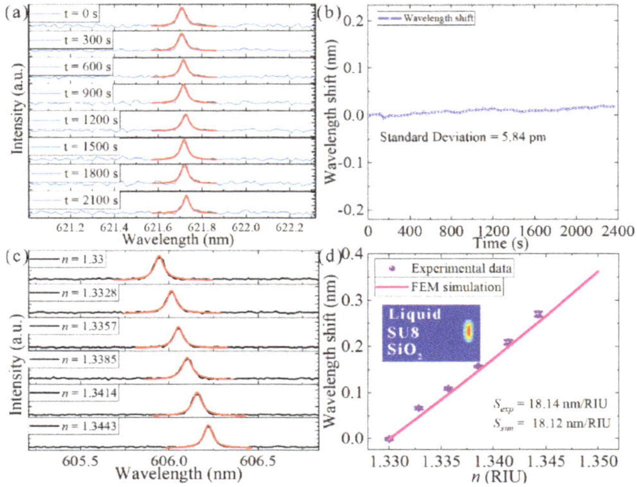

Figure 4. (**a**) Laser spectra of optofluidic microdisk laser at different times. (**b**) Measured wavelengths of the lasers as a function of time. Standard deviation was 5.84 pm. (**c**) Laser spectra shifts when the refractive index of the DMSO solution is slightly increased. (**d**) Mean wavelength shifts of the lasers as a function of the refractive indices of DMSO solutions. Violet dots and pink line represent the experimental data and FEM simulation data, respectively. Insert: fundamental-order-radial mode with TE polarization.

The spectral modulation of the optofluidic microdisk laser can be achieved by changing the refractive index of the external environment, similar to the method for measuring the bulk refractive index sensitivity (*S*) of a sensor. Because this is an optofluidic device, it is essential to first characterize the sensitivity according to the changes in the refractive index of the solvent on the sensing surface [36]. In Figure 4c,d, the optofluidic microdisk laser was tested by flowing progressively higher concentrations of dimethyl sulfoxide (DMSO, CAS: 67-68-5, Lingfeng Chemical Reagent Co. Ltd., Shanghai, China) solutions over the sensing surface. The DMSO was diluted with deionized water (DI water) into concentrations of 2%, 4%, 6%, 8%, and 10%. The refractive indices of the DMSO solutions were proportional to the volume ratios of the DMSO (details can be found in Supplementary Materials) [37]. Prior to use, the DI water and DMSO–water mixtures were kept at room temperature for a longer period, to minimize temperature-induced spectrum changes. A redshift of the wavelength in response to an increasing refractive index of the solution was observed from the laser spectra measurement and can be explained by $m\lambda = 2\pi n_{eff} R$ (see details in Supplementary Materials) [38]. S_{exp} and S_{sim} represent the *S* from the experiment and from the FEM simulation, respectively. S_{exp} was obtained by directly measuring the wavelength shift, which was 18.14 nm/RIU. The process of detecting the *S* of the microdisk was also simulated using FEM, based on a 2-D axisymmetric model. Results show that the S_{sim} of the fundamental and second-order radial modes with TE polarization were 18.12 and 28.90 nm/RIU, respectively, and that the S_{sim} of the fundamental and second-order radial modes with transverse magnetic (TM) polarization were 22.98 and 37.65 nm/RIU, respectively. When the GTP was used to observe the experiment, the polarization state of the output laser was determined to be TE polarization. Therefore, the *S* of the measured microdisk was determined to be fundamental-order-radial mode with TE polarization.

3.3. Mode Splitting from the Coupled-Microdisk Laser in Aqueous Media

The method of purifying and adjusting the microcavity laser spectrum using the coupled microcavity structure is compact and effective. Normally, a typical WGM propagates along a spherical equatorial plane, whose evanescent field component allows interaction with the surrounding environment. Therefore, when two WGM microcavities are close to each other, their WGMs can be evanescently coupled and form mode splitting [31]. A large array of dye-doped polymer microlasers can be fabricated in parallel via a single step of deep-ultraviolet (DUV) lithography [39]. The coupled microdisks were investigated by immersing in water. At this time, the two modes experience the same gain or loss ($g_a = g_b$), and the frequency difference equals 2κ. Using the gap between the coupled microdisk of 0.3 μm, the diameters of these two coupled microdisks were 19.71 and 19.63 μm, and their SEM image is plotted in Figure 5a. Under evenly pumping schemes, as shown in the insert of Figure 5b, the mode splitting of the laser spectrum can be observed. $\Delta\lambda$ represents the distance between two laser peaks under mode splitting and is 0.183 nm in this case. Figure 5c shows the field distributions of WGMs with the fundamental-order-radial mode in a symmetric coupled-microdisk system. The wavelengths of the supermodes are 620.619 and 620.437 nm, respectively. $\Delta\lambda$ of the FEM simulation is 0.182 nm, which is very close to the experimental result of 0.183 nm. Different gaps between coupled microdisks were investigated under the same pumping conditions. Because the optical field decays exponentially outside the microdisk, the coupling strengths κ between the two cavities will decrease. $\Delta\lambda$ will also decrease, which is shown in Figure 5d. The experimental data and FEM simulation results were well-fitted by the exponential function. As the gap exceeded 0.6 μm, the coupling strengths K of the supermodes became weak and the $\Delta\lambda$ value stayed at a level of no more than a few dozen nanometers. Because of the limited resolution of the monochromator, narrower mode splitting cannot be observed in experiments.

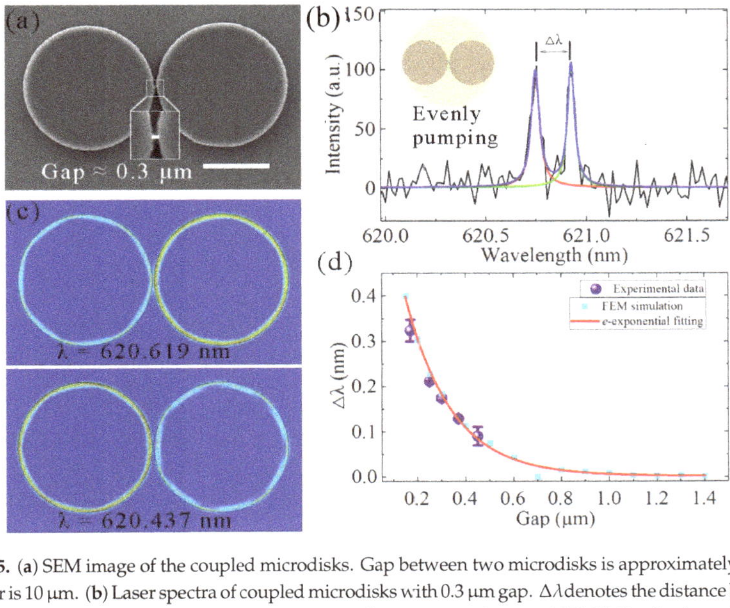

Figure 5. (**a**) SEM image of the coupled microdisks. Gap between two microdisks is approximately 0.3 μm. Scale bar is 10 μm. (**b**) Laser spectra of coupled microdisks with 0.3 μm gap. $\Delta\lambda$ denotes the distance between two peaks of mode splitting. Insert: schematic of evenly pumping schemes. (**c**) Field distributions of WGMs

with fundamental-order-radial mode. (**d**) $\Delta\lambda$ of the experimental results and FEM simulations as a function of the gap. Purple dots represent the experimentally detected data; cerulean dots and red curves refer to the FEM simulation results and exponential fitting, respectively. FEM simulations were performed with the same parameters for the experimental data.

Furthermore, when we change the refractive index of the aqueous medium n, it will also affect the coupling strengths of the supermodes K. Here, we used coupled microdisks with a gap of 0.4 μm. As shown in Figure 6a, when the refractive index of the DMSO solution was changed, the resonance detuning of the mode splitting was affected. Figure 6b shows that $\Delta\lambda$ of the experimental data was increased, which agreed well with the results of the FEM simulation. The increased background refractive index n will extract a greater part of the optical field into the coupling gap, leading to enhanced coupling strength and a larger difference in resonance frequencies of the coupled modes. The coupling coefficient K is as a function of n_{eff}, which is given in Equation (5), and the experimental phenomenon is consistent with the FEM results in Figure 1b.

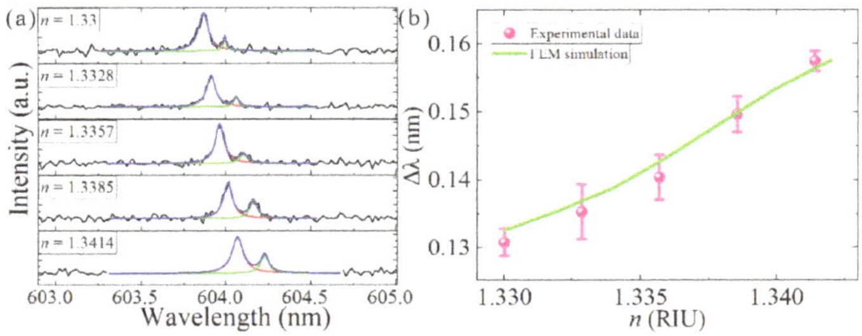

Figure 6. (**a**) Mode splitting in a coupled-microdisk resonator laser with resonance detuning. (**b**) $\Delta\lambda$ of the experimental data and FEM simulation, shown as a function of refractive index. Pink dots represent the experimentally detected data, while the green curves represent the FEM simulation results. FEM simulations were performed using the same parameters for the experimental data.

In order to increase the difference in the gains/losses between the two microcavities, we used selective pumping schemes in which we changed the position of the pump beam spot gradually from the left side of the coupled microdisks to the right side by 8 μm per step. Figure 6 demonstrates the laser spectra of coupled microdisks with a 0.3 μm gap, with the coupled microdisks immersed in water. We focused on the wavelengths of the spectra from 612 to 615 nm, which are marked in red in Figure 7a,c. The laser spectra in Figure 7a,c correspond to the pump positions plotted in Figure 7b,d. The laser emerged when one of the coupled microdisks was pumped, and the splitting mode arose when the coupled microdisks were evenly pumped. When the pump beam spot gradually moved to the other side of the coupled microdisks, the splitting mode disappeared, and the laser finally vanished. This result is due to the selective pumping process, which changed the gain/loss status between the two coupled microdisks. The modulation gain branches of various supermodes will lead to a switchable single-frequency laser or mode splitting [9]. When only one of the coupled microdisks is pumped, the eigenfrequencies difference will become purely imaginary because in this case $\Delta\tilde{g}$ is large enough that the value under the square root in Equation (4) is smaller than zero. This will lead to the emission of a single-mode laser when the other laser modes are also below the threshold. Under even pumping, the gain difference between the two microcavities is very small and $\Delta\omega$ is real, so it shows doublet peaks in the laser spectrum, i.e., mode splitting.

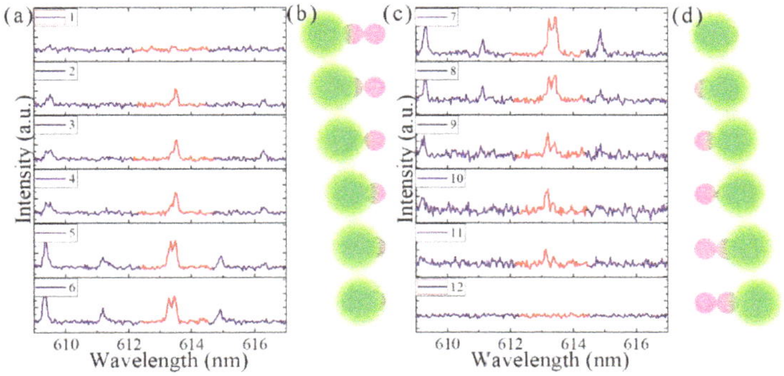

Figure 7. Laser spectra from different positions of the pump beam spot. (**a**) and (**c**) spectra during the moving of the pump laser from left to right. (**b**) and (**d**) schematics of the position of the pump beam spot.

3.4. Single-Frequency Lasing from the Coupled Microdisk in Aqueous Media

Because of their high beam quality and spectral purity, lasers with a single-frequency laser emission feature are indispensable to many scientific and industrial applications, such as laser spectroscopy, laser metrology, and biomolecular sensing [5,6,39]. Normally, WGM lasers are usually multimode because of the lack of mode selection [9,40]. Methods such as microcavity size reduction, Vernier effect, and PT symmetry effect of coupled microcavities are possible strategies to realize single-frequency laser [9,14,40–42]. These methods need to control the gap size or thermal modulation of the two microcavities in order to control the coupling strength between them, which often requires a complex implementation system and results in low accuracy.

The results of the previous section indicate that selective pumping schemes would change the spectrum of the coupled microdisks. In order to overcome the major drawback of a single-microcavity resonator laser, which essentially involves multimode laser emission, we used selective pumping schemes in a coupled-microdisk system, in order to realize a single-frequency laser emission. The coupled-microdisk resonator was dipped in water. In this case, the gap of the coupled microdisks was 0.5 μm. The position of the pump laser was "left pumping", and the simulated mode field of the single-frequency laser is plotted in the inserts of Figure 8b. In the range 580–610 nm (much larger than one FSR and covers the tuning range of RhB in the ethanol solvent), only one laser peak has been collected by the spectrometer. Therefore, it is determined to be a single-frequency laser. The single-frequency laser spectra of the coupled microdisks at different pumping intensities are plotted in Figure 8a. The central wavelength of the single-frequency laser was 594.39 nm. When the pump laser energy was below the threshold, the spectrum was actually the fluorescence, which increases slowly with the pump energy. Moreover, the fluorescence efficiency of the gain media was very low. When the pump laser energy was higher than the laser threshold, the laser intensity will increase with the pump energy with a higher slope efficiency. On the curve in Figure 8b, there is a sudden change point (laser threshold) between the fluorescence and laser as the pump energy increases. The corresponding laser threshold curve shows that the laser threshold was 83.88 μJ/mm^2, as shown in Figure 8b. Selective breaking of PT symmetry can systematically improve the effective amplification of single-mode operation [23].

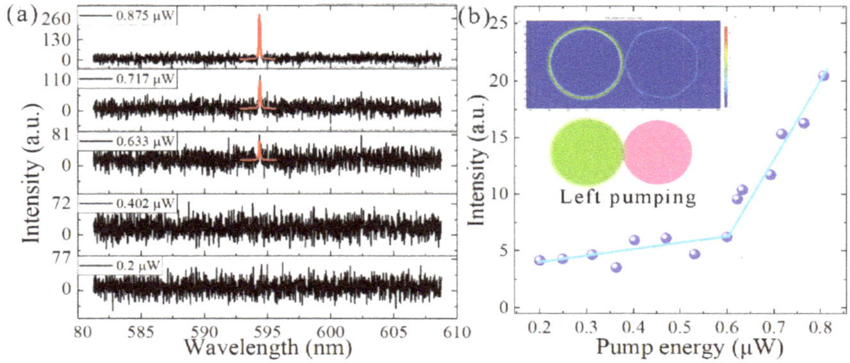

Figure 8. (**a**) Single-frequency emissions of the coupled microdisks under different pump energies. Insert: the energy of the pump laser. (**b**) Lasing intensity as a function of the pump energy intensity extracted from (**a**). Lasing threshold is approximately 0.60 μW (84.88 μJ/mm^2) in a water environment. Insert: the position of the pump laser and the mode field of the single-frequency laser.

An interesting phenomenon was observed when we changed the surrounding refractive index of the single-frequency laser. Another coupled-microdisk resonator was dipped in water under left pumping, with the gap at 0.6 μm. As shown in Figure 9a, the wavelength of the original single-frequency laser was 606.994 nm. With the increase in the liquid refractive index, the wavelength of the single-frequency laser gradually shifted to longer wavelengths, and the intensity decreased. Another hopped laser mode occurred at 603.131 nm. If the liquid refractive index is further increased, the original laser will disappear and only one laser peak can be seen in the spectrum. Figure 9c shows that $ln(I_{hopped}/I_{original})$ changed linearly with the liquid refractive index. Here, $I_{original}$ and I_{hopped} are the light intensity of the original laser line and the hopped laser line, respectively. The hopped laser mode resulted from the slightly asymmetric structure of the coupled microdisks, which responded slightly differently to the change in the surrounding refractive index [30]. Therefore, the wavelength shifts of the coupled-microdisk resonances did not shift synchronously. Furthermore, the slight change of refractive index will also change the coupling efficiency of two coupled microdisks. As shown in Figure 9b,c, the wavelength shift sensitivities ($S_{original}$ and S_{hopped}) were 14.04 and 14.65 nm/RIU, respectively, and the sensitivity of the intensity change (S_I) was 149.87 RIU^{-1}. Considering standard deviation (σ = 10 a.u.) of the background signal intensity as the lowest detectable lasing intensity, the detection limit of S_I was 5.34 × 10^{-5} RIU, based on the equation in a previous article [30]. The resolution of the monochromator can reach 0.01 nm. The detection limits of $S_{original}$ and S_{hopped} were 7.12 × 10^{-4} and 6.83 × 10^{-4} RIU, respectively. These results indicate that S_I was 13 times more sensitive compared with $S_{original}$ and S_{hopped}. Because of the Vernier effect, the coupling of the WGMs in the two resonators can produce a large amplification sensitivity. It was noted that the attenuation of light intensity was due to the bleaching of dye, caused by long-time laser pumping. The relative intensities of the original and hopped laser peaks were not affected.

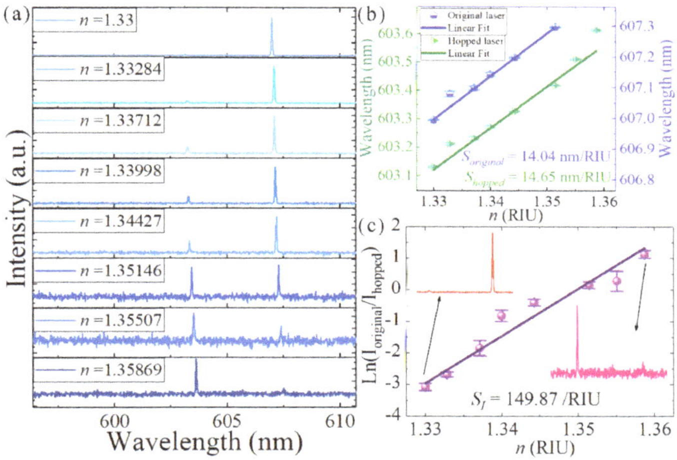

Figure 9. (**a**) Single-frequency emissions of the coupled-microdisk resonator laser. (**b**) Wavelength shifts of single-frequency lasers as a function of refractive index. (**c**) Intensity ratio $[ln(I_{hopped}/I_{original})]$ of the two lasing modes, as a function of refractive index. Insert: spectra at the two refractive indices values are given.

4. Conclusions

In this study, the spectral characteristics of coupled microdisks in aqueous media were examined. The spectral characteristics and modulation mechanism of mode splitting and single-frequency lasers from coupled microdisks were investigated. The resonance detuning of mode splitting was studied by changing the gap between two microdisks, the refractive index of the aqueous media, and the position of the pump laser, namely, selective pumping schemes. The variation of the single-frequency laser was studied by changing the intensity of the pump light and the refractive index of the aqueous media. For single-frequency lasers, the sensitivity can be amplified by changing the light intensity for sensing. The results of this study will help to deepen the understanding of the spectral characteristics and the modulation mechanism of microcavity lasers. Such optofluidic microcavity lasers have broad application prospects in tunable single-mode on-chip lasers and biosensors.

Supplementary Materials: The following are available online at http://www.mdpi.com/2079-4991/9/10/1439/s1, Figure S1. 3-D schematic of the solid-state dye-doped polymer microdisks. Figure S2. 3-D schematic diagram shows the fabrication process of the optofluidic microdisk device.

Author Contributions: X.W. originally conceived the idea. Z.G., C.Z. and L.C. fabricated the devices. Z.G., S.L. and J.H. finished the experiment. Z.G., H.W. and X.W. wrote the manuscript. Z.G. and Y.Z. carried out SEM measurements. Z.G. and H.W. carried out the COMSOL Multiphysics. All authors discussed the progress of the research and reviewed the manuscript.

Funding: This research was funded by Special Project of National Key Technology R & D Program of the Ministry of Science and Technology of China, grant number 2016YFC0201401; Shanghai Science and Technology Committee Grant, grant number 18JC1411500; National Natural Science Foundation of China (NSFC), grant number 61805112; National Key Basic Research Program of China (Program 973), grant number 2015CB352006; Natural Science Foundation of Jiangsu Province, grant number BK20181003.

Conflicts of Interest: The authors declare no conflict of interest.

References

1. Yang, S.; Wang, Y.; Sun, H. Advances and Prospects for Whispering Gallery Mode Microcavities. *Adv. Opt. Mater.* **2015**, *3*, 1136–1162. [CrossRef]
2. Helbo, B.; Kristensen, A.; Menon, A. A micro-cavity fluidic dye laser. *J. Micromech. Microeng.* **2003**, *13*, 307–311. [CrossRef]

3. Shopova, S.I.; Zhou, H.; Fan, X.; Zhang, P. Optofluidic ring resonator based dye laser. *Appl. Phys. Lett.* **2007**, *90*, 221101. [CrossRef]
4. Vahala, K.J. Optical microcavities. *Nature* **2003**, *424*, 839–846. [CrossRef] [PubMed]
5. Gu, F.; Xie, F.; Lin, X.; Linghu, S.; Fang, W.; Zeng, H.; Tong, L.; Zhuang, S. Single whispering-gallery mode lasing in polymer bottle microresonators via spatial pump engineering. *Light. Sci. Appl.* **2017**, *6*, e17061. [CrossRef] [PubMed]
6. Shang, L.; Liu, L.; Xu, L. Single-frequency coupled asymmetric microcavity laser. *Opt. Lett.* **2008**, *33*, 1150–1152. [CrossRef] [PubMed]
7. Arbabi, A.; Kang, Y.M.; Lu, C.-Y.; Chow, E.; Goddard, L.L. Realization of a narrowband single wavelength microring mirror. *Appl. Phys. Lett.* **2011**, *99*, 91105. [CrossRef]
8. Li, M.; Zhang, N.; Wang, K.; Li, J.; Xiao, S.; Song, Q. Inversed Vernier effect based single-mode laser emission in coupled microdisks. *Sci. Rep.* **2015**, *5*, 13682. [CrossRef]
9. Wang, H.; Liu, S.; Chen, L.; Shen, D.; Wu, X. Dual-wavelength single-frequency laser emission in asymmetric coupled microdisks. *Sci. Rep.* **2016**, *6*, 38053. [CrossRef]
10. Zhang, X.; Li, H.; Tu, X.; Wu, X.; Liu, L.; Xu, L. Suppression and hopping of whispering gallery modes in multiple-ring-coupled microcavity lasers. *J. Opt. Soc. Am. B* **2011**, *28*, 483. [CrossRef]
11. Wu, X.; Sun, Y.; Suter, J.D.; Fan, X. Single mode coupled optofluidic ring resonator dye lasers. *Appl. Phys. Lett.* **2009**, *94*, 241109. [CrossRef]
12. Miri, M.-A.; Alù, A. Exceptional points in optics and photonics. *Science* **2019**, *363*, eaar7709. [CrossRef] [PubMed]
13. Brandstetter, M.; Liertzer, M.; Deutsch, C.; Klang, P.; Schöberl, J.; Türeci, H.E.; Strasser, G.; Unterrainer, K.; Rotter, S. Reversing the pump dependence of a laser at an exceptional point. *Nat. Commun.* **2014**, *5*, 4034. [CrossRef] [PubMed]
14. Chang, L.; Jiang, X.; Hua, S.; Yang, C.; Wen, J.; Jiang, L.; Li, G.; Wang, G.; Xiao, M. Parity–time symmetry and variable optical isolation in active–passive-coupled microresonators. *Nat. Photon.* **2014**, *8*, 524–529. [CrossRef]
15. Peng, B.; Özdemir, Ş.K.; Lei, F.; Monifi, F.; Gianfreda, M.; Long, G.L.; Fan, S.; Nori, F.; Bender, C.M.; Yang, L. Parity–time-symmetric whispering-gallery microcavities. *Nat. Phys.* **2014**, *10*, 394–398. [CrossRef]
16. Gao, Z.; Fryslie, S.T.M.; Thompson, B.J.; Carney, P.S.; Choquette, K.D. Parity-time symmetry in coherently coupled vertical cavity laser arrays. *Optica* **2017**, *4*, 323. [CrossRef]
17. Regensburger, A.; Bersch, C.; Miri, M.-A.; Onishchukov, G.; Christodoulides, D.N.; Peschel, U. Parity–time synthetic photonic lattices. *Nature* **2012**, *488*, 167–171. [CrossRef]
18. Guo, A.; Salamo, G.J.; Duchesne, D.; Morandotti, R.; Volatier-Ravat, M.; Aimez, V.; Siviloglou, G.A.; Christodoulides, D.N. Observation ofPT-Symmetry Breaking in Complex Optical Potentials. *Phys. Rev. Lett.* **2009**, *103*, 093902. [CrossRef]
19. Peng, B.; Özdemir, Ş.K.; Liertzer, M.; Chen, W.; Kramer, J.; Yılmaz, H.; Wiersig, J.; Rotter, S.; Yang, L. Chiral modes and directional lasing at exceptional points. *Proc. Natl. Acad. Sci. USA* **2016**, *113*, 6845–6850. [CrossRef]
20. Hodaei, H.; Hassan, A.U.; Wittek, S.; Garcia-Gracia, H.; El-Ganainy, R.; Christodoulides, D.N.; Khajavikhan, M. Enhanced sensitivity at higher-order exceptional points. *Nature* **2017**, *548*, 187–191. [CrossRef]
21. Chen, W.; Özdemir, Ş.K.; Zhao, G.; Wiersig, J.; Yang, L. Exceptional points enhance sensing in an optical microcavity. *Nature* **2017**, *548*, 192–196. [CrossRef] [PubMed]
22. Wu, X.; Li, H.; Liu, L.; Xu, L. Unidirectional single-frequency lasing from a ring-spiral coupled microcavity laser. *Appl. Phys. Lett.* **2008**, *93*, 81105. [CrossRef]
23. Hodaei, H.; Miri, M.-A.; Heinrich, M.; Christodoulides, D.N.; Khajavikhan, M. Parity-time-symmetric microring lasers. *Science* **2014**, *346*, 975–978. [CrossRef] [PubMed]
24. Feng, L.; Wong, Z.J.; Ma, R.-M.; Wang, Y.; Zhang, X. Single-mode laser by parity-time symmetry breaking. *Science* **2014**, *346*, 972–975. [CrossRef] [PubMed]
25. Yang, C.; Jiang, X.; Hua, Q.; Hua, S.; Chen, Y.; Ma, J.; Xiao, M. Realization of controllable photonic molecule based on three ultrahigh-Q microtoroid cavities. *Laser Photon. Rev.* **2017**, *11*, 1600178. [CrossRef]
26. Chen, Y.; Lei, L.; Zhang, K.; Shi, J.; Wang, L.; Li, H.; Zhang, X.M.; Wang, Y.; Chan, H.L.W. Optofluidic microcavities: Dye-lasers and biosensors. *Biomicrofluidics* **2010**, *4*, 043002. [CrossRef]
27. Lu, Q.; Chen, X.; Fu, L.; Xie, S.; Wu, X. On-Chip Real-Time Chemical Sensors Based on Water-Immersion-Objective Pumped Whispering-Gallery-Mode Microdisk Laser. *Nanomaterials* **2019**, *9*, 479. [CrossRef]

28. Ren, L.; Wu, X.; Li, M.; Zhang, X.; Liu, L.; Xu, L. Ultrasensitive label-free coupled optofluidic ring laser sensor. *Opt. Lett.* **2012**, *37*, 3873–3875. [CrossRef]
29. Zhang, X.; Ren, L.; Wu, X.; Li, H.; Liu, L.; Xu, L. Coupled optofluidic ring laser for ultrahigh-sensitive sensing. *Opt. Express* **2011**, *19*, 22242–22247. [CrossRef]
30. Li, H.; Shang, L.; Tu, X.; Liu, L.; Xu, L. Coupling Variation Induced Ultrasensitive Label-Free Biosensing by Using Single Mode Coupled Microcavity Laser. *J. Am. Chem. Soc.* **2009**, *131*, 16612–16613. [CrossRef]
31. Li, M.; Wu, X.; Liu, L.; Fan, X.; Xu, L. Self-Referencing Optofluidic Ring Resonator Sensor for Highly Sensitive Biomolecular Detection. *Anal. Chem.* **2013**, *85*, 9328–9332. [CrossRef] [PubMed]
32. Testa, G.; Persichetti, G.; Bernini, R. Liquid Core ARROW Waveguides: A Promising Photonic Structure for Integrated Optofluidic Microsensors. *Micromachines* **2016**, *7*, 47. [CrossRef]
33. Wu, X.; Oo, M.K.K.; Reddy, K.; Chen, Q.; Sun, Y.; Fan, X. Optofluidic laser for dual-mode sensitive biomolecular detection with a large dynamic range. *Nat. Commun.* **2014**, *5*, 3779. [CrossRef] [PubMed]
34. Wu, X.; Chen, Q.; Sun, Y.; Fan, X. Bio-inspired optofluidic lasers with luciferin. *Appl. Phys. Lett.* **2013**, *102*, 203706. [CrossRef]
35. Oxborrow, M. Traceable 2-D finite element simulation of the whispering gallery modes of axisymmetric elec- tromagnetic resonators. *IEEE Trans. Microw. Theory Tech.* **2007**, *55*, 1209–1218. [CrossRef]
36. Guo, Y.; Ye, J.Y.; Divin, C.; Huang, B.; Thomas, T.P.; Baker, J.J.R ; Norris, T.B.; Baker, J.R. Real-Time Biomolecular Binding Detection Using a Sensitive Photonic Crystal Biosensor. *Anal. Chem.* **2010**, *82*, 5211–5218. [CrossRef] [PubMed]
37. Aminabhavi, T.M. Use of mixing rules in the analysis of data for binary liquid mixtures. *J. Chem. Eng. Data* **1984**, *29*, 54–55. [CrossRef]
38. Chandrahalim, H.; Fan, X. Reconfigurable Solid-state Dye-doped Polymer Ring Resonator Lasers. *Sci. Rep.* **2015**, *5*, 18310. [CrossRef]
39. Zhu, H.; White, I.M.; Suter, J.D.; Dale, P.S.; Fan, X. Analysis of biomolecule detection with optofluidic ring resonator sensors. *Opt. Express* **2007**, *15*, 9139–9146. [CrossRef] [PubMed]
40. Xie, F.; Yao, N.; Fang, W.; Wang, H.; Gu, F.; Zhuang, S. Single-mode lasing via loss engineering in fiber-taper-coupled polymer bottle microresonators. *Photon Res.* **2017**, *5*, B29. [CrossRef]
41. Grossmann, T.; Wienhold, T.; Bog, U.; Beck, T.; Friedmann, C.; Kalt, H.; Mappes, T. Polymeric photonic molecule super-mode lasers on silicon. *Light. Sci. Appl.* **2013**, *2*, e82. [CrossRef]
42. Ta, V.D.; Chen, R.; Sun, H. Coupled polymer microfiber lasers for single mode operation and enhanced refractive index sensing. *Adv. Opt. Mater.* **2014**, *2*, 220–225. [CrossRef]

nanomaterials

MDPI

Article

Guided Mode Resonance Sensors with Optimized Figure of Merit

Yi Zhou, Bowen Wang, Zhihe Guo and Xiang Wu *

Key Laboratory of Micro and Nano Photonic Structures (Ministry of Education), Department of Optical Science and Engineering, Shanghai Engineering Research Center of Ultra Precision Optical Manufacturing, Fudan University, Shanghai 200433, China; 18110720008@fudan.edu.cn (Y.Z.); 16210720012@fudan.edu.cn (B.W.); 17110720004@fudan.edu.cn (Z.G.)

* Correspondence: wuxiang@fudan.edu.cn

Received: 6 May 2019; Accepted: 22 May 2019; Published: 1 June 2019

Abstract: The guided mode resonance (GMR) effect is widely used in biosensing due to its advantages of narrow linewidth and high efficiency. However, the optimization of a figure of merit (FOM) has not been considered for most GMR sensors. Aimed at obtaining a higher FOM of GMR sensors, we proposed an effective design method for the optimization of FOM. Combining the analytical model and numerical simulations, the FOM of "grating–waveguide" GMR sensors for the wavelength and angular shift detection schemes were investigated systematically. In contrast with previously reported values, higher FOM values were obtained using this method. For the "waveguide–grating" GMR sensors, a linear relationship between the grating period and groove depth was obtained, which leads to excellent FOM values for both the angular and wavelength resonance. Such higher performance GMR sensors will pave the way to lower detection limits in biosensing.

Keywords: guided mode resonance; figure of merit; biosensor; detection limits

1. Introduction

Guided mode resonance (GMR) is a physical effect that occurs in thin-film structures that contain diffractive elements and a waveguide layer ("grating–waveguide" GMR structure) [1] or a planar dielectric-layer diffraction grating ("waveguide–grating" GMR structure) [2]. Light can be coupled into the waveguide modes by different grating diffraction orders. Due to periodic modulation in the waveguide, part of the guided light will leak, and thus, guided modes cannot be sustained on the waveguide grating and will interfere with the noncoupled reflected or transmitted waves [3]. The GMR effect occurs within a narrow spectral band at a particular wavelength, angle, and polarization [4].

Since Magnusson and Wang suggested the application of the GMR effect for sensing purposes due to its narrow, controllable linewidth and high efficiency [5], many researchers have shown great interest in GMR sensors, especially in biosensing [6–13]. To date, there are four main detection schemes for GMR sensors, including wavelength detection [14–17], angular shift detection [18–20], intensity shift detection [21–25], and phase shift detection [26–29]. The wavelength and angular shift detection schemes are used most often as the incident light is totally reflected with highly angular and spectral selectivity at resonance [30]. To reduce the cost of detection systems, for example, by eliminating the need for an expensive spectrometer and laser source, several intensity detection schemes have been proposed [21–24]. In addition, the sensing performance can also be enhanced by implementing phase detection schemes [26–29].

Although different detection methods result in different detection effects, the structural parameters and optical properties of GMR sensors are also important because these factors directly determine the sensors' performance [31–36]. In a label-free biosensing system, a figure of merit (FOM), which combines the sensitivity (S) and the full width at half maximum (FWHM) of the resonance, is the accepted marker to compare the performance of biosensors, which can be defined as FOM = S/FWHM [37]. This value can simultaneously reflect the effects of the magnitude of a measured quantity change and the ability to measure small wavelength shifts [38]. In addition, a FOM can be considered as a good indicator for the expected performance of a sensor [39]. However, a narrow resonant linewidth will result in a sacrifice in the sensitivity, as resonant modes are mostly confined within the solid dielectric medium, rather than the sensing medium [40]. In contrast, highly resonant sensitivity may result in a broadened resonant linewidth, which limits the capability to measure small resonant wavelength shifts with higher accuracy [37]. To solve this problem, Ge et al. proposed an external cavity laser (ECL) GMR sensor using an active resonator laser that maintains the high resolution of laser biosensors without sacrificing sensitivity [41–43]. However, although the properties of sensitivity and linewidth have been investigated widely by varying structural parameters [31–36], the optimization of a FOM has not been considered for most GMR sensors. Most recently, Lan et al. investigated the FOM of a "waveguide-grating" GMR sensor by varying the geometric parameters and incident angle [44]. However, their work only investigated one GMR sensor for wavelength resonance rather than providing a design guideline that is suitable for all GMR sensors. Therefore, a convenient and effective design method to achieve an optimized FOM of GMR sensors for both angular and wavelength resonance is necessary.

In this work, by combining an analytical model and numerical simulation, we provide a convenient and efficient method as a guideline for parametric design to achieve the optimized FOM of "grating–waveguide" GMR sensors for both angular and wavelength resonance. Focusing on "waveguide–grating" GMR sensors, a linear relationship between the grating period and grating groove depth was found as a guideline to achieve optimized FOM.

2. Analytical Model for GMR Sensors

Lin et al. proposed a model for calculating the wavelength sensitivity of GMR sensors [35]. Briefly, as the incident light λ passes through the grating with an incident angle θ_i, the diffraction grating equation can be expressed as follows [17]:

$$\Lambda[n_{wg}\sin(\theta_d) - n_c\sin(\theta_i)] = m_g\lambda, m_g = 0, \pm1, \pm2, \ldots, \tag{1}$$

where Λ is the grating period, n_{wg} is the refractive index of the waveguide layer, n_c is the refractive index of the surrounding medium, θ_d is the diffraction angle and m_g is the order of the diffracted wave. The grating diffracted wave will couple into the waveguide layer, once the diffracted wave is phase-matched to the waveguide mode, and the diffraction angle θ_d is the propagation angle in the waveguide layer because the grating groove depth (d_g) is assumed to be extremely thin, and thus, other influences are ignored. The guided wave condition of the planar waveguide can be defined as follows:

$$k_0 n_{wg} d_{wg}\cos(\theta_d) - m\pi = \phi_t + \phi_d, m = 0, 1, 2, \ldots, \tag{2}$$

where $k_0 = 2\pi/\lambda$ is the wavenumber in a vacuum, d_{wg} is the thickness of the waveguide layer, m is a positive integer number that stands for the mode number of a waveguide, φ_t and φ_d represent phase shifts that occur due to the total internal Fresnel reflection at the waveguide grating interface and the waveguide substrate interface, respectively. For a given waveguide structure, we can use Equation (2) to calculate the θ_d, and thus the incident angle θ_i in Equation (1) will be obtained. When the surrounding refractive index n_c is changed, the different incident angles can be solved using Equations (1) and (2), and thus the angular sensitivity $S_a = \Delta\theta_i/\Delta n_c$ can be solved, where $\Delta\theta_i$ is the shift in the resonant angle induced by a change in the refractive index of the surrounding medium Δn_c.

As Equation (2) only stands for a single wavelength, it needs to be modified as follows to be suitable for multiple wavelengths (suitable for normal incidence only) [35]:

$$\frac{2\pi}{\Lambda} m_g d_{wg} \cot(\theta_d) - m\pi = \phi_t + \phi_d, m = 0, 1, 2, \ldots, \tag{3}$$

Using Equations (1) and (3), the wavelength sensitivity $S_w = \Delta\lambda/\Delta n_c$ can be solved, where $\Delta\lambda$ is the shift in the resonant wavelength induced by a change in the refractive index of the surrounding medium Δn_c.

3. Simulation Results and Analysis

The RSoft 7.1 (RSoft Design Group, Inc., Ossining, NY, USA) based on rigorous coupled-wave analysis (RCWA) was used to simulate the GMR effect [36]. To calculate the field distributions, we used COMSOL Multiphysics 5.2a (COMSOL Inc., Stockholm, Sweden), which is based on the finite element (FEA) analysis method [45]. In our case, we simulated the GMR sensors made of a high-index material, such as silicon nitride (Si_3N_4, $n_g = n_{wg} = 2.00$) immersed in water ($n_c = 1.333$) and with silicon dioxide (SiO_2, $n_s = 1.45$) as a substrate. The filling factor (FF) of the grating is fixed at 0.5 and the total thickness of the grating and waveguide layer d was also kept constant.

3.1. Angular Shift Detection Scheme for "Grating–Waveguide" GMR Sensors

For an angular interrogation technique, a monochromatic light source, such as a He–Ne laser, is typically used to measure the GMR effect in this case. The incident wavelength λ is set at 633 nm, and Λ is set at 280 nm. The total thickness of the grating and waveguide layer d is fixed at 100 nm. d_g and d_{wg} are the thicknesses of the grating and waveguide layer, respectively.

The schematic geometries of the angular GMR sensors are depicted in Figure 1a. The incident light can excite transverse electric (TE) or transverse magnetic (TM) modes, depending on the electric field E_z or magnetic field H_z perpendicular to the plane of incidence (x–y plane). When n_c varies, the resonant angle θ_i will change according to Equation (1). The angular position of the resonant peak will shift as illustrated in Figure 1b. Figure 1c shows the sensitivity and FWHM versus d_g for TE polarization. S_a increases and the resonant linewidth broadens as d_g increases for TE polarization. Consequently, the angular FOM cannot achieve a higher value with higher sensitivity, as shown in Figure 1d. However, S_a increases and the resonant linewidth decreases as d_g decreases for TM polarization when d_g is below 60 nm, as shown in Figure 1e. Figure 1f shows that, for TM polarization, the angular FOM increases as the d_g decreases. A maximum FOM of 5709 was achieved when d_g was 10 nm and d_{wg} was 90 nm, which is higher than previously reported values [18]. The higher FOM not only arises from a lower d_g resulting in a lower resonance linewidth, but also generates a higher sensitivity. The maximum FOM is 16.3 times higher than the minimum FOM of 350, as shown in Figure 1f.

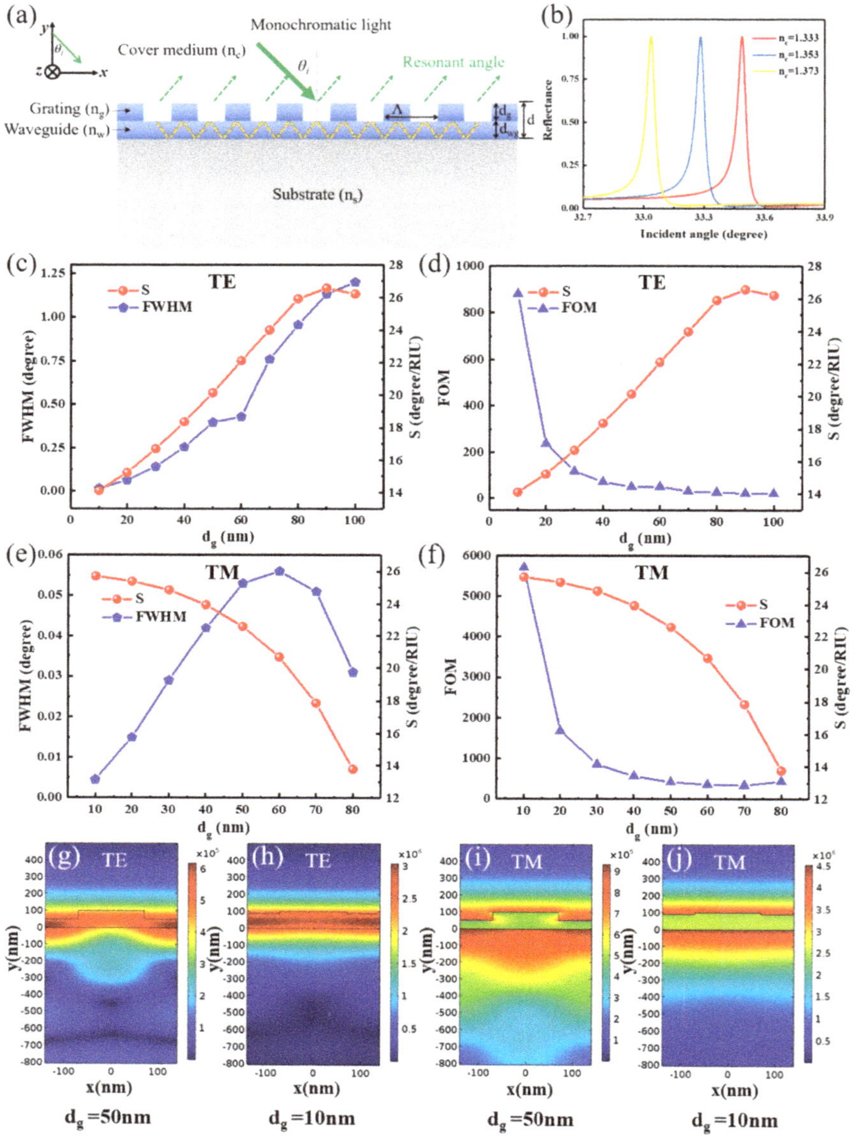

Figure 1. (**a**) Schematic structure of the "grating–waveguide" guided mode resonance (GMR) structure for angular resonance. (**b**) An example of reflection spectra for oblique incidence at the resonance of a monochromatic light ($d_g = d_{wg} = 50$ nm for transverse magnetic (TM) polarization). Calculated sensitivity, resonant linewidth and FOM versus d_g for transverse electric (TE) polarization (**c**,**d**) and TM polarization (**e**,**f**). Electric field distribution of the angular resonance of the GMR structure with different d_g values of 50 nm (d_{wg} of 50 nm), 10 nm (d_{wg} of 90 nm) and for TE (**g**,**h**) and TM (**i**,**j**). (d is 100 nm, Λ is 280 nm, and the filling factor (FF) is 0.5).

TE and TM polarization have different performances when d is fixed at 100 nm. To investigate the different performances between TE and TM polarization, the COMSOL Multiphysics 5.2a (COMSOL Inc., Stockholm, Sweden) software was used to calculate the field distributions of resonant positions. Figure 1g,h, respectively, show the electric field distributions of TE polarization when d_g is 50 nm and 10 nm. The electric fields are mainly stored in the waveguide layer. Therefore, a thick waveguide layer confines more electric energy, which results in a narrower resonant linewidth but reduces the sensitivity. Therefore, the value is mainly attributed to a narrow linewidth of 0.016 degrees at the cost of sensitivity, although a higher FOM of 881 was obtained for TE polarization. For the TM case, most electric energy is not confined in the waveguide layer and is distributed in the cover medium and substrate, as shown in Figure 1i,j. In the case where d_g = 50 nm, up to 70.65% of the total resonance energy is distributed in the substrate, which is 82.56% of the total evanescent energy, and only 19.55% of the total resonance energy is distributed in the cover medium, which is 17.44% of the total evanescent energy. In the d_g = 10 nm case, 62% and 25% of the total resonance energy is distributed in the substrate and cover medium, respectively, which respectively contain 69.72% and 30.38% of the total evanescent energy. Therefore, the evanescent energy increases by almost 13%, which leads to a higher angular sensitivity. Meanwhile, an obvious increase in the maximum electric intensity from 9.27×10^5 V/m (d_g = 50 nm) to 4.44×10^6 V/m (d_g = 10 nm) is observed for TM polarization, as shown in Figure 1i,j. This indicates that more energy was confined in the structure, thus decreasing the resonant linewidth. This result can also be proven by referencing [46], because the grating groove depth controls the coupling-loss coefficient in an exponential manner, thus the coupling-loss coefficient tends to be smaller as d_g decreases. Therefore, when d is fixed at 100 nm for TM polarization, a shallow d_g facilitates a narrow linewidth, and a relatively thick d_{wg} (near 90 nm) also results in a higher sensitivity. A much higher FOM will be obtained thanks to the higher sensitivity and narrower linewidth.

To explain this phenomenon, we used the previous angular model to calculate S_a. The red curve in Figure 2a depicts the S_a of TE polarization, and a higher sensitivity region is located near d_{wg} = 50 nm. The curved sensitivity line of TM polarization has the largest sensitivity region near d_{wg} = 100 nm, as shown in Figure 2b. The highest sensitivity is 25.9 degrees/RIU (refractive index unit), which is very close to the highest sensitivity value in Figure 1f for TM polarization. Moreover, this largest sensitivity region d_{wg} = 100 nm is the same as the value of the total depth of the grating and waveguide layer d that we set in Section 3.1. Combined with these results, the extremely high FOM accompanied by a high sensitivity and narrow linewidth shown in Figure 1f can be explained as follows: a shallower d_g facilitates a narrower linewidth and a thicker d_{wg}, approaching the highest sensitivity region (100 nm) and resulting in higher sensitivity.

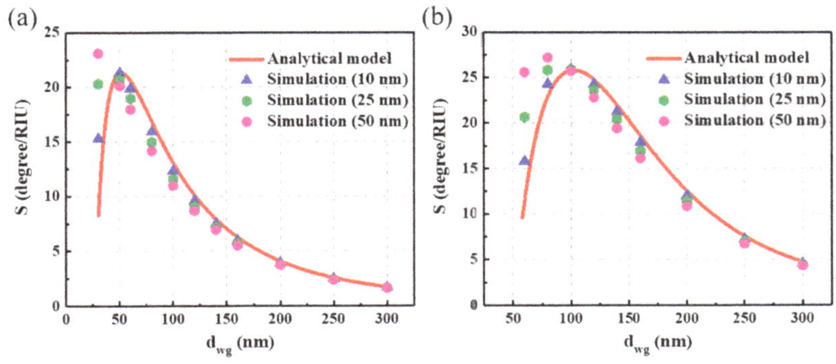

Figure 2. The red curved line represents the analytical model of the S_a calculation for the transverse electric (TE) (**a**) and transverse magnetic (TM) (**b**) modes. Blue, green, and magenta marked symbols respectively represent calculation results of 10 nm, 25 nm and 50 nm d_g of the rigorous coupled-wave analysis (RCWA) simulation method.

To further verify the analytical result, an RCWA method was used to compare the results. In Figure 2, blue, green, and magenta marked symbols represent the numerical results for cases with a d_g of 10 nm, 25 nm, and 50 nm, respectively. The results display evident differences when d_g is relatively high because the influence of the grating on both the phase shift and total internal reflection cannot be ignored. However, this model is still useful for calculating the angular sensitivity when the grating depth is small.

For optimizing the FOM of "grating–waveguide" GMR sensors, a high value region of sensitivity should be evaluated first and then d should be maintained near this value. Finally, by fabricating a shallower grating depth, higher sensitivity and narrower linewidth will be achieved, resulting in a higher FOM. This is a design guideline for "grating–waveguide" GMR sensors to achieve optimized FOM.

However, the TE polarization has its highest sensitivity region near d_{wg} = 50 nm, as shown in Figure 2a. Based on the previous deduction, we set d = 50 nm and simultaneously varied d_g and d_{wg} (more details are shown in Table 1). A maximum FOM (with a narrower linewidth and higher sensitivity) of 1158 was achieved when d_g = 10 nm and d_{wg} = 40 nm. The missing values in Table 1 indicate the instances when the phase-matched condition was not satisfied [17,35].

Table 1. Linewidth (FWHM), sensitivity (S) and figure of merit (FOM) of angular resonance for d = 50 nm in transverse electric (TE) mode.

d_g (nm)	d_{wg} (nm)	FWHM (Degree)	S (Degree/RIU)	FOM
10	40	0.018	20.854	1158.567
20	30	0.058	19.138	329.959
30	20	0.083	14.934	179.923
40	10	Not a value	Not a value	Not a value

3.2. *Wavelength Shift Detection Scheme for "Grating–Waveguide" GMR Sensors*

The most common detection technique for GMR sensors is the wavelength shift-tracking method. A broadband light source is used, such as an LED light and white light source. To investigate the wavelength detection scheme, a normal incidence light was used and Λ was set at 410 nm. Similarly, we set d at 100 nm and d_g and d_{wg} were simultaneously varied. The schematic geometry of the GMR sensors is depicted in Figure 3a. When n_c is varied, the resonant wavelength λ changes according to Equation (1). Thus, the resonant peak position of the wavelength will shift, as illustrated in Figure 3b.

Figure 3c,d demonstrates the relationship among resonant linewidth, sensitivity, and FOM for TE polarization. As d_g decreases, a narrow linewidth and low S_w is obtained, and thus the FOM cannot achieve a relatively high value combined with high sensitivity. On the other hand, for TM polarization, S_w increases and the resonant linewidth decreases when d_g decreases (d_{wg} increases), as shown in Figure 3e. A maximum FOM of 2154 was achieved when the d_g was 10 nm (d_{wg} was 90 nm), as shown in Figure 3f, which is higher than those of common "grating–waveguide" GMR sensors [39].

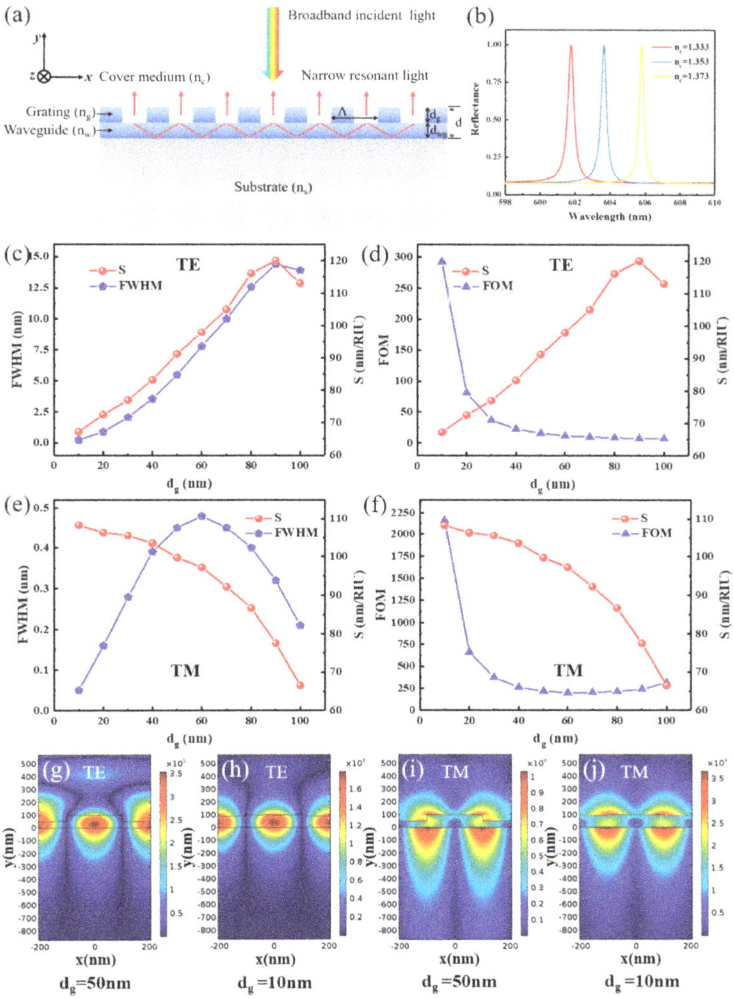

Figure 3. (**a**) Schematic representation of the "grating–waveguide" structure for wavelength resonance. (**b**) An example of reflection spectra for normal incidence at the resonance of a broadband light ($d_g = d_{wg}$ = 50 nm for transverse magnetic (TM) polarization). Calculated sensitivity, resonant linewidth and figure of merit (FOM) versus dg for transverse electric (TE) polarization (**c**,**d**) and TM polarization (**e**,**f**). Electric field distribution of the wavelength resonance of the guided mode resonance (GMR) structure with different d_g of 50 nm (d_{wg} of 50 nm), 10 nm (d_{wg} of 90 nm) and for TE (**g**,**h**) and TM (**i**,**j**), for which the resonance wavelengths are 628.91, 651.51, 601.76 and 611.26 nm, respectively. d is set at 100 nm, Λ is 410 nm, and filling factor (FF) is 0.5.

The electric field distributions for both TE and TM polarization were simulated to further explain the previous results and are shown in Figure 3g–j. Compared to Figure 1g–j, similar results can be achieved here. For the TE mode, most of the evanescent field energy exists in the waveguide layer, and thus the sensing performances were affected. TM cases have a larger light-matter interaction region in the cover medium, and more light is strongly confined (the maximum electric intensity up to 3.61×10^6 V/m), making it more suitable to achieve better performances in biosensing. The sensitivity curve of the TM polarization has the largest sensitivity region near d_{wg} = 100 nm, which is the same as the value of d that we set here. On the other side, the TE polarization has the largest sensitivity region near the 50 nm of d_{wg}, and d = 50 nm is maintained. Table 2 shows the difference in performance caused by varying d_g and d_{wg}. As shown in Table 2, a higher FOM is obtained with a narrower linewidth and higher sensitivity.

Table 2. Linewidth (FWHM), sensitivity (S) and figure of merit (FOM) of wavelength resonance for d = 50 nm in transverse electric (TE) mode.

d_g (nm)	d_{wg} (nm)	FWHM (nm)	S (nm/RIU)	FOM
10	40	0.25	89.00	356
20	30	0.90	83.25	92.5
30	20	1.49	70.00	46.98
40	10	0.79	36.75	46.52

Combining these results with those of Section 3.1, TM polarization has better results for both angular and wavelength resonance, although different detection methods were used. This phenomenon can be explained in two respects: first, TM polarized modes have a larger phase shift length at the resonance point compared to TE modes, resulting in higher sensitivity, as shown in Figure 4. The red and blue curves represent the TM polarization phase shift in the waveguide layer at different surrounding n_c. Yellow and green curves represent the TE polarization phase shift at different surrounding n_c. The label d_{wg} = 60 nm (purple curve) and d_{wg} = 90 nm (gray curve) represent the calculated results coming from the left side of Equations (2) and (3) (called structure relation [35]). The intersection points of the structure relation curve and phase shift curve represent the points of resonant condition, where the GMR effect occurs. The length between two intersection points is proportional to resonant angle or wavelength shift. Therefore, a larger phase change length at different n_c, results in greater sensitivity. In Figure 4, the TM polarization has a larger phase change length compared to the TE mode, when the d_{wg} was 90 nm. Second, the coupling-loss coefficient for a TM polarization was smaller than that for the TE polarization, resulting in a narrow linewidth [46]. Furthermore, the phase shift length of d_{wg} = 90 nm (purple curve) exceeds that of the d_{wg} = 60 nm (gray curve) condition for TM polarization. This also explains why a higher sensitivity was obtained for the case where d_{wg} = 90 nm, as shown in Figures 1f and 3f.

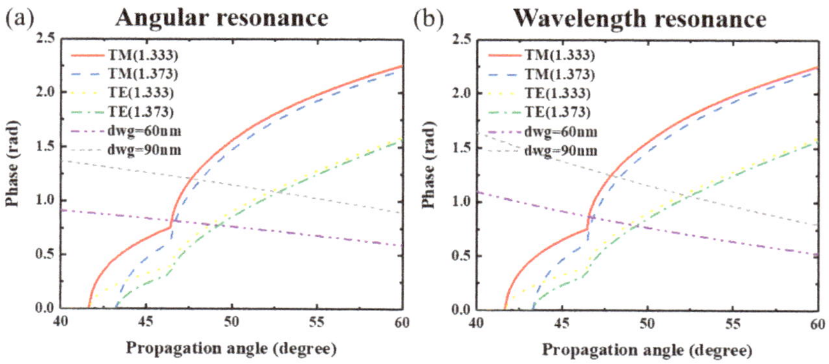

Figure 4. Phase shift curve in waveguide layer at different propagation angle, (**a**) angular resonance, (**b**) wavelength resonance.

3.3. Optimized FOM for "Waveguide–Grating" GMR Sensors

The schematic geometries of angular and wavelength resonance of the "waveguide–grating" GMR sensors are depicted in Figure 5a,b, respectively. Aimed at "waveguide–grating" GMR sensors, different d_g values were swept by a "MOST Optimize/Scanner" (DiffractMod, RSoft 7.1, RSoft Design Group, Inc., Ossining, NY, USA) under certain Λ values to investigate whether a minimum resonant linewidth and a higher sensitivity exist for both the TE and TM modes. The chosen parameters are similar to those presented in Section 3.1 for the angular shift and Section 3.2 for the wavelength shift detection schemes, except d (d_g and d_{wg}).

For angular resonance, a better result was obtained under TM polarization; the series of results for different d_g (from 150 nm to 400 nm) are shown in Figure 5c,d. Figure 5c shows the resonant linewidth and S_a versus d_g for TM polarization. A minimum resonant linewidth and a higher sensitivity occur for a value of d_g of 300 nm. A higher FOM, almost up to 10^6, was achieved, as shown in Figure 5d, which is 168 times greater than the highest FOM value presented in Figure 1f. For wavelength resonance, a superior result was obtained under TE polarization, and the series of results for different d_g values (from 100 nm to 500 nm) are shown in Figure 5e,f. Figure 5c shows the resonant linewidth and S_w versus d_g for TE polarization. A minimum resonant linewidth and a higher sensitivity will occur under 380 nm of d_g. Thus, a high FOM of 1618 will be achieved as shown in Figure 5d. The electric energy distribution helps us to further understand the mechanism, as shown in Figure 5a,b. For both angular and wavelength resonance, the electric field will ascend from the substrate to the grating and the surrounding medium as d_g increases. Therefore, the total evanescent energy will increase in the surrounding medium, thus increasing the sensitivity. Meanwhile, an obvious increase in the maximum electric intensity from 7.44×10^5 V/m (d_g = 200 nm) to 3.89×10^7 V/m (d_g = 300 nm) was observed for angular resonance under TM polarization, as shown in Figure 4g,h. For wavelength resonance, the maximum electric intensity increased from 1.84×10^5 V/m (d_g = 200 nm) to 2.31×10^6 V/m (d_g = 380 nm) for TE polarization, as shown in Figure 5i,j. These results indicate that more energy was confined in the GMR structure, and thus, the linewidth was decreased. Combining these two points, a higher FOM value was achieved, accompanied by high sensitivity and a narrow linewidth.

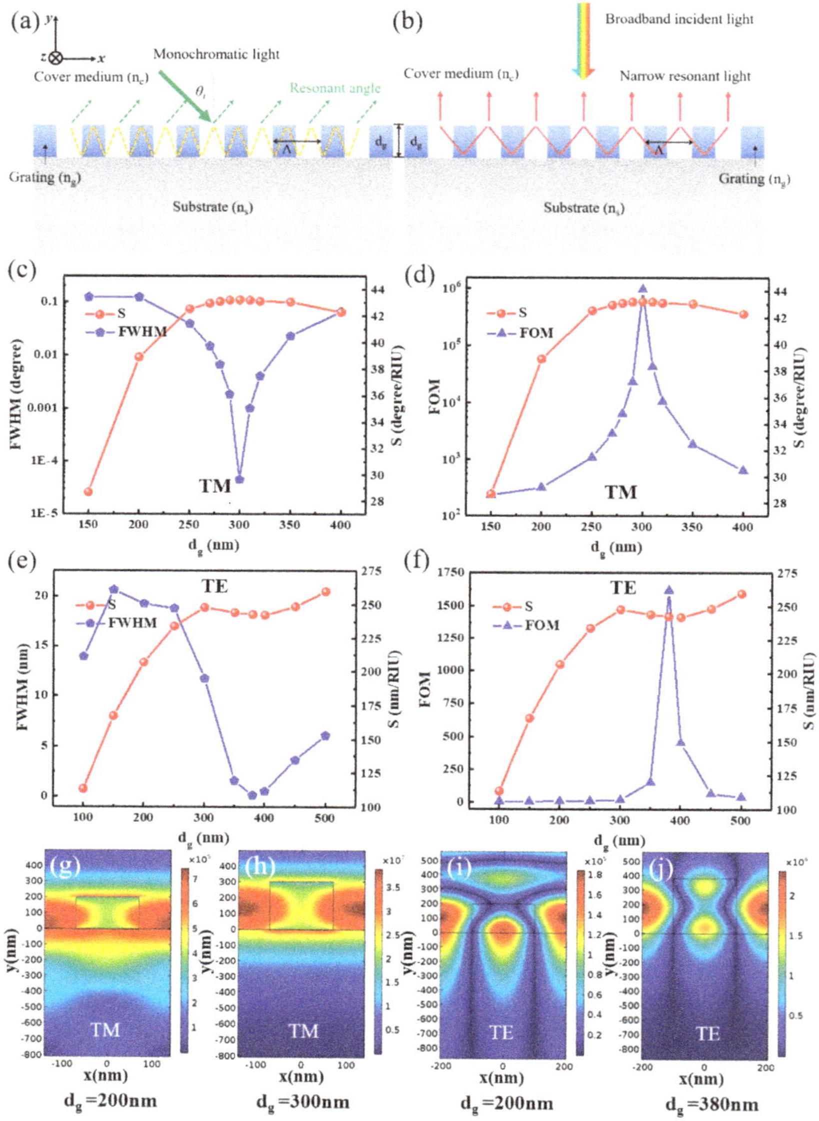

Figure 5. Schematic structure of the "waveguide–grating" structure for angular (**a**) and wavelength (**b**) resonance. Calculated sensitivity, resonant linewidth and figure of merit (FOM) versus d_g for angular resonance under transverse magnetic (TM) polarization (**c**,**d**), and for wavelength resonance under transverse electric (TE) polarization (**e**,**f**). Electric field distribution of the "waveguide–grating" guided mode resonance (GMR) structure of the wavelength resonance with d_g values of 200 nm and 300 nm for TM (**g**,**h**) and angular resonance with d_g values of 200 nm, 380 nm for TE (**i**,**j**). Λ is 280 and 410 nm for the angular and wavelength resonance respectively and filling factor (FF) is 0.5.

Different Λ were chosen to determine the d_g value that lead to a superior FOM. For wavelength resonance, different Λ varying from 330 nm to 550 nm, were chosen to calculate the position at which there is a minimum resonant linewidth and higher sensitivity for TE polarization, as shown in Table 3. The relationship among Λ, d_g, and the corresponding FOM is depicted in Figure 6a.

Table 3. Linewidth (FWHM), sensitivity (S) and figure of merit (FOM) of the wavelength resonance under different Λ and corresponding d_g for transverse electric (TE) modes.

Λ (nm)	d_g (nm)	FWHM (nm)	*S* (nm/RIU)	FOM
330	306.44	0.12	195	1625
410	380.00	0.15	242.75	1618.333
450	417.50	0.16	266.00	1662.500
550	509.74	0.20	325.25	1626.250

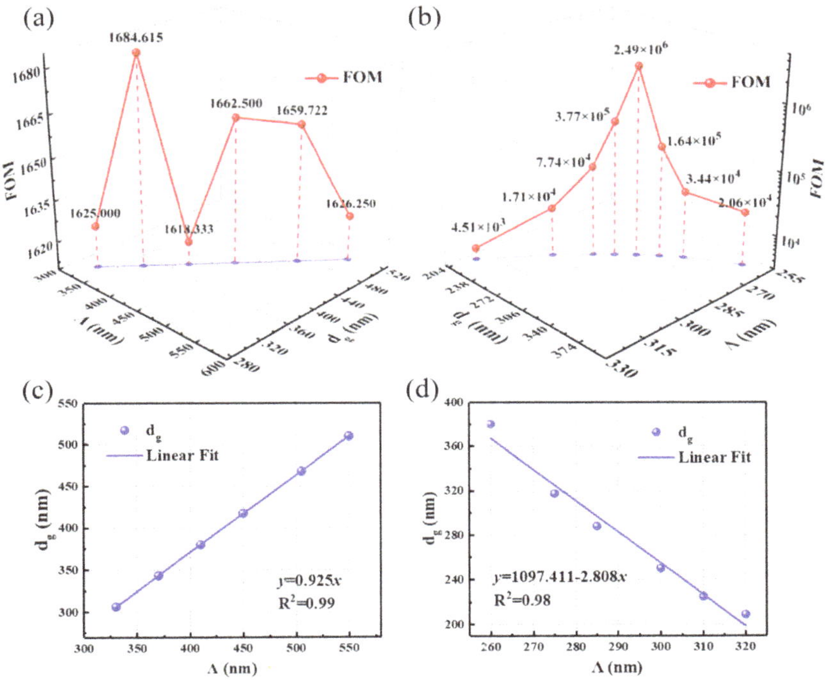

Figure 6. The best figure of merit (FOM) value for wavelength (**a**) and angular resonance (**b**) under different Λ and corresponding d_g values. Λ as a function of d_g for wavelength (**c**) and angular (**d**) resonance.

For angular resonance, different Λ, varying from 260 nm to 320 nm, were chosen to determine the best FOM under TM polarization, as shown in Table 4. The angular sensitivity is inversely proportional to Λ. The relationship between Λ, d_g, and the corresponding FOM is depicted in a three-dimensional schematic diagram in Figure 6b. All FOM values are higher than those of already reported values [18,39]. Λ as a function of d_g for the wavelength and angular resonance are respectively shown in Figure 6c,d. The ratio of Λ to d_g for the wavelength resonance was 0.925. A linear relationship was obtained between Λ and d_g for the angular resonance, as shown in Figure 6d, and the formula of the best-fit line is $y = 1097.411 - 2.808x$. We can apply the linear relationship to design "waveguide–grating" GMR sensors with optimized FOM.

Table 4. Linewidth (FWHM), sensitivity (S) and figure of merit (FOM) of the angular resonance under different Λ and corresponding d_g for transverse magnetic (TM) modes.

Λ (nm)	d_g (nm)	FWHM (Degree)	S (Degree/RIU)	FOM
260	380	3.3×10^{-3}	67.921	2.058×10^{4}
280	302	1.7×10^{-5}	48.322	2.486×10^{6}
290	274	4.9×10^{-4}	37.902	7.735×10^{4}
310	225	3.9×10^{-3}	30.927	7.930×10^{3}

The point of these optimized FOM results is that a narrow resonant linewidth occurred abruptly, as shown in Figure 5c,e. Norton et al. investigated linewidth of "waveguide–grating" GMR structure through coupled-mode theory [47]. Briefly, the angular and wavelength linewidth is proportional to the coupling loss coefficient, and this coefficient is determined by the overlap of the bound mode and the radiation mode. At higher grating depths, the orthogonality of the two modes decrease the overall magnitude of the coupling loss, resulting in a narrow resonant linewidth. In our work, similar resonant linewidth phenomenon can be found in Figure 5c,e. Although Norton et al. just investigated one grating period, it can be predicted that different grating periods have their own grating depths, where the narrow resonant linewidth occurs abruptly.

4. Conclusions

In summary, with the aim of achieving a higher angular and wavelength FOM for GMR sensors, we systematically presented a parametric analysis elucidating the influence of structural design factors on the performance of GMR sensors. Combining an analytical model and numerical algorithm, we determine an effective and convenient method to achieve higher angular and wavelength FOM of "grating–waveguide" GMR sensors. A suitable fixed depth of the grating and waveguide will facilitate a higher FOM. This method is suitable for both angular and wavelength resonance. A linear relationship to design "waveguide–grating" GMR sensors with optimized FOMs was determined using numerical stimulations. These high FOM values can facilitate the performance of GMR sensors to achieve lower detection limits.

Author Contributions: Conceptualization, Y.Z. and X.W.; methodology, Y.Z. and X.W.; software, Y.Z., Z.G. and B.W.; validation, X.W., Z.G. and B.W.; writing—original draft preparation, Y.Z.; writing—review and editing, X.W.; funding acquisition, X.W.; All authors discussed the results and commented on the manuscript.

Funding: This work was supported by the Special Project of National Key Technology R&D Program of the Ministry of Science and Technology of China (2016YFC0201401), the National Key Basic Research Program of China (973 program-No. 2015CB352006) and the National Natural Science Foundation of China (NSFC) (61378080, 61327008, 61505032, and 61705039).

Conflicts of Interest: The authors declare no conflict of interest.

References

1. Sharon, A.; Rosenblatt, D.; Friesem, A.A.; Weber, H.G.; Engel, H.; Steingrueber, R. Light modulation with resonant grating-waveguide structures. *Opt. Lett.* **1996**, *21*, 1564–1566. [CrossRef]
2. Liu, Z.S.; Magnusson, R. Concept of multiorder multimode resonant optical filters. *IEEE Photonics Technol. Lett.* **2002**, *14*, 1091–1093. [CrossRef]
3. Quaranta, G.; Basset, G.; Martin, O.J.F.; Gallinet, B. Recent Advances in Resonant Waveguide Gratings. *Laser Photonics Rev.* **2018**, *12*, 1800017. [CrossRef]
4. Wawro, D.; Tibuleac, S.; Magnusson, R. Optical waveguide-mode resonant biosensors. In *Optical Imaging Sensors and Systems for Homeland Security Applications*; Javidi, B., Ed.; Springer: New York, NY, USA, 2006; pp. 367–384.
5. Magnusson, R.; Wang, S.S. New Principle for Optical Filters. *Appl. Phys. Lett.* **1992**, *61*, 1022–1024. [CrossRef]
6. Cunningham, B.; Li, P.; Lin, B.; Pepper, J. Colorimetric resonant reflection as a direct biochemical assay technique. *Sens. Actuators B-Chem.* **2002**, *81*, 316–328. [CrossRef]

7. Cunningham, B.; Lin, B.; Qiu, J.; Li, P.; Pepper, J.; Hugh, B. A plastic colorimetric resonant optical biosensor for multiparallel detection of label-free biochemical interactions. *Sens. Actuators B-Chem.* **2002**, *85*, 219–226. [CrossRef]
8. Block, I.D.; Chan, L.L.; Cunningham, B.T. Photonic crystal optical biosensor incorporating structured low-index porous dielectric. *Sensor Actuators B-Chem.* **2006**, *120*, 187–193. [CrossRef]
9. Zhang, W.; Ganesh, N.; Block, I.D.; Cunningham, B.T. High sensitivity photonic crystal biosensor incorporating nanorod structures for enhanced surface area. *Sens. Actuators B-Chem.* **2008**, *131*, 279–284. [CrossRef]
10. Shafiee, H.; Lidstone, E.A.; Jahangir, M.; Inci, F.; Hanhauser, E.; Henrich, T.J.; Kuritzkes, D.R.; Cunningham, B.T.; Demirci, U. Nanostructured Optical Photonic Crystal Biosensor for HIV Viral Load Measurement. *Sci. Rep.* **2014**, *4*, 4116. [CrossRef]
11. Tu, Y.K.; Tsai, M.Z.; Lee, I.C.; Hsu, H.Y.; Huang, C.S. Integration of a guided-mode resonance filter with microposts for in-cell protein detection. *Analyst* **2016**, *141*, 4189–4195. [CrossRef]
12. Tsai, M.Z.; Hsiung, C.T.; Chen, Y.; Huang, C.S.; Hsu, H.Y.; Hsieh, P.Y. Real-time CRP detection from whole blood using micropost-embedded microfluidic chip incorporated with label-free biosensor. *Analyst* **2018**, *143*, 503–510. [CrossRef]
13. Cunningham, B.T.; Zhang, M.; Zhuo, Y.; Kwon, L.; Race, C. Recent Advances in Biosensing With Photonic Crystal Surfaces: A Review. *IEEE Sens. J.* **2016**, *16*, 3349–3366. [CrossRef]
14. Ku, Y.F.; Li, H.Y.; Hsieh, W.H.; Chau, L.K.; Chang, G.E. Enhanced sensitivity in injection-molded guided-mode-resonance sensors via low-index cavity layers. *Opt. Express* **2015**, *23*, 14850–14859. [CrossRef]
15. Wan, Y.H.; Krueger, N.A.; Ocier, C.R.; Su, P.; Braun, P.V.; Cunningham, B.T. Resonant Mode Engineering of Photonic Crystal Sensors Clad with Ultralow Refractive Index Porous Silicon Dioxide. *Adv. Opt. Mater.* **2017**, *5*, 1700605. [CrossRef]
16. Hermannsson, P.G.; Sorensen, K.T.; Vannahme, C.; Smith, C.L.C.; Klein, J.J.; Russew, M.M.; Grutzner, G.; Kristensen, A. All-polymer photonic crystal slab sensor. *Opt. Express* **2015**, *23*, 16529–16539. [CrossRef] [PubMed]
17. Lin, S.F.; Wang, C.M.; Ding, T.J.; Tsai, Y.L.; Yang, T.H.; Chen, W.Y.; Chang, J.Y. Sensitive metal layer assisted guided mode resonance biosensor with a spectrum inversed response and strong asymmetric resonance field distribution. *Opt. Express* **2012**, *20*, 14584–14595. [CrossRef] [PubMed]
18. Mizutani, A.; Urakawa, S.; Kikuta, H. Highly sensitive refractive index sensor using a resonant grating waveguide on a metal substrate. *Appl. Opt.* **2015**, *54*, 4161–4166. [CrossRef]
19. Sader, E.; Sayyed-Ahmad, A. Design of an optical water pollution sensor using a single-layer guided-mode resonance filter. *Photonic Sens.* **2013**, *3*, 224–230. [CrossRef]
20. Wei, X.; Weiss, S.M. Guided mode biosensor based on grating coupled porous silicon waveguide. *Opt. Express* **2011**, *19*, 11330–11339. [CrossRef]
21. Li, H.Y.; Hsu, W.C.; Liu, K.C.; Chen, Y.L.; Chau, L.K.; Hsieh, S.; Hsieh, W.H. A low cost, label-free biosensor based on a novel double-sided grating waveguide coupler with sub-surface cavities. *Sens. Actuators B-Chem.* **2015**, *206*, 371–380. [CrossRef]
22. Nazirizadeh, Y.; Bog, U.; Sekula, S.; Mappes, T.; Lemmer, U.; Gerken, M. Low-cost label-free biosensors using photonic crystals embedded between crossed polarizers. *Opt. Express* **2010**, *18*, 19120–19128. [CrossRef] [PubMed]
23. Lin, Y.C.; Hsieh, W.H.; Chau, L.K.; Chang, G.E. Intensity-detection-based guided-mode-resonance optofluidic biosensing system for rapid, low-cost, label-free detection. *Sens. Actuators B-Chem.* **2017**, *250*, 659–666. [CrossRef]
24. Jahns, S.; Brau, M.; Meyer, B.O.; Karrock, T.; Gutekunst, S.B.; Blohm, L.; Selhuber-Unkel, C.; Buhmann, R.; Nazirizadeh, Y.; Gerken, M. Handheld imaging photonic crystal biosensor for multiplexed, label-free protein detection. *Biomed. Opt. Express* **2015**, *6*, 3724–3736. [CrossRef] [PubMed]
25. Takashima, Y.; Haraguchi, M.; Naoi, Y. High-sensitivity refractive index sensor with normal incident geometry using a subwavelength grating operating near the ultraviolet wavelength. *Sens. Actuators B-Chem.* **2018**, *255*, 1711–1715. [CrossRef]
26. Magnusson, R.; Lee, K.J.; Wawro, D. Guided-mode resonance biosensors employing phase detection. In Proceedings of the Diffractive Optics and Micro-Optics, Rochester, NY, USA, 10–13 October 2004; pp. 52–57.

27. Kuo, W.K.; Huang, N.C.; Weng, H.P.; Yu, H.H. Tunable phase detection sensitivity of transmitted-type guided-mode resonance sensor in a heterodyne interferometer. *Opt. Express* **2014**, *22*, 22968–22973. [CrossRef] [PubMed]
28. Sahoo, P.K.; Sarkar, S.; Joseph, J. High sensitivity guided-mode-resonance optical sensor employing phase detection. *Sci. Rep.* **2017**, *7*, 7607. [CrossRef] [PubMed]
29. Kuo, W.K.; Syu, S.H.; Lin, P.Z.; Yu, H.H. Tunable sensitivity phase detection of transmitted-type dual-channel guided-mode resonance sensor based on phase-shift interferometry. *Appl. Opt.* **2016**, *55*, 903–907. [CrossRef] [PubMed]
30. Boonruang, S.; Mohammed, W.S. Multiwavelength guided mode resonance sensor array. *Appl. Phys. Express* **2015**, *8*, 092004. [CrossRef]
31. Wang, Z.S.; Sang, T.; Zhu, J.T.; Wang, L.; Wu, Y.G.; Chen, L.Y. Double-layer resonant Brewster filters consisting of a homogeneous layer and a grating with equal refractive index. *Appl. Phys. Lett.* **2006**, *89*, 241119. [CrossRef]
32. El Beheiry, M.; Liu, V.; Fan, S.H.; Levi, O. Sensitivity enhancement in photonic crystal slab biosensors. *Opt. Express* **2010**, *18*, 22702–22714. [CrossRef] [PubMed]
33. Szeghalmi, A.; Kley, E.B.; Knez, M. Theoretical and Experimental Analysis of the Sensitivity of Guided Mode Resonance Sensors. *J. Phys. Chem. C* **2010**, *114*, 21150–21157. [CrossRef]
34. Sang, T.; Wang, Z.S.; Zhu, J.T.; Wang, L.; Wu, Y.G.; Chen, L.Y. Linewidth properties of double-layer surfacerelief resonant Brewster filters with equal refractive index. *Opt. Express* **2007**, *15*, 9659–9665. [CrossRef]
35. Lin, S.F.; Wang, C.M.; Tsai, Y.L.; Ding, T.J.; Yang, T.H.; Chen, W.Y.; Yeh, S.F.; Chang, J.Y. A model for fast predicting and optimizing the sensitivity of surface-relief guided mode resonance sensors. *Sens. Actuators B-Chem.* **2013**, *176*, 1197–1203. [CrossRef]
36. Ju, J.; Han, Y.A.; Kim, S.M. Design Optimization of Structural Parameters for Highly Sensitive Photonic Crystal Label-Free Biosensors. *Sensors* **2013**, *13*, 3232–3241. [CrossRef]
37. Lu, H.; Huang, M.; Kang, X.B.; Liu, W.X.; Dong, C.; Zhang, J.; Xia, S.Q.; Zhang, X.Z. Improving the sensitivity of compound waveguide grating biosensor via modulated wavevector. *Appl. Phys. Express* **2018**, *11*, 082202. [CrossRef]
38. Zhang, M.; Lu, M.; Ge, C.; Cunningham, B.T. Plasmonic external cavity laser refractometric sensor. *Opt. Express* **2014**, *22*, 20347–20357. [CrossRef]
39. Pitruzzello, G.; Krauss, T.F. Photonic crystal resonances for sensing and imaging. *J. Opt.* **2018**, *20*, 073004. [CrossRef]
40. Huang, Q.L.; Peh, J.; Hergenrother, P.J.; Cunningham, B.T. Porous photonic crystal external cavity laser biosensor. *Appl. Phys. Lett.* **2016**, *109*, 071103. [CrossRef]
41. Ge, C.; Lu, M.; George, S.; Flood, T.A.; Wagner, C.; Zheng, J.; Pokhriyal, A.; Eden, J.G.; Hergenrother, P.J.; Cunningham, B.T. External cavity laser biosensor. *Lab Chip* **2013**, *13*, 1247–1256. [CrossRef]
42. Zhang, M.; Ge, C.; Lu, M.; Zhang, Z.X.; Cunningham, B.T. A self-referencing biosensor based upon a dual-mode external cavity laser. *Appl. Phys. Lett.* **2013**, *102*, 213701. [CrossRef]
43. Zhang, M.; Peh, J.; Hergenrother, P.J.; Cunningham, B.T. Detection of Protein-Small Molecule Binding Using a Self-Referencing External Cavity Laser Biosensor. *J. Am. Chem. Soc.* **2014**, *136*, 5840–5843. [CrossRef] [PubMed]
44. Lan, G.L.; Zhang, S.; Zhang, H.; Zhu, Y.H.; Qing, L.Y.; Li, D.M.; Nong, J.P.; Wang, W.; Chen, L.; Wei, W. High-performance refractive index sensor based on guided-mode resonance in all-dielectric nano-silt array. *Phys. Lett. A* **2019**, *383*, 1478–1482. [CrossRef]
45. Abutoama, M.; Abdulhalim, I. Self-referenced biosensor based on thin dielectric grating combined with thin metal film. *Opt. Express* **2015**, *23*, 28667–28682. [CrossRef] [PubMed]
46. Priambodo, P.S. Theoretical analysis, design, fabrication and characterization of dielectric and nonlinear guided-mode resonance optical filters. *Diss. Abstr. Intern.* **2003**, *26*, 203–208.
47. Norton, S.M.; Erdogan, T.; Morris, G.M. Coupled-mode theory of resonant-grating filters. *J. Opt. Soc. Am. A* **1997**, *14*, 629–639. [CrossRef]

nanomaterials

MDPI

Article

High-Performance Ultraviolet Light Detection Using Nano-Scale-Fin Isolation AlGaN/GaN Heterostructures with ZnO Nanorods

Fasihullah Khan, Waqar Khan and Sam-Dong Kim *

Division of Electronics and Electrical Engineering, Dongguk University, Seoul 100-715, Korea; fasihullah.khan@dongguk.edu (F.K.); waqarkyz@gmail.com (W.K.)
* Correspondence: samdong@dongguk.edu; Tel.: +822-2260-3800

Received: 24 February 2019; Accepted: 8 March 2019; Published: 15 March 2019

Abstract: Owing to their intrinsic wide bandgap properties ZnO and GaN materials are widely used for fabricating passive-type visible-blind ultraviolet (UV) photodetectors (PDs). However, most of these PDs have a very low spectral responsivity $\boldsymbol{R}$, which is not sufficient for detecting very low-level UV signals. We demonstrate an active type UV PD with a ZnO nanorod (NR) structure for the floating gate of AlGaN/GaN high electron mobility transistor (HEMT), where the AlGaN/GaN epitaxial layers are isolated by the nano-scale fins (NFIs) of two different fin widths (70 and 80 nm). In the dark condition, oxygen adsorbed at the surface of the ZnO NRs generates negative gate potential. Upon UV light illumination, the negative charge on the ZnO NRs is reduced due to desorption of oxygen, and this reversible process controls the source-drain carrier transport property of HEMT based PDs. The NFI PDs of a 70 nm fin width show the highest $\boldsymbol{R}$ of a ~3.2×10^7 A/W at 340 nm wavelength among the solid-state UV PDs reported to date. We also compare the performances of NFI PDs with those of conventional mesa isolation (MI, 40×100 μm^2). NFI devices show ~100 times enhanced $\boldsymbol{R}$ and on-off current ratio than those of MI devices. Due to the volume effect of the small active region, a much faster response speed (rise-up and fall-off times of 0.21 and 1.05 s) is also obtained from the NFI PDs with a 70 nm fin width upon the UV on-off transient.

Keywords: high-responsivity; ultraviolet photodetectors; nano-scale fin isolation; wide-band gap semiconductors; ZnO nanorods; two-dimensional electron gas; visible-blind

---

1. Introduction

Ultraviolet (UV) light detection with high responsivity is of great interest, due to its promising applications in pathology [1], water treatment, safety, and defense-technologies [2]. In most commercialized systems, UV detection has been realized by photomultiplier tubes (PMTs), thermal detectors, Si or Ge based photodetectors (PDs), and charge-coupled devices (CCDs). However, these detection methods pose various problems for practical applications. For example, PMTs exhibit fragile vacuum-tube construction and require a high-voltage bias. Moreover, PMTs are sensitive to the magnetic-field; therefore, they need to be shielded from ambient light. Si based PDs or CCDs are one of the most commonly used solutions, due to their highly matured process technology [3]. Recently, n-Si/p$^+$-B photodiodes of a very high sensitivity and stability have been demonstrated, but they still exhibit responsivity in the order of ~10^{-1} A/W [4]. Many attempts have been made to improve the limited responsivity of Si PDs working in linear mode. Among them, avalanche-mode PDs are now very popular, and single-photon avalanche diodes are also used to obtain high-gain responsivity [5]. However, the use of stop-band filters for visible light reduces the quantum efficiency of these devices in the UV range.

In recent years, the nanostructures of wide bandgap (WBG) materials, such as diamond, ZnO, III-nitride, and SiC based UV PDs have attracted tremendous research interest for their many advantages [6]. For instance, they are capable of room temperature (RT) operation and have intrinsic optical transparency (visibly-blindness) in the visible spectral range. Furthermore, they have a low thermal conductivity, high breakdown field, and stability at elevated temperatures. In spite of these promising advantages, the WBG semiconductors have very low electron mobility. Even though noble heterostructures with a very high electron mobility of ~10^6 $cm^2/V{\cdot}s$ using a material system such as MgZnO/ZnO [7] have been reported, most of the passive PDs fabricated using conventional WBG semiconductors have very low spectral responsivity [8]. Moreover, the response speeds of PDs based on ZnO or GaN are very slow in general because the photoresponse characteristics depend on the well-known bottle-neck chemisorption process of oxygen at the surface of such materials [9].

Zinc oxide (ZnO) nanorod (NR) based UV PDs [10–15] have shown promising results in terms of response speed and spectral responsivity [16]. Significant research effort has been made to improve the performance of the PDs, either by improving the crystalline quality of NRs or by utilizing the composite coaxial structure of ZnO with other materials, such as copper oxide [17] and graphene [18]. However, most of the passive type PDs reported to date exhibit very slow response speed (tens of seconds) and low responsivity (~hundreds of A/W). Recently, an active-type UV PD using the ZnO NR-gated AlGaN/GaN high electron mobility transistor (HEMT) structure has been attempted as a part of an effort to attain a breakthrough in responsivity (~10^5 A/W) and to obtain a relatively fast response speed [19]. However, due to the very high dark current of the device, the HEMT based PD showed a relatively low on-off current ratio of ~3.

In this study, we demonstrate high-responsivity UV PDs based on the ZnO NR-gated AlGaN/GaN HEMT structure with nano-scale fin isolation (NFI). ZnO NRs act as the floating gate while the UV driven chemisorption process of oxygen at the surface of ZnO NRs [20] controls the conduction of the underlying two-dimensional electron gas (2-DEG) channel. The 2-DEG present at the interface of the AlGaN (barrier) and GaN (channel) layers is due to polarization induced electric potential in the heterostructure [21]. In the NFI PD structure, ZnO NRs surround the channel in the gate area. Consequently, the carriers in the 2-DEG channel are confined along the channel and in perpendicular direction to the interface. Therefore, under dark conditions, the 2-DEG channel is fully depleted due to negative surface potential generated via oxygen adsorption at the surface of ZnO NRs [9]. This structure resembles enhancement-mode (normally-off) AlGaN/GaN fin-shaped field-effect transistors (FINFETs) [22], demonstrating extremely broad transconductance and excellent off-state characteristics. In this study, the performance of the NFI PDs is compared to conventional mesa isolation (MI) AlGaN/GaN HEMT based PD of ~100×40 μm^2 active area. We examine various photoresponse characteristics of NFI and MI devices and investigate how structural differences influence their performance.

2. Device Fabrication and Characterization

Figure 1 shows schematic illustrations of two different HEMT based PD structures examined in this study. The interface of the GaN (channel) and AlGaN (barrier) layers have a confinement of 2-DEG, which acts as channel of very high electron concentration and mobility.

Figure 2 illustrates the essential process steps for our gateless NFI and MI HEMT photodetectors (PDs). The epitaxial layers of AlGaN/GaN were deposited on a 6 inch Si (111) by a metal-organic chemical vapor deposition system at NTT-AT (NTT-Advanced Technology Corporation, Kanagawa, Japan). As shown in the schematics, the undoped GaN buffers and channel layers of 3000–4000 nm thickness were grown first, followed by subsequent growths of the barrier layer of ~20 nm $Al_{0.25}GaN_{0.75}$ and the GaN cap layer of ~1.2 nm. The measured electron sheet carrier concentration and Hall mobility of the epitaxial layer were ~$5 \times 10^{12}/cm^2$ and ~1750 $cm^2/V{\cdot}s$, respectively. The active area (100×40 μm^2) for the gateless HEMT was defined by a mesa etching of 100 nm depth in the case of conventional MI structures, as shown in Figure 1. The NFI structure, on the other hand, was fabricated with 10 nano-fin-shaped isolations (10 gate fingers) with the same mesa depth of 100 nm and two different fin widths (W_{fin}) of 70 and 80 nm. By opening the 100 nm Si_3N_4 passivation layers deposited on the active region by plasma-enhanced chemical vapor deposition (PECVD), the gate area of a 2 μm gate length was defined. The ZnO NRs were then selectively grown as an active element for the UV light detection.

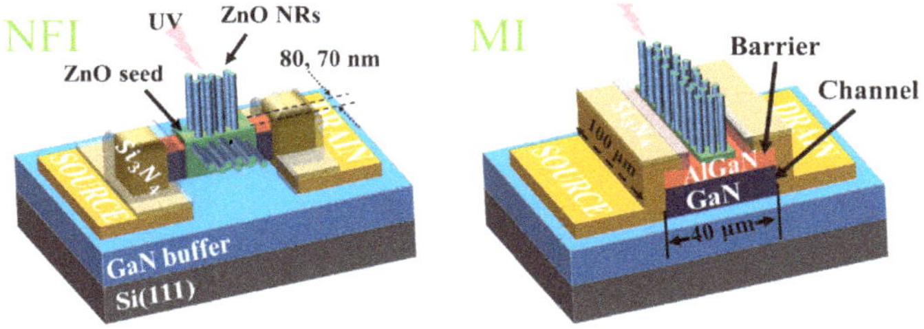

Figure 1. Schematics of the NFI (nanoscale fin isolation) and MI (mesa isolation) structures for the AlGaN/GaN HEMT (high electron mobility transistor)-based UV photodetectors.

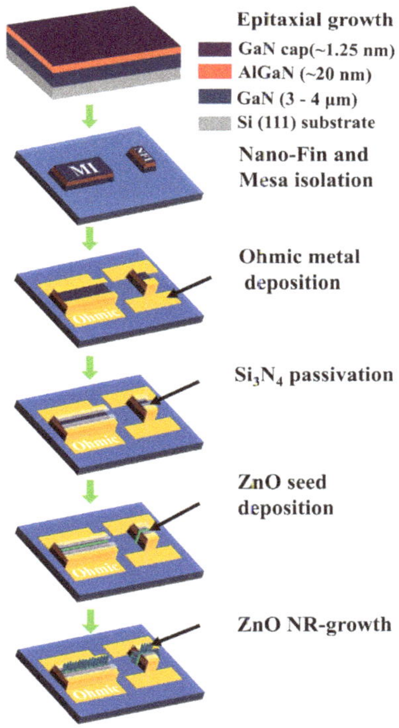

Figure 2. Schematic illustration of the major process steps used for the fabrication of the NFI and MI devices.

The epitaxial layers were cleaned and agitated with acetone and ethanol in bath sonication to remove dust and surface contamination, followed by rinsing in deionized (DI) water, then drying with nitrogen (N_2) gas. Optical lithography (Karl Suss, Garching, Germany, MA6 mask aligner, 365 nm) was used to define the active MI regions, including all the device patterns used in this experiment. However, patterning for the NFIs was performed by an electron beam lithography tool (Jeol, Tokyo,

Japan, JBX-9300FS, 100 keV) with a 70 nm PECVD SiO_2 hard mask to avoid the mask pattern erosion during the mesa etching. After the pattern development of the active regions, mesa etching was performed by removing 100 nm of thickness from the peripheral areas in a reactive ion etching system (RIE, STS Multiplex ICP) using BCl_3 and Cl_2 gases. Ohmic contacts were achieved by depositing the metal stack of Ti/Ni/Au (20/30/80 nm), by using an electron beam evaporation system, and a pattern lift-off method, using image reversal photoresist. Ohmic metals were then subjected to a subsequent rapid thermal alloy process at 900 °C for 35 s in N_2. The samples were then passivated by a 100 nm silicon nitride (Si_3N_4) layer deposited in a PECVD system at 200 °C and RF power of 1 kW using a NH_3/SiH_4 gas flow rate ratio of 1.5. The gate areas (2 × 100 µm) of two different structures were opened by etching the Si_3N_4 passivation using RIE in CF_4 plasma at a gas flow rate of 110 sccm and a chamber pressure of 40 mTorr (at an etching rate of ~9.4 nm/min) under the RF power of 100 W.

ZnO NRs were then grown in the gate area by using the hydrothermal synthesis method. Prior to the growth of NRs, a 20 nm thick seed layer (SL) (as shown in the top-right inset of Figure 3c) was deposited by spin coating the seed solution (3000 rpm and prebaking at 120 °C for 60 s) repeatedly 15 times. After that, the crystalline quality of the SL was improved by annealing the samples at ~350 °C on a hotplate for 1 h. The seed solution used in this study was prepared by dispersing 0.66 g of zinc acetate-dehydrate ($C_4H_6O_4Zn{\cdot}2H_2O$) salt in 30 mL of 1-propanol ($CH_3CH_2CH_2OH$). In this work, a 20 mM zinc-acetate-dehydrate concentration for seed solution was selected for the desired SL film quality, assuring the ZnO NR crystalline characteristics to be used for this PD application [23]. Various attempts, such as vacuum annealing and O_2 plasma post-treatment methods [10,24] to improve the SL film quality, which is the key to the high-quality NR crystallites, are also underway in our laboratory. The growth of ZnO NRs was limited to the gate area, which was done by etching the SL from all other areas except the gate region. The samples were then placed in a growth solution for 6 h on a hotplate at ~90 °C to grow ZnO NRs. The growth solution was prepared by mixing 0.25 mole equimolar concentration of zinc-nitrate-hexahydrate ($Zn(NO_3)_2{\cdot}6H_2O$, 99%) and hexamethylenetetramine (HMTA) ($C_6H_{12}N_4$, 99.5%) in deionized (DI) water. After growth of the the NRs, the samples were carefully cleaned with acetone, ethanol, and DI water, sequentially.

Sloped etching profiles for the NFI structure, the growth morphology of NRs, and the processed device structures were examined by plane-view scanning electron microscopy (SEM, 10 kV S-4800S-Hitachi, Tokyo, Japan). As shown in the top-right inset of Figure 3a, the measured bottom dimensions of the trapezoidal fins were 170 nm after RIE for the $\boldsymbol{W_{fin}}$ of 70 nm. The transmission electron microscope (TEM, 9500-Hitachi) was used to characterize the cross-sectional view and crystalline quality of the as-grown NRs. The RT photo-luminescence (PL) emission spectra from the as grown NRs was obtained by using a He-Cd laser illumination source with a 325 nm wavelength. Gate areas opened with a trench-shaped pattern inside the dark-shaded Si_3N_4 passivation and the NRs grown on them are shown in the plane-view SEM micrographs in Figure 3a,b. An inset (bottom-right) of Figure 3a shows the morphology of the ZnO NRs grown around the underlying nano-fins in the gate area of the NFI structures at a magnification of ×3000. Across the gate region, 10 fingers of NFIs are running in parallel, whereas the NRs are grown all the way vertically along the gate area in the MI structure, as shown in Figure 3b. The SEM image in Figure 3c shows that the NRs exhibit an average diameter of ~85 nm and a length of ~1.4 µm. Figure 3d shows a PL spectrum measured at RT from the as-grown ZnO NRs employed as a light absorbing structure in this work. PL characterization at RT is one of the most efficient tools to evaluate the crystalline quality for the WBG ZnO NRs of a direct band gap property. The spectrum exhibits a strong near band edge emission peak at a wavelength of ~380 nm, which is mainly associated with the band-to-band excitonic recombination of ZnO [25]. The near band edge emission intensity was increased about two times with the increase of the NR aspect ratio (AR, length/diameter) from ~8 to ~16. For this reason, NRs with an AR of ~16 were used in this experiment. Emissions in a visible range (420–650 nm) are due to various form of intrinsic defects, such as oxygen vacancy, zinc vacancy, and hydrogen and oxygen interstitials [10]. Despite the inevitable intrinsic defects and consequent visible emissions, as observed from most of the nano-crystallites grown through the aqueous solution based growth methods, our PL spectrum reveals fairly good optical properties in our ZnO crystallites compared to those of the NRs grown using similar methods [2,26,27].

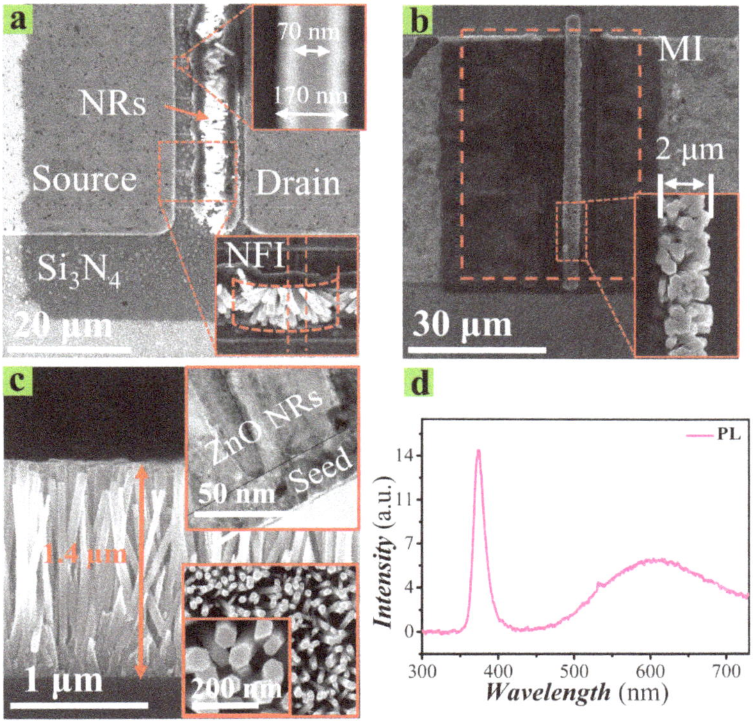

Figure 3. Plane-view SEM images of (**a**) 70 nm NFI PDs (photodetectors) (×1000) with an enlarged top-view of the NFI (W_{fin} = 70 nm) (top-right inset, ×5000) and ZnO NR (nanorod) profiles (bottom-right inset, ×3000) and (**b**) MI PDs (×500) with enlarged ZnO NRs profile (bottom-right inset, ×3000). (**c**) Cross-sectional SEM image of the as-grown NRs. Cross-sectional TEM image of the NRs and SL (seed layer) (top-right inset). Top-view of the NRs (bottom-right inset). (**d**) RT (room temperature) PL (photo-luminescence) spectrum of the as-grown ZnO NRs.

A wide band (300–700 nm) Xenon (Xe) lamp was employed as a light source to measure the photoresponse characteristics of the fabricated devices under optical intensities ranging from 0.5 to 16.5 μW/cm^2 (100–300 W lamp power). The transient characteristic measurements, according to the UV light (370 nm) on-and-off transient, were carried out by a programmable light shutter controlled in our measurement set-up. The transients of the drain current (I_{ds}) for the HEMT PDs as a function of time were recorded in a Keithley source measurement unit with a floating gate configuration at a drain voltage (V_{ds}) of 4 V. The spectral response of the PDs was measured by a focused illumination of a monochromatic light from a wide band (300–1100 nm) Xe-lamp (Ushio UXL-75XE, Ushio Inc., Tokyo, Japan) light source of a 16.5 μW incident optical power. In this measurement set-up, the responsivity was measured at a light chopping frequency of 30 Hz using a lock-in amplifier in a series configuration with drain-source probes for the detection of an amplified change in current. The change of I_{ds} as a function of incident light wavelength varying from 300 to 800 nm was recorded under a floating gate configuration at a V_{ds} of 4 V. To obtain a monochromatic light from the Xe-lamp, a spectral optics monochromator (CM110 1/8 m) with 2400 lines/mm grating was used.

3. Sensing Mechanism

As illustrated in the schematics of Figure 4, the UV sensing mechanism of the HEMT based PDs depends on the chemisorption of oxygen at the surface of NRs and the consequent change in carrier concentration in underlying 2-DEG. The as-grown crystalline ZnO-NRs are n-type in nature due to a large number of donor defects, such as hydrogen interstitial and Zn interstitial [28]. Under dark conditions (no UV light illumination), oxygen molecules (O_2) transported from the ambient air to the ZnO NR surface tend to trap electrons from the conduction band of ZnO and leave behind positively charged ionized donors in the surface space charge region, while the negatively charged O_2 molecules are fixed to the surface of NRs as adsorbed oxygen ions ($O^-_{2_{ads}}$):

$$O_2 + e^- \rightarrow O^-_{2_{ads}}. \quad (1)$$

This reaction leads to the expansion of the space charge region near the surface of the NRs due to the depletion of the surface electron states by $O^-_{2_{ads}}$, as depicted in the left of Figure 4a. As a consequence, the adsorption process gives rise to a negative potential at the gate of our HEMT based PDs (as shown in the left of Figure 4b,c), thereby dropping the carrier concentration in the 2-DEG channel at the interface of AlGaN/GaN. This process will eventually reduce the conductance of channel and drain-to-source current ($\boldsymbol{I_{ds}}$) in dark conditions. The Schottky gate (Ni/Au) HEMTs (2 μm gate length, 100 μm gate width) fabricated in our group using the same epitaxial structure and Si_3N_4 passivation showed a threshold voltage ($\boldsymbol{V_{TH}}$) of ~−3 V [29]. This result demonstrates that our HEMT devices with an Au/Ni gate electrode are normally on at zero gate voltage $\boldsymbol{V_{gs}}$ (depletion-mode). We suppose that the $\boldsymbol{V_{TH}}$ of the HEMTs is given by [30]

$$\varnothing_b - \Delta E_c - \left(en_s d^2\right)/2\varepsilon \quad (2)$$

where $\varnothing_b$ is the Schottky barrier height, ΔE_c is the conduction band offset, $\boldsymbol{e}$ is the electron charge, $\boldsymbol{n_s}$ is the 2-DEG sheet carrier density, $\boldsymbol{d}$ is the barrier layer thickness, and $\boldsymbol{\varepsilon}$ is the dielectric constant. Because the difference between the estimated $\varnothing_b$ of Ni at the interface with GaN (~1.1 eV) [31] and the GaN surface band bending (~1.0 eV) [32] in ambient air (fully saturated by $O^-_{2_{ads}}$ in the dark state) is quite small, it can be reasonably assumed that a significant amount of dark current ($\boldsymbol{I_{dark}}$) through the 2-DEG channel is unavoidable from our gateless MI HEMT devices under a floating gate condition due to their normally-on characteristics as shown in Figure 4c.

On the other hand, a different mode of operation is expected for the NFI structure because the 2-DEG channel region inside the nano-fins is three-dimensionally surrounded by a free surface, whereon the ZnO NRs are grown with many $O^-_{2_{ads}}$ around the surface in dark conditions, as shown in Figure 4b. This makes the NFI PDs operate more closely to a normally-off mode (enhancement mode), thereby exhibiting a much lower $\boldsymbol{I_{dark}}$, because the carriers in the 2-DEG region can be highly depleted, even under the floating gate conditions, due to the surface depletion of the NRs by the $O^-_{2_{ads}}$ surrounding the extremely small volume of the active region. The same phenomenon was also observed in the AlGaN/GaN heterojunction FinFETs [33]. With the decrease of $\boldsymbol{W_{fin}}$ from 200 to 60 nm, it was observed from the HEMT devices that electron density in the 2-DEG channel at zero $\boldsymbol{V_{gs}}$ rapidly drops with a positive shift in $\boldsymbol{V_{TH}}$ because of the fringing-field from the side gates depleting the 2-DEG channel. As a result, the fin-HEMT showed a change in the conduction mechanism from normally-on to normally-off modes.

Due to light absorption, UV light illumination generates electron-hole pairs near the surface of the ZnO NRs [34]. The generated holes recombine with the electrons trapped by $O^-_{2_{ads}}$ at the surface. In this way, the O_2 molecules start to desorb from the surface, as shown in the right of Figure 4a. This phenomenon gives rise to a reduction of negative charge in the gate region, thereby increasing the carrier concentration in the 2-DEG channel and the drain to the source current ($\boldsymbol{I_{photo}}$) under UV illumination, as shown in the right parts of Figure 4b,c.

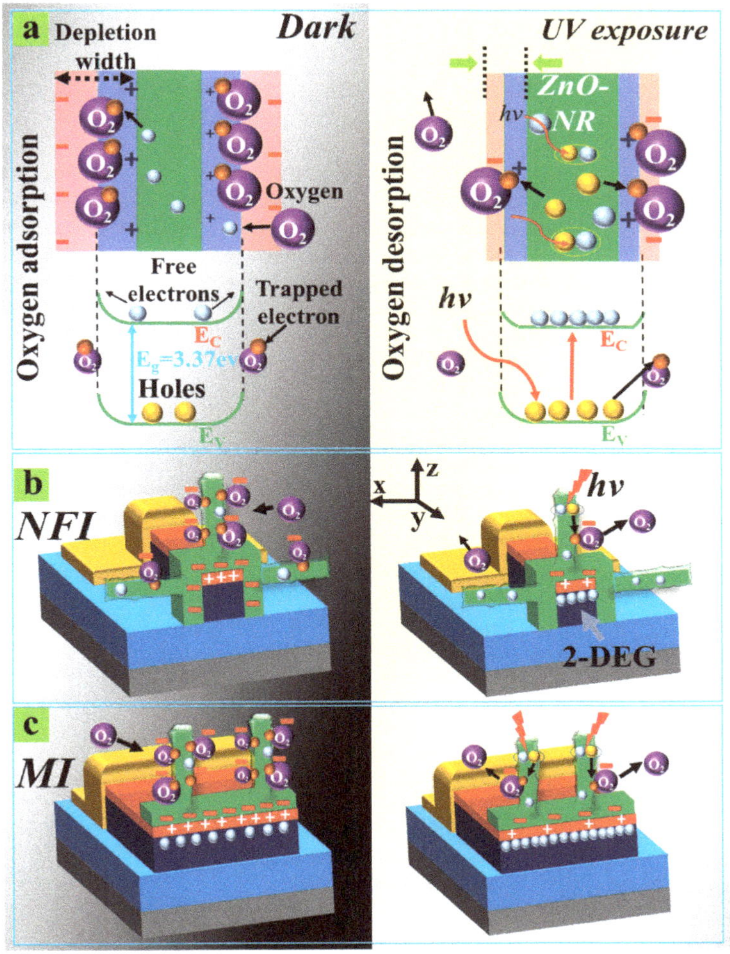

Figure 4. Schematic illustrations of (**a**) oxygen adsorption (left: in the dark) and desorption (right: under UV illumination) at the surface of ZnO-NRs. Oxygen adsorption (left: in the dark) and desorption (right: under UV illumination) processes are taking place at the surface of ZnO NRs grown on the gate areas of (**b**) NFI and (**c**) MI PDs. Cross-sectional schematics of each PD structure viewed in y-axis direction are illustrated; 2-Deg (two-dimensional electron gas).

As long as the response speed of PDs is accounted for, it can be assumed that the response kinetics upon the UV light on-and-off transient are not controlled by the drift motion of electrons in the 2-DEG channel but by the adsorption and desorption reaction of O_2 on the ZnO NR surface. This result occurs because the mobility of the electrons confined two-dimensionally in the channel region (~1750 $cm^2\ V^{-1}s^{-1}$) [33,35] is so fast that the carrier channel transit time cannot be a bottleneck parameter controlling the whole sensor response speed. As proposed in our previous study, the rate of charge change dQ/dt can be given by [35]:

$$dQ/dt = -\alpha e^{\beta Q} \tag{3}$$

where α and β are the constants. By numerically solving Equation (3), it can be determined that one critical parameter affecting the response time is the gate area. From the calculations based on this model, the response (or recovery) time increases with the increase of gate area caused by the consequent increase of the total gate charge. We have much smaller gate capacitance in the NFI PD structure than in the MI structure; therefore, a faster response speed upon UV transient illumination can be expected from the NFI structure.

Spectral responsivity $\boldsymbol{R}$, which can be defined as a ratio of $I_{photo} - I_{dark}$ to the incident optical power P_i, is one of the key measures to evaluate the performance of PDs. As reported in our previous studies [19,35], a significant enhancement in $\boldsymbol{R}$ was shown from the NR-gated PDs due to a vast surface area of the ZnO nanostructure and a much higher surface-to-volume ratio than that of the planar ZnO thin film gate structure. In our HEMT-based PDs, the $\boldsymbol{R}$ can be also significantly influenced by the gain characteristic for the HEMT, which can be expressed by the change of I_{photo} according to the change of light power irradiating on the gate area.

4. Results and Discussion

Figure 5a shows the equivalent electrical circuit diagram of both MI and NFI devices, where MI PD is illustrated as a normally-on transistor while the NFI PD is a normally-off transistor. To examine the characteristic difference of UV responses from the two different PD structures, we first measured I_{ds} in dark conditions (I_{dark}) and under UV illumination (I_{photo}) at drain voltage (V_{ds}), ranging from 0–5 V. UV light exposure was provided from an Xe lamp, operating at 300 W power, with a monochromatic light filter at 370 nm. This UV source produced an incident light intensity of 16 μW/cm^2, as measured by a power meter. Figure 5b demonstrates that our MI PD exhibits a high I_{dark} of ~10 mA/mm at a V_{ds} of 4 V due to the enhancement-mode operation, as discussed in the previous section. In the case of NFI PDs, significantly reduced values of I_{dark} (~0.19 and ~0.27 mA/mm for W_{fin} of 70 and 80 nm, respectively) were measured at a V_{ds} of 4 V.

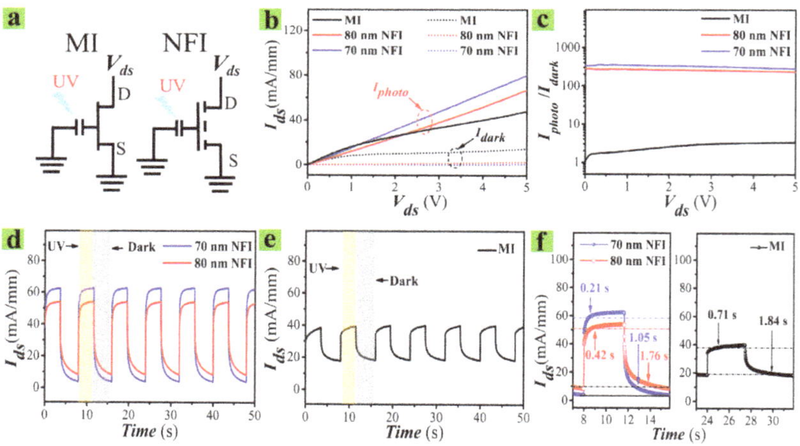

Figure 5. Photoresponse of MI and NFI PDs. (**a**) Equivalent electrical circuits of MI (left) and NFI devices (right). (**b**) Measured I_{ds} versus V_{ds} and (**c**) on-off current ratio (I_{photo}/I_{dark}) of MI and NFI PDs. Time resolved photoresponse characteristics of (**d**) NFI and (**e**) MI PDs (V_{ds} = 4 V). (**f**) Magnified views of transient characteristics of NFI (left) and MI (right) PDs.

The performance of PDs can be evaluated by few important parameters, such as spectral responsivity $\boldsymbol{R}$, photoconductive gain $\boldsymbol{G}$, specific detectivity $\boldsymbol{D^*}$, and sensitivity $\boldsymbol{S}$ [36]. Despite the high $\boldsymbol{R}$ nature of our MI HEMT-based PDs, this high I_{dark} of the MI structure can critically deteriorate the photo-sensitivity performance associated with $\boldsymbol{S}$, which is given by ($I_{photo} - I_{dark})/I_{dark}$), or the on-off current ratio (I_{photo}/I_{dark}), as well as D^*. As shown in Figure 5c, much improved on-off current ratios (290~340) were recorded from the NFI PDs compared to those of the MI devices (~4). This result is mainly due to the suppression of I_{dark} caused by the fully depleted 2-DEG channel of the NFI devices in dark conditions.

The high-speed transient characteristic of the UV PD is one of the key factors for real-time application. Specific cases, such as the non-invasive assessment of cancer cells by optical biopsy [37], require ultra-fast PDs with response and recovery time in the order of a few milliseconds. To assess the photocurrent transient of the fabricated devices, the change in I_{ds} as a function of the on-and-off of the UV light exposure time was measured using a programmable shutter. Figure 5d,e shows that both the NFI and MI PDs produce sharp increases in I_{ds} upon UV illumination and a slower fall-off upon termination of UV exposure. The NFI device with a 70 nm W_{fin} showed the fastest rise-up time (or UV response time) and fall-off time (or recovery time) of 0.21 and 1.05 s, respectively. We hereafter define the rise-up and fall-off time as the time intervals for I_{ds} to ramp up to 90% of the maximum saturation value after UV turn-on and to ramp down by 90% from the maximum value after UV-off. On the other hand, the MI device showed much slower response and recovery times of ~0.71 s and ~1.84 s, respectively. This significant improvement in response speed of the NFI device is due to the minimized dimension of gate area where the light absorption takes place; therefore, less time is required to complete the O_2 adsorption-desorption process in the smaller area of the NFI PDs than in the larger area of the MI device, as discussed in the previous section. The ZnO NR based photoconductive PDs [26,38] still show a long recovery time on a seconds scale, even though a great amount of research effort has been made on enhancing the response speed of these passive PDs. UV PDs based on ZnO nanowire networks with Pt contacts have been fabricated on glass substrates by exhibiting a fast recovery time of 0.2 s with a high photosensitivity (~5 × 10^3) at 365 nm [39]. The fastest UV PDs of GaN-based metal-semiconductor-metal, p-i-n, or metal Schottky barrier devices [6] have shown extremely high speed (from microseconds to picoseconds) and low-noise capabilities. However, PDs of these GaN-based structures, developed specially to improve the UV response characteristics, exhibit a very low spectral responsivity of less than 1 A/W [8].

Figure 6a shows the schematic illustration of PD before and after the growth of the ZnO NRs. Shown in Figure 6b,c are the measured I_{dark} values of the two different device structures before and after NR growths on the gate region. For the MI PDs, the dark currents were significantly reduced by the attachment of NRs on the gate region. However, the I_{dark} values measured over the entire drain voltage range were still very high (~10 mA/mm) due to the incomplete depletion of the 2-DEG channel in a large volume of the active region. The measured I_{dark} of the NFI PDs before the NR growth were very high, reaching up to ~560 mA/mm, but they rapidly dropped to a very low level of ~hundreds of µA/mm by the attachment of NRs. This result reveals that both the surface depletion effect in the small volume of an active region and the attachment of the NRs with many $O_{2\,ads}^-$ on their surfaces lead to the formation of a fully-depleted 2-DEG channel in the NFI structure.

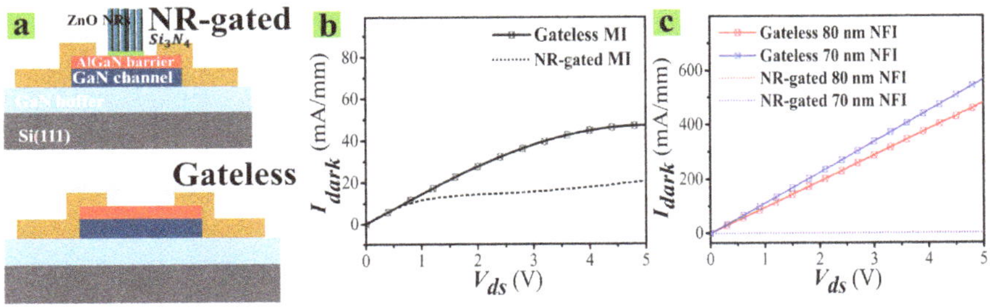

Figure 6. (**a**) Schematic illustration of NR-gated PD and Gateless PD. Measured I_{dark} of MI (**b**) and NFI (**c**) device structure before (gateless) and after NR growths (NR-gated).

The performance of the PDs can be also assessed by spectral responsivity $\boldsymbol{R}$ and specific detectivity $\boldsymbol{D}^*$, which are expressed in the following equations [36]:

$$R = \frac{I_{photo} - I_{dark}}{P_i} \tag{4}$$

$$D* = \sqrt{A}R/\sqrt{2qI_{dark}} \tag{5}$$

where $\boldsymbol{A}$ is the active area of device, $\boldsymbol{q}$ is the electron charge, and $\boldsymbol{P_i}$ is the radiant light power incident on the active area of the device. The responsivities of our PDs were measured using an incident light intensity of 16 $\mu W/cm^2$ with an effective area of 2.4×2.4 mm^2. NFI PDs showed an extremely high responsivity of ~3.2×10^7 A/W at 340 nm, which is the best performance among any solid-state PDs reported to date [40–44], and even ~100 times higher than that (~2×10^5 A/W at 340 nm) of the MI device, as shown in the right of Figure 7a. This improvement in responsivity of the NFI PD is attributed to a higher photonic transconductance (hereafter, we call it $g_{m,photo}$) characteristic of our HEMT-based PDs, which represents the ratio of the photocurrent change at the drain terminal to the change in incident optical power at the gate terminal. We define the $g_{m,photo}$ as following.

$$g_{m,photo} = \frac{\Delta I_{photo}}{\Delta P_i}(A/W). \tag{6}$$

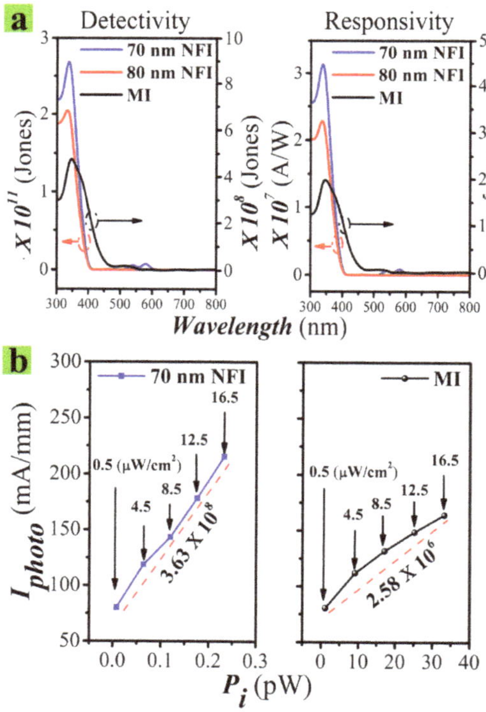

Figure 7. (**a**) Specific detectivity (left) and spectral responsivity (right) as functions of radiant light wavelength measured from NFI and MI UV detectors. (**b**) I_{photo} versus P_i of the NFI (left) and MI (right) PDs measured at V_{ds} of 5 V.

To calculate the $g_{m,photo}$ values of the two different device structures, the I_{photo} values were measured at various optical intensities, varying from 0.5 to 16 μW/cm^2 at a V_{ds} of 5 V. Shown in Figure 7b is the measured I_{photo} versus P_i from the MI and NFI PDs. The estimated $g_{m,photo}$ values for the NFI and MI PDs were 3.63 × 10^8 and 2.58 × 10^6 A/W, respectively. Even though this $g_{m,photo}$ parameter does not directly represent the gain ($g_m = \Delta I_{ds}/\Delta V_{gs}$) characteristics of the field-effect transistors, it can be used as a performance measure to estimate the gain characteristic of our gateless HEMT PDs, because the change of incident optical power is directly associated with the virtual ΔV_{gs} induced by the change in numbers of $O_2^-{}_{ads}$ on the ZnO NR surface under the illumination of UV light.

FinFET technology has recently shown a major increase in adoption of use within Si integrated circuits. The advantages of a FINFET structure, even though there are a number of subtly different forms, can be numerous, but they are basically based on "channel controllability" in a nano-scale channel-length regime of FETs [45]. Furthermore, Si-based FINFETs have shown a significant enhancement in the dependence of I_{ds} on V_{gs} at any applied bias in the sub- and near-threshold regimes by the superior electrostatics of the device architecture [46]. Normally-off Al_2O_3/GaN metal-insulator FINFETs (W_{fin} = 50 nm, 1 μm gate length) also showed very high maximum I_{ds}, g_m, and maximum field-effect mobility of 1.51 A/mm, 580 mS/mm, and 293 cm^{2}V^{-1}s^{-1}, respectively, due to the more effective increase of 2-DEG electron concentration and higher electron mobility by enhanced gate controllability than the planar devices [47,48]. The superior performance of R in our NFI PDs is most likely attributed to this enhancement of "gate controllability" screening the field lines effectively from the interface traps or the defects near the 2-DEG channel, thereby reducing the electron scattering in the GaN channel.

The D^* of a PD, as defined in Equation (5), is also an important figure of merits used to describe performance. The NFI PDs of 70 nm W_{fin} exhibited a maximum D^* of ~3.2 × 10^{12} Jones (cm·Hz$^{-1/2}$/W) at 340 nm, which is four orders of magnitude higher than that of MI PDs (~4 × 10^8 Jones), as shown in the left of Figure 7a, which is mainly due to the very low I_{dark} of the NFI structure. A very high D^* up to a value of 1.4 × 10^{15} Jones [14] was reported from the ZnO NRs based UV PDs. However, the R achieved by this device was much lower (~10^3 A/W) than that of PDs.

5. Conclusions

Most of the performance parameters, such as R, D^*, on-off current ratio, and response speed, were all significantly improved by employing the NFI structure for the AlGaN/GaN HEMT based UV PDs with a ZnO NR UV-absorbing structure. The NFI PD, especially, exhibited an extremely high R of ~3.2 × 10^7 A/W. This performance enhancement was due to the subsequent characteristic change of the gateless HEMTs, induced by the reduction of W_{fin} to 70–80 nm. The NFI structure significantly improved the gain characteristics caused by enhanced gate controllability in nano-fin channels beyond the inherent high performance in conversion efficiency of the photon to electron-hole pair generation due to the large surface-to-volume ratio of the ZnO NRs grown in the active region. As the width of the 2-DEG channel is reduced by the NFI profile, the side-wall surface depletion in nano-fins and the attachment of NRs with numerous $O_{2\,ads}^-$ on the surface of the NRs lead to the formation of a fully-depleted 2-DEG channel and pushed the V_{TH} to a positive value. The measured high on-off current ratio and D^* are mainly due to this normally-off operation of the NFI PD structure. An improvement in response speed of the PDs is associated with the minimized dimension of the gate area and the resulting gate capacitance of the NFI structure, where much less time for charging or discharging is required for the O_2 adsorption-desorption process. The fabricated PDs also showed a linear dependence of photocurrent on the input light intensity in a range of 0.5–16.5 μW/cm^2, regardless of device structure. The measured $g_{m,photo}$ value for the NFI PDs of 70 nm W_{fin} was 3.63 × 10^8 A/W, which was ~100 times greater than that of the MI PDs.

Author Contributions: Conceptualization, F.K.; methodology, F.K.; validation, F.K.; formal analysis, F.K.; investigation, F.K.; resources, S.-D.K.; data curation, F.K.; writing—original draft preparation, F.K.; writing—review and editing, F.K. and S.-D.K.; visualization, S.-D.K.; supervision, S.-D.K. and W.K.; project administration, S.-D.K.; funding acquisition, S.-D.K.

Funding: All the experiments performed in this work are financially sponsored by National Research Foundation (NRF) grant (2016006533) under the program: "Mid-career Researchers" by the Ministry of Science and Technology, Republic of Korea. Similarly, the materials characterization was achieved with the financial support from NRF, Republic of Korea under the program: "Nano-materials Technology Development" (2009-0082580) managed by Ministry of Science, ICT and Future Planning.

Acknowledgments: The authors would like to thank Jihyun Kim, Hafiz Muhammad Salman Ajmal, Kiyun Nam and Noor ul Huda from Dongguk University for the continued encouragement, help, and support in carrying out the experiments and preparation of results.

Conflicts of Interest: The authors declare no conflicts of interest.

References

1. Bigio, I.J.; Mourant, J.R. Ultraviolet and visible spectroscopies for tissue diagnostics: Fluorescence spectroscopy and elastic-scattering spectroscopy. *Phys. Med. Biol.* **1997**, *42*, 803–814. [CrossRef] [PubMed]
2. Monroy, E.; Omns, F.; Calle, F. Wide-bandgap semiconductor ultraviolet photodetectors. *Semicond. Sci. Technol.* **2003**, *18*, R33–R51. [CrossRef]
3. Li, Z. Radiation damage effects in Si materials and detectors and rad-hard Si detectors for SLHC. *J. Instrum.* **2009**, *4*, P03011. [CrossRef]
4. Nanver, L.K.; Qi, L.; Mohammadi, V.; Mok, K.R.M.; Boer, W.B.D.; Golshani, N.; Sammak, A.; Scholtes, T.L.M.; Gottwald, A.; Kroth, U.; et al. Robust UV/VUV/EUV PureB Photodiode Detector. *IEEE J. Sel. Top. Quantum Electron.* **2014**, *20*, 306–316. [CrossRef]
5. Qi, L.; Mok, K.R.C.; Aminian, M.; Charbon, E.; Member, S.; Nanver, L.K. UV-sensitive low dark-count pureb single-photon avalanche diode. *IEEE Trans. Electron Devices* **2014**, *61*, 3768–3774. [CrossRef]
6. Omnès, F.; Monroy, E.; Muñoz, E.; Reverchon, J.-L. Wide bandgap UV photodetectors: A short review of devices and applications. In *Gallium Nitride Materials and Devices*; Integrated Optoelectronic Devices: San Jose, CA, USA, 2007; p. 64730E.
7. Falson, J.; Kozuka, Y.; Uchida, M.; Smet, J.H.; Arima, T.; Tsukazaki, A.; Kawasaki, M. MgZnO/ZnO heterostructures with electron mobility exceeding 1×106 cm^2/Vs. *Sci. Rep.* **2016**, *6*, 26598. [CrossRef] [PubMed]
8. Al-Heuseen, K.; Hashim, M.R. Enhancing responsivity of porous GaN metal-semiconductor-metal ultraviolet photodiodes by using photoelectrochemical etching. *Acta Phys. Pol. A* **2012**, *121*, 71–73. [CrossRef]
9. Melnick, D.A. Zinc Oxide Photoconduction, an Oxygen Adsorption Process. *J. Chem. Phys.* **1957**, *26*, 1136–1146. [CrossRef]
10. Khan, W.; Khan, F.; Ajmal, H.; Huda, N.; Kim, J.; Kim, S.-D. Evolution of structural and optical properties of zno nanorods grown on vacuum annealed seed crystallites. *Nanomaterials* **2018**, *8*, 68. [CrossRef] [PubMed]
11. Humayun, Q.; Kashif, M.; Hashim, U.; Qurashi, A. Selective growth of ZnO nanorods on microgap electrodes and their applications in UV sensors. *Nanoscale Res. Lett.* **2014**, *9*, 29. [CrossRef] [PubMed]
12. Mohammad, S.M.; Hassan, Z.; Ahmed, N.M.; Al-Hardan, N.H.; Bououdina, M. Fabrication of low cost UV photo detector using ZnO nanorods grown onto nylon substrate. *J. Mater. Sci. Mater. Electron.* **2015**, *26*, 1322–1331. [CrossRef]
13. Khan, W.; Kim, S.-D. Ultra-violet photo-response characteristics of p-Si/i-SiO_2/n-ZnO heterojunctions based on hydrothermal ZnO nanorods. *Mater. Sci. Semicond. Process.* **2017**, *66*, 232–240. [CrossRef]
14. Chen, C.H.; Lee, C.T. High detectivity mechanism of ZnO-based nanorod ultraviolet photodetectors. *IEEE Photonics Technol. Lett.* **2013**, *25*, 348–351. [CrossRef]
15. Farhat, O.F.; Halim, M.M.; Ahmed, N.M.; Oglat, A.A.; Abuelsamen, A.A.; Bououdina, M.; Qaeed, M.A. A study of the effects of aligned vertically growth time on ZnO nanorods deposited for the first time on Teflon substrate. *Appl. Surf. Sci.* **2017**, *426*, 906–912. [CrossRef]
16. Zhang, H.; Babichev, A.V.; Jacopin, G.; Lavenus, P.; Julien, F.H.; Yu Egorov, A.; Zhang, J.; Pauporté, T.; Tchernycheva, M. Characterization and modeling of a ZnO nanowire ultraviolet photodetector with graphene transparent contact. *J. Appl. Phys.* **2013**, *114*, 234505. [CrossRef]

17. Wu, J.-K.; Chen, W.-J.; Chang, Y.H.; Chen, Y.F.; Hang, D.-R.; Liang, C.-T.; Lu, J.-Y. Fabrication and photoresponse of ZnO nanowires/CuO coaxial heterojunction. *Nanoscale Res. Lett.* **2013**, *8*, 387. [CrossRef] [PubMed]
18. Dang, V.Q.; Trung, T.Q.; Duy, L.T.; Kim, B.-Y.; Siddiqui, S.; Lee, W.; Lee, N.-E. High-performance flexible ultraviolet (UV) phototransistor using hybrid channel of vertical ZnO nanorods and graphene. *ACS Appl. Mater. Interfaces* **2015**, *7*, 11032–11040. [CrossRef] [PubMed]
19. Dogar, S.; Khan, W.; Kim, S.-D. Ultraviolet photoresponse of ZnO nanostructured AlGaN/GaN HEMTs. *Mater. Sci. Semicond. Process.* **2016**, *44*, 71–77. [CrossRef]
20. Li, Q.H.; Gao, T.; Wang, Y.G.; Wang, T.H. Adsorption and desorption of oxygen probed from ZnO nanowire films by photocurrent measurements. *Appl. Phys. Lett.* **2005**, *86*, 123117. [CrossRef]
21. Ambacher, O.; Smart, J.; Shealy, J.R.; Weimann, N.G.; Chu, K.; Murphy, M.; Schaff, W.J.; Eastman, L.F.; Dimitrov, R.; Wittmer, L.; et al. Two-dimensional electron gases induced by spontaneous and piezoelectric polarization charges in N- and Ga-face AlGaN/GaN heterostructures. *J. Appl. Phys.* **1999**, *85*, 3222–3233. [CrossRef]
22. Ture, E.; Brückner, P.; Quay, R.; Ambacher, O. Enhancement-Mode AlGaN/GaN FinFETs With High On/Off Performance in 100 nm Gate Length. In Proceedings of the 11th European Microwave Integrated Circuits Conference Enhancement-Mode; EuMA, London, UK, 3–4 October 2016; pp. 61–64.
23. Chang, S.-B.; Chae, H.U.; Kim, H.-S. Structural, optical, electrical and morphological properties of different concentration sol-gel ZnO seeds and consanguineous ZnO nanostructured growth dependence on seeds. *J. Alloys Compd.* **2017**, *729*, 571–582.
24. Khan, W.; Ajmal, H.; Khan, F.; Huda, N.; Kim, S.-D. Induced photonic response of ZnO Nanorods Grown on oxygen plasma-treated seed crystallites. *Nanomaterials* **2018**, *8*, 371. [CrossRef] [PubMed]
25. Yoon, Y.C.; Park, K.S.; Kim, S.-D. Effects of low preheating temperature for ZnO seed layer deposited by sol-gel spin coating on the structural properties of hydrothermal ZnO nanorods. *Thin Solid Films* **2015**, *597*, 125–130. [CrossRef]
26. Alenezi, M.R.; Henley, S.J.; Silva, S.R.P. On-chip fabrication of high performance nanostructured ZnO UV detectors. *Sci. Rep.* **2015**, *5*, 8516. [CrossRef] [PubMed]
27. Talib, R.A.; Abdullah, M.J.; Al-Salman, H.S.; Mohammad, S.M.; Allam, N.K. ZnO nanorods/polyaniline heterojunctions for low-power flexible light sensors. *Mater. Chem. Phys.* **2016**, *181*, 7–11. [CrossRef]
28. Willander, M.; Nur, O.; Sadaf, J.R.; Qadir, M.I.; Zaman, S.; Zainelabdin, A.; Bano, N.; Hussain, I. Luminescence from zinc oxide nanostructures and polymers and their hybrid devices. *Materials* **2010**, *3*, 2643–2667. [CrossRef]
29. Iqbal, M.; Ko, P.-S.; Kim, S.-D. Hot-electron reliability improvement using perhydropolysilazane spin-on-dielectric passivation buffer layers for AlGaN/GaN HEMTs. *Curr. Appl. Phys.* **2014**, *14*, 1099–1104. [CrossRef]
30. Alim, M.A.; Rezazadeh, A.A.; Gaquiere, C. Anomaly and threshold voltage shifts in GaN and GaAs HEMTs over temperature. In Proceedings of the 12th European Microwave Integrated Circuits Conference (EuMIC), Nuernberg, Germany, 8 October 2017; pp. 33–36.
31. Lee, M.L.; Sheu, J.K.; Lin, S.W. Schottky barrier heights of metal contacts to n-type gallium nitride with low-temperature-grown cap layer. *Appl. Phys. Lett.* **2006**, *88*, 032103. [CrossRef]
32. Cho, S.J.; Dogan, S.; Sabuktagin, S.; Reshchikov, M.A.; Johnstone, D.K.; Morkoc, H. Surface band bending in as-grown and plasma-treated n-type GaN films using surface potential electric force microscopy. *Appl. Phys. Lett.* **2004**, *84*, 3070–3072. [CrossRef]
33. Im, K.S.; Won, C.H.; Jo, Y.W.; Lee, J.H.; Bawedin, M.; Cristoloveanu, S.; Lee, J.H. High-performance GaN-based nanochannel FinFETs With/Without AlGaN/GaN heterostructure. *IEEE Trans. Electron Devices* **2013**, *60*, 3012–3018. [CrossRef]
34. Janotti, A.; van de Walle, C.G. Fundamentals of zinc oxide as a semiconductor. *Rep. Prog. Phys.* **2009**, *72*, 126501. [CrossRef]
35. Khan, F.; Khan, W.; Kim, J.H.; Huda, N.U.; Salman Ajmal, H.M.; Kim, S.-D. Oxygen desorption kinetics of ZnO nanorod–gated AlGaN/GaN HEMT-based UV photodetectors. *AIP Adv.* **2018**, *8*, 075225. [CrossRef]
36. An, Q.; Meng, X.; Xiong, K.; Qiu, Y. Self-powered ZnS nanotubes/Ag nanowires MSM UV photodetector with high on/off ratio and fast response speed. *Sci. Rep.* **2017**, *7*, 4885. [CrossRef] [PubMed]

37. Brancaleon, L.; Lin, G.; Kollias, N. The in vivo fluorescence of tryptophan moieties in human skin increases with UV exposure and is a marker for epidermal proliferation. *J. Investig. Dermatol.* **1999**, *113*, 977–982. [CrossRef] [PubMed]
38. Al-Hardan, N.H.; Abdul Hamid, M.A.; Ahmed, N.M.; Jalar, A.; Shamsudin, R.; Othman, N.K.; Kar Keng, L.; Mohammed, S.M. A study on the UV photoresponse of hydrothermally grown zinc oxide nanorods with different aspect ratios. *IEEE Sens. J.* **2015**, *15*, 6811–6818. [CrossRef]
39. Bai, Z.; Yan, X.; Chen, X.; Zhao, K.; Lin, P.; Zhang, Y. High sensitivity, fast speed and self-powered ultraviolet photodetectors based on ZnO micro/nanowire networks. *Prog. Nat. Sci. Mater. Int.* **2014**, *24*, 1–5. [CrossRef]
40. Agrawal, J.; Dixit, T.; Palani, I.A.; Ramachandra Rao, M.S.; Singh, V. Fabrication of high responsivity deep UV photo-detector based on Na doped ZnO nanocolumns. *J. Phys. D Appl. Phys.* **2018**, *51*, 185106. [CrossRef]
41. Hen, H.A.O.C.; Inzhou, X.M.A.; Hang, J.I.Z.; Iuguo, Q.L.I.; Iu, H.U.L.; Hen, Z.U.C.; Hu, G.U.C.; Hu, S.H.C. Avalanche solar blind photodetectors with high responsivity based on MgO/MgZnO heterostructures. *Opt. Mater. Express* **2018**, *8*, 785–793.
42. Khan, M.A.; Kuznia, J.N.; Olson, D.T.; Van Hove, J.M.; Blasingame, M.; Reitz, L.F. High-responsivity photoconductive ultraviolet sensors based on insulating single-crystal GaN epilayers. *Appl. Phys. Lett.* **1992**, *60*, 2917–2919. [CrossRef]
43. Bhardwaj, R.; Sharma, P.; Singh, R.; Gupta, M.; Mukherjee, S. High responsivity Mg $\times$ $Zn_{1-x}O$ based ultraviolet photodetector fabricated by dual ion beam sputtering. *IEEE Sens. J.* **2018**, *18*, 2744–2750. [CrossRef]
44. Weng, W.Y.; Hsueh, T.J.; Chang, S.J.; Wang, S.B.; Hsueh, H.T.; Huang, G.J. A high-responsivity GaN nanowire UV photodetector. *IEEE J. Sel. Top. Quantum Electron.* **2011**, *17*, 996–1001. [CrossRef]
45. Bhattacharya, D.; Jha, N.K. FinFETs: From devices to architectures. In *Digitally-Assisted Analog and Analog-Assisted Digital IC Design*; Jiang, X., Ed.; Cambridge University Press: Cambridge, UK, 2015; Volume 2014, pp. 21–55. ISBN 9781316156148.
46. Crupi, F.; Alioto, M.; Franco, J.; Magnone, P.; Togo, M.; Horiguchi, N.; Groeseneken, G. Understanding the basic advantages of bulk FinFETs for sub- and near-threshold logic circuits from device measurements. *IEEE Trans. Circuits Syst. II Express Briefs* **2012**, *59*, 439–442. [CrossRef]
47. Wakejima, A.; Ando, A.; Watanabe, A.; Inoue, K.; Kubo, T.; Osada, Y.; Kamimura, R.; Egawa, T. Normally off AlGaN/GaN HEMT on Si substrate with selectively dry-etched recessed gate and polarization-charge-compensation δ-doped GaN cap layer. *Appl. Phys. Express* **2015**, *8*, 026502. [CrossRef]
48. Im, K.-S.; Kim, R.-H.; Kim, K.-W.; Kim, D.-S.; Lee, C.S.; Cristoloveanu, S.; Lee, J.-H. Normally off single-nanoribbon Al_2O_3 GaN MISFET. *IEEE Electron Device Lett.* **2013**, *34*, 2012–2014. [CrossRef]

nanomaterials

MDPI

Article

Porous Si-SiO_2 UV Microcavities to Modulate the Responsivity of a Broadband Photodetector

María R. Jimenéz-Vivanco [1], Godofredo García [1], Jesús Carrillo [1], Francisco Morales-Morales [2], Antonio Coyopol [1], Miguel Gracia [3], Rafael Doti [4], Jocelyn Faubert [4] and J. Eduardo Lugo [4,*]

[1] Semiconductor Devices Research Center, ICUAP, BUAP, Ciudad Universitaria, Puebla Pue. C.P. 72570, Mexico; jimenezvmr10@gmail.com (M.R.J.-V.); godgarcia@yahoo.com (G.G.); jecarril@siu.buap.mx (J.C.); acoyopol@gmail.com (A.C.)

[2] Optics Research Center, A.C., Loma del Bosque 115, Col. Lomas del Campestre León, León C.P. 37150, Gto, Mexico; fcomm9@gmail.com

[3] IFUAP, Benemérita Universidad Autónoma de Puebla, Ed. IF2, Col. San Manuel, Puebla C.P. 72570, Mexico; gracia@ifuap.buap.mx

[4] Faubert Lab, School of Optometry, University de Montreal, Montreal, QC H3T 1P1, Canada; rafael.doti@gmail.com (R.D.); faubert.umontreal@gmail.com (J.F.)

* Correspondence: eduardo.lugo@gmail.com; Tel.: +1-514-343-6111 (ext. 1685)

Received: 28 November 2019; Accepted: 9 January 2020; Published: 28 January 2020

Abstract: Porous Si-SiO_2 UV microcavities are used to modulate a broad responsivity photodetector (GVGR-T10GD) with a detection range from 300 to 510 nm. The UV microcavity filters modified the responsivity at short wavelengths, while in the visible range the filters only attenuated the responsivity. All microcavities had a localized mode close to 360 nm in the UV-A range, and this meant that porous Si-SiO_2 filters cut off the photodetection range of the photodetector from 300 to 350 nm, where microcavities showed low transmission. In the short-wavelength range, the photons were absorbed and did not contribute to the photocurrent. Therefore, the density of recombination centers was very high, and the photodetector sensitivity with a filter was lower than the photodetector without a filter. The maximum transmission measured at the localized mode (between 356 and 364 nm) was dominant in the UV-A range and enabled the flow of high energy photons. Moreover, the filters favored light transmission with a wavelength from 390 nm to 510 nm, where photons contributed to the photocurrent. Our filters made the photodetector more selective inside the specific UV range of wavelengths. This was a novel result to the best of our knowledge.

Keywords: photonic nanoscience; nanotechnology; porous Si-SiO_2; UV filters; responsivity

1. Introduction

Porous silicon (PS) is a promising material for many different applications, such as solar cells, specifically as anti-reflection coating [1], chemical sensing [2,3], biomedical applications [4], biosensing [5,6], as a photodetector [7], or light-emitting diode [8]. The anti-reflection coating (ARC) reduces reflection losses thereby increasing the photocurrent and efficiency in solar cells [9–12]. A microcavity filter (MF) obtained in the IR range has been used as the ARC, where the maximum transmission at the localized mode wavelength was 95.2%, with a minimum reflectance of 4.8% at the same wavelength. The PS ARC was deposited onto a solar cell, where the entire surface of the solar cell was covered, and its external quantum efficiency (EQE) was measured, with and without the MF. The EQE at different wavelengths matched the shape of the MF transmission spectra [9]. ARCs are used to convert down the higher energy solar radiation into a wide range of light spectra, which is absorbed more efficiently into bulk Si [10]. In another interesting application, silver nanoparticles were infiltrated within a porous silicon photonic crystal to detect the trace of explosives (Rhodamine 6G dye and Picric acid explosives) using surface-enhanced Raman scattering [13]. Recently, porous silicon nanoparticles (PSN) have been used for the tunable delivery of camptothecin, a small therapeutic molecule, where PSN acted as a therapeutics nanocarrier into the electrospun composite of poly fibers [14]. Additionally, mesoporous Si has been investigated for applications in biology and medicine, and an example is a porous silicon container employed to enclose the drug and release it in a controlled manner [15]. Previously, PS MFs have been fabricated to work as red-infrared filters coupled with a silicon photodetector, where the localized mode can be tuned by changing the optical path of the defect layer [16]. The reflectivity at the localized mode wavelength also increases. This effect occurs because PS has less absorption at long wavelengths [16,17]. This kind of filter can be lifted off from the Si substrate creating a free-standing PS MF. This MF can be transferred to another substrate such as quartz and glass [18,19]. Infrared long-wave pass and short-wave pass filters based in macroporous Si have been investigated [20]; they can be used together to suppress the radiation from shorter wavelengths. Consequently, the signal-to-noise performance of detectors may improve [18]. Moreover, luminescent silicon quantum dots embedded in free-standing PS and PS MFs infiltered with CdSe/ZnS and $AgInS_2$/ZnS quantum dots have been reported [21–23]. The localized mode of the PS MF was tuned experimentally to match the emission spectra of CdSe/Zn and $AgInS_2$/ZnS quantum dots to achieve enhanced photoluminescence. The authors mentioned that the localized mode of the MF modulated the photoluminescence of the quantum dots [22,23]; where electrical and thermal tuning of localized modes could also be achieved by infiltrating liquid crystals in PS MFs [24]. PS was also utilized as an anode for fast charge-discharge in lithium-ion batteries [25,26], and it was found that laser carbonization and wet oxidation PS structures had memristive properties. The first memristor showed properties of plasticity and short/long term memory, whereas the second exhibited strong filamentary-type resistance switching; and they have been used as two terminal resistive memory cells [27].

As the reader already noticed, PS is a versatile material, and herein, we will focus on the optical properties of PS, which allow the design of various kinds of interference filters in a broad range of wavelengths. Among these designs, we found rugate filters [28,29], Fibonacci filters [30], Bragg reflectors filters (BRF) [31,32], and MFs [33]. However, many drawbacks have been found in these filters, such as high chemical instability, high photon losses due to light absorption, and scattering in the visible and UV ranges. Some solutions to these problems have been identified by carrying out dry oxidation in PS structures [32,34], and recently it has been possible to manufacture porous Si-SiO_2 filters (BRF and MF) in the UV range [34–36]. Our specific goal is integrating a PS MF in a photodetector to enlarge the responsivity spectrum bandwidth. In the past, the integration of different PS bandpass filters in Si photodetectors was attainable to achieve different filtering effects. It was possible to tune the responsivity spectrum from green to the near-infrared range obtaining all-Si color-sensitive photodetectors in that wavelength interval but with no below the green range [16,31,37]. It has been reported that UV filters based in multilayer stacks show substantial drawbacks in the deep UV spectral range due to the limited number of transparent materials within that range. They have been proposed as an alternative to extending optoelectronic technology towards the UV spectral range; for instance, PS filters based on macroporous silicon and low-pressure chemical vapor deposition were tested for many solar-blind applications such as electrical spark imaging and non-line of sight (NLOS) UV optical communications [18,38]. Following the same line of thought of extending silicon-based selective filters towards the UV spectral range; in this work, porous Si-SiO_2 microcavities were used to modulate the responsivity of a broad photodetector with a detection range from 300 to 510 nm. The porous Si-SiO_2 microcavities made the photodetector more selective in the UV range; and all microcavities had a localized mode close to 360 nm in the UV-A range. The maximum transmission at the localized mode wavelength (360 nm) was dominant in the UV-A range, allowing high energy photons to pass and produce a maximum peak in the photodetector.

This work is divided as follows: In Section 2, we give some technical details for materials and methods. Section 3 describes the results and their discussion. Finally, in Section 4, we present some conclusions.

2. Materials and Methods

2.1. Porous Silicon Microcavities Filters

PS microcavities filters were fabricated using p-type Si wafers, with (100) orientation and electrical resistivity between the range 0.01–0.02 Ω cm. The microcavities were etched anodically in an aqueous electrolyte based in HF and ethanol with a volume ratio of 1:1. Before etching, the Si wafers were cleaned with HF and ethanol for 5 min to remove the native oxide. A Keithley 2460 current source controlled by a laptop was employed to deliver a current profile for the microcavities formation, where the current profile consisted of switching two different current pulses, with each current pulse producing low/high porosity layers that respectively corresponded to the low/high refractive index layers. This process produced a stack of layers with specific refractive indexes, while anodization time determined the thickness of the layers. A pause of 3 s was introduced between each current pulse to prevent any porosity gradient.

Anodization etching included low/high current pulses of 5 mA/cm^2 and 80 mA/cm^2 to obtain layers with porosities of 39% and 64%, respectively. Meanwhile, the anodization times of 4.1 and 1.1 s were applied to form the first and second porous layers, and finally a third porous layer (defect) was built-in using a current pulse of 80 mA/cm^2 for 2.2 s. The PS microcavities filters were detached from the Si substrates using the same aqueous electrolyte before mentioned, but a high current density (450 mA/cm^2) was applied for 2 s. This process created a free-standing PS microcavity filter, which was transferred to a quartz substrate. All samples were rinsed with ethanol and dried at room temperature after the anodization process.

2.2. Dry Oxidation in PS Microcavities Filters

Microcavities filters were systematically oxidized using two stages of dry oxidation. In the first stage, a temperature of 350 °C for 30 min was used. In the second stage the oxidation temperature was increased to 900 °C for 1 h. During the oxidation process, the oxygen flow was changed from 1.15 to 4.52 SLPM (standard liter per minute) to observe any change in the microcavity optical response. The reasons behind the two dry oxidation steps were: the first step was a low-temperature pre-oxidation needed to equilibrate the silicon structure. That is, this step prevented the aggregation of the pores during further treatments at higher temperatures. The second step was applied to grow an oxide layer of greater thickness than that obtained in the last oxidation step. The layer thickness is higher than the natural native oxide grown in the environment resulting in the consolidation of SiO_2.

Transmittance and reflectance spectra measurements were carried out before and after dry oxidation with a Varian (Agilent Technologies, CA, USA) UV-Vis-NIR spectrophotometer at normal incidence, and 20° from 200 to 800 nm. The cross-section SEM images of an oxidized microcavity was obtained using a JEOL-JSM7600F (Jeol, MA, USA).

2.3. Photocurrent Modulation by Porous Si-SiO_2 Microcavities Filters

The photodetector employed to modulate its responsivity was a GVGR-T10GD (Electro Optical Components Inc., CA, USA) based in indium gallium nitride with a spectral detection range from 300 to 510 nm. In the experimental process, an optical chopper SR540 (Stanford Research Systems Inc., CA, USA) with a frequency of 287 Hz was used to cut off the light coming from the Xenon lamp 6254 (Oriel Corporation, CT, USA), concentrated at the input of a monochromator. A lens was used to focus the output light from the chopper, on the photodetector. The PS microcavity filter was placed before the photodetector, which was polarized with 5 volts using a Keithley 2460 source (Keithley, OHIO, USA), and then the photodetector photocurrent inputted a Look-in amplifier SR-530 (Stanford

Research Systems, CA, USA)). The filtered and amplified photocurrent versus light wavelength (from 276 to 536 nm) was displayed in a PC. The experimental set-up is shown in Figure 1.

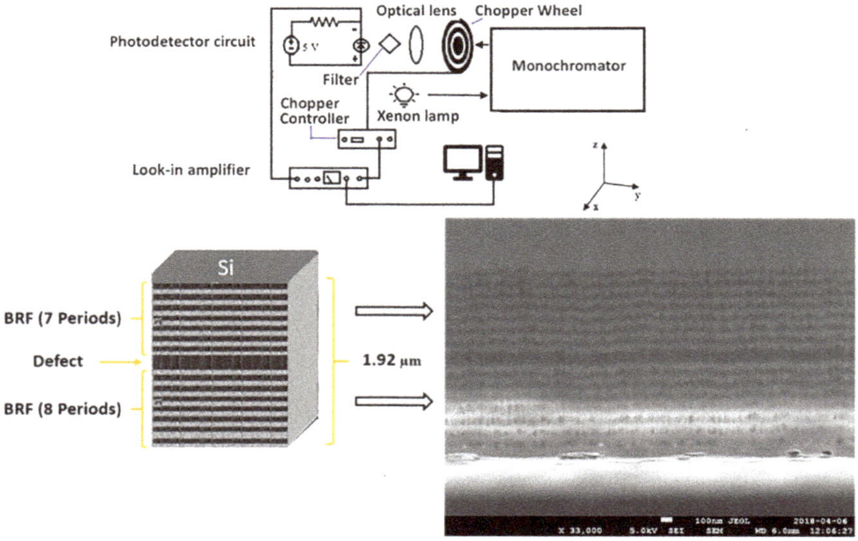

Figure 1. Experimental set-up used to obtain the responsivity of a commercial photodetector with different porous silicon (PS) microcavity filters (up panel). The bottom panel shows the scheme and cross-section SEM images of an oxidized microcavity filter with a total thickness of 1.92 μm.

2.4. Theoretical Mechanism to Modulate the Responsivity of a Broadband Photodetector in the UV Optical Range with Porous Si-SiO$_2$ Microcavities

In an ideal photodetector, its responsivity is proportional to the input light power being the proportional factor, the so-called quantum efficiency η, which is a function of the light reflection R at the surface, the fraction of electron-hole pairs that contribute effectively to the photocurrent ζ, the light absorption in the bulk of the material α, the photodetector depth d, and the light wavelength.

Herein, we proposed to modulate the responsivity of a commercial photodetector by changing R. This was done by adding a microcavity filter, where the spectral reflection response of the microcavity filter modifies the original reflection spectral response at the surface of the commercial photodetector completely. The microcavity filter reflection depends on the high and low refractive indexes and thickness values, as well as the refractive index and thickness of the defect layer values. Initially, the microcavities are made of PS to filter out blue light. This means that the layers' refractive index values, controlled by the porosities and in which the crystalline silicon nanostructures remains, were chosen to respond in that particular region of the light spectrum. However, our primary goal was to achieve responsivity modulation within the UV region using a silicon-based filter. We cannot use the sensitive blue light microcavities for that purpose because the crystalline silicon nanostructures strongly absorb light within the UV and blue regions. This absorption process is equivalent to having a 100% reflection at the photodetector surface. Thus, the quantum efficiency, and consequently, the responsivity, tends to zero. The oxidation process helps to avoid this problem in two ways: first, since there is a phase change, from crystalline silicon to silicon dioxide, the layers' refractive index values decrease, pushing the filter response towards the UV region. Second, there is much less light absorption in the UV band. The oxygen flow value determines the concentration of silicon dioxide, followed by the layer's refractive index values, and consequently, the spectral response of the filter. More specifically, the filters spectral response contains a unique photonic state known as a localized state represented as a resonant peak in the reflection spectrum. It is a state that drastically modifies the responsivity, and therefore, it is the flow of oxygen that determines the final position of the localized state along with its reflection amplitude.

3. Results and Discussion

3.1. Porous Silicon Microcavities Filters in the Blue Range

The averaged theoretical (red line) and experimental (black line) reflection (broken line) and transmission (solid) spectra from the PS microcavities filters are depicted in Figure 2a. This corresponded to the average measurements of transmittance and reflectance spectrum from four microcavities. All four MCs were designed to show a localized mode in the same wavelength (590 nm). Experimental transmission and reflection measurements of the microcavities were taken using a UV-VIS-NIR spectrophotometer in the wavelength range from 200 to 800 nm. The transmission spectrum was measured at normal incidence and reflectance spectrum of the same microcavities was obtained at 20°. Meanwhile, theoretical transmission and reflection spectrum were obtained using the matrix method. This method is well known (Supplementary Materials) for obtaining the theoretical spectrums that we considered the angle measurements taken with the UV-VIS-NIR spectrophotometer.

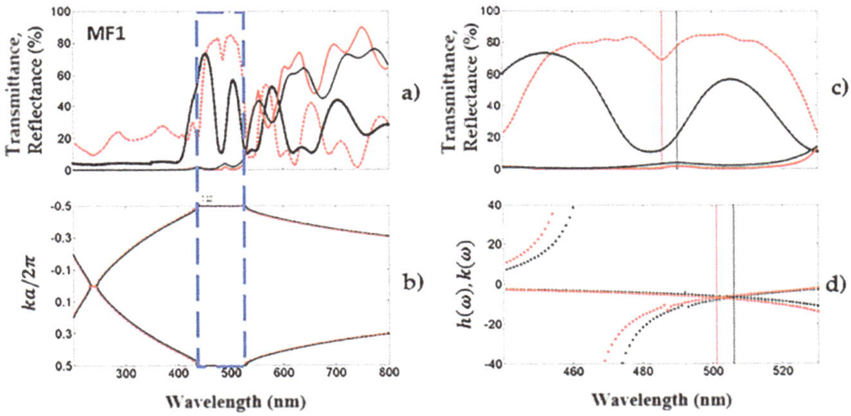

Figure 2. Experimental averaged transmission and reflection spectra of four microcavities compared to its theoretical spectra. (**a**) The theoretical (red line) and experimental (black line) averaged reflection and transmission spectra for four unoxidized microcavities obtained in the blue optical range; (**b**) the averaged band structure of the same microcavities. (**c**,**d**) are the comparison of the averaged theoretical and experimental defect mode locations. In (**d**), two angles of incidence (zero and 20 degrees) were considered.

The theoretical reflection spectrum of the microcavity predicts more reflection in the UV range than its experimental counterpart, while the theoretical and experimental transmission spectra matched very well. Moreover, the light transmission was low in the UV-VIS range due to strong absorption of Si. The localized mode of the microcavity was located at 490 nm and exhibited a maximum transmission peak of 4%, while that in the visible range from 555 to 800 nm showed high light transmission. This result was because PS had small absorption losses in this range of the electromagnetic spectrum, where the primary photon loss was light scattering. When the reflection spectrum of the microcavity was taken at 20°, the position of the localized mode changed, and it shifted towards high energies (483 nm). It could also be observed in the theoretical reflection spectrum. Figure 2b shows the averaged photonic bandgap (PBG) of an unoxidized microcavity. It was obtained using the dispersion relation inside the first Broullin zone. We calculated it only in one dimension and both measurement angles were considered, which were mentioned before.

We obtained the defect frequency can be obtained using a combination of the transfer matrix technique and variational methods. The defect modes are represented by the maximum peak of transmission in the microcavities [39,40]. When the symmetry of the photonic structure is broken up by a defect layer, it is possible to have a confined state, where photons are trapped and cannot escape, being confined inside the bandgap. We calculated the defect mode frequency for an antisymmetric state because our photonic structures were built as follows:

$$(HL)_8L(HL)_7$$

Our antisymmetric photonic structures based on PS and porous Si-SiO_2 consisted of 31 layers, alternating between high refractive and low refractive indexes, with a low refractive index defect layer, and where the structure was surrounded by air.

The averaged theoretical defect mode frequency location is shown in Figure 2d, where it is compared with the averaged transmission and reflection spectra (Figure 2c). The defect mode is shown as a transmission maximum or as a reflection minimum at a specific wavelength between the photonic bandgap (PBG) edges. The variational method defect mode location was 18 nm off, concerning the experimental result from the reflection measurements, which was taken at 20 degrees. In this case, the defect mode location is depicted in Figure 2d (red broken line). Using the same variational method, the defect mode location was 16 nm off for the experimental transmission spectrum, which was taken at normal incidence. The localized mode is shown in Figure 2d (black broken line). The transfer matrix method predicted the same position as the experimental result. Both theoretical methods showed good agreement with the experiments.

3.2. Porous Si-SO_2 Microcavities Filters in the UV

Averaged transmission (solid line) and reflection (broken line) spectra from porous Si-SiO_2 microcavities filters are depicted in Figure 3a. The theoretical (red line) and experimental (black line) spectrum showed a good fit. The theoretical spectra were obtained by the transfer matrix (Supplementary Materials) considering both measurement angles, where one was 0°, which corresponded to the transmission measurements, and the other was 20°, considered for the reflection measurements. The microcavity to the left was first oxidized at 350 °C for 30 min and then at 900 °C for 1 hr, and the oxygen flow was 1.15 SLPM (standard liter per minute). The microcavity shown to the right was oxidized using the same temperature and oxidation time mentioned above. In this case, an oxidation flow of 2.21 SLPM was applied. The microcavity MF2 exhibited a maximum transmission amplitude of 70%, and the microcavity MF3 showed a maximum transmission amplitude of 57% in the localized mode wavelength, which indicated that the microcavity MF3 had more absorption losses than the microcavity MF3. Porous Si-SiO_2 microcavities filters in Figure 3a show a transmission maximum peak (356 and 364 nm) and a reflection minimum peak (349 nm) in the UV range. We observed a mismatch between both peaks because the measurements were taken at different incidence angles.

Figure 3c shows the averaged transmission and reflection spectrum from UV filters compared to the defect mode location obtained with the variational method (Figure 3d). The defect mode location corresponding to the reflection spectrum was 4 nm off using the variational method, which is shown in Figure 3d (red broken line). The position of the defect mode obtained from the variational method (black broken line, Figure 3d) was 9 nm off for the transmission spectrum, as compared to what was predicted using the transfer matrix method.

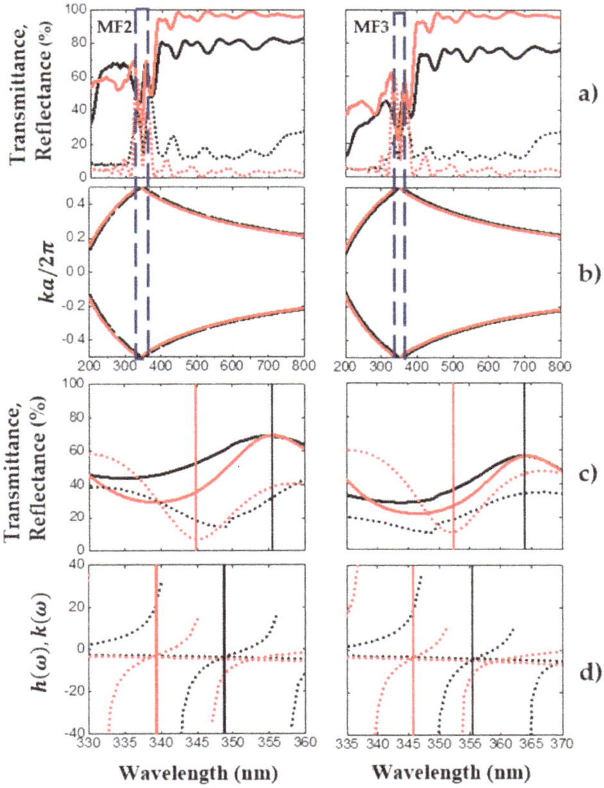

Figure 3. The averaged theoretical and experimental transmission and reflection spectra of two UV microcavities filters, with the averaged photonic bandgap and defect mode locations. (**a**) The averaged theoretical (red line) and experimental (black line) reflection (broken line) and transmission (solid line) spectra of oxidized microcavities obtained in the UV range; (**b**) the averaged photonic bandgap structure (empty box) of the microcavities. (**c**,**d**) show the comparison of the averaged theoretical and experimental defect mode locations. In (**d**), two angles of incidence (zero and 20 degrees) were considered.

Figure 4a depicts the averaged theoretical (red line) and experimental (black line) transmission (solid line) and reflection (broken line) spectra of two oxidized MF. In this case, the porous Si-SiO_2 microcavity filter MF4 showed a maximum transmission peak (368 nm) in the UV range, with an amplitude of 42%. Meanwhile, the microcavity MF5 displayed a maximum transmission peak at 369 nm, with an amplitude of 21% (red and black solid lines). A reflection minimum peak (349 nm) is also shown in both microcavities (red and black broken lines). The maximum transmission peaks were less intense than the microcavities shown in Figure 3. Here, the same temperature and oxidation time were employed to obtain the UV filters, but the oxygen flow applied for sample MF4 was 3.39 SLPM, while for sample MF5, it was 4.52 SLPM. The oxygen flow was increased gradually.

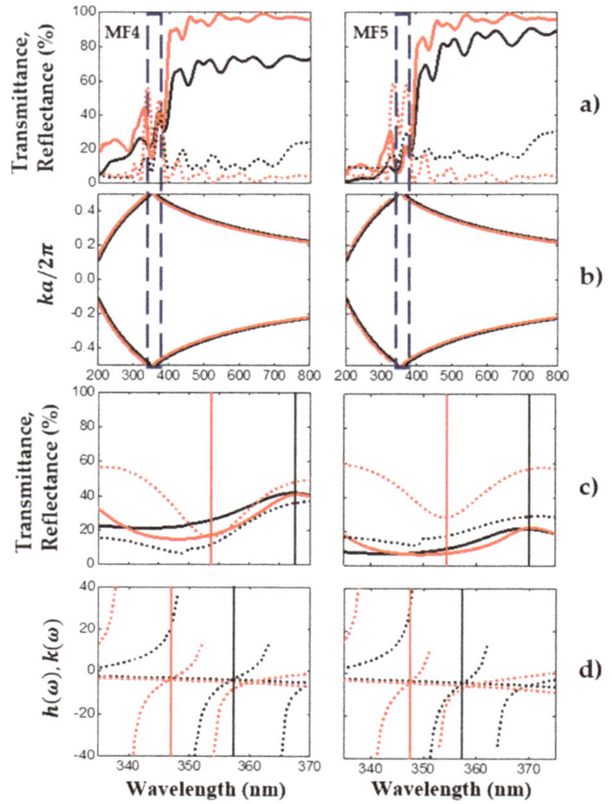

Figure 4. The averaged theoretical and experimental reflection and transmission spectrum of another two UV microcavities filters. Additionally, its averaged photonic bandgap and the defect mode locations are depicted. (**a**) The averaged theoretical (red line) and experimental (black line) reflection (broken line) and transmission (solid line) spectra of oxidized microcavities obtained in the UV range; (**b**) the averaged photonic bandgap structure (empty box) of the microcavities. (**c**,**d**) are the comparison between the averaged theoretical and experimental defect mode locations. In (**d**), two angles of incidence (zero and 20 degrees) were considered.

Figure 4c,d shows the averaged theoretical (red line) and experimental (black line) reflection (broken line) and transmission (solid line) spectra near to the defect mode location. The variational method was employed to obtain the defect mode location between the PBG edges. The broken red plots (Figure 4d) corresponded to the defect mode position of the microcavities measured at 20°. The defect mode locations of microcavities measured at normal incidence are depicted as broken black lines in Figure 4d. There was a mismatch between the defect mode position calculated using the transfer matrix and the variational method.

3.3. Refractive Index of Porous Silicon and Porous Si-SiO_2

The complex refractive index (refractive index and extinction coefficient) of the PS was obtained using the effective medium approximation Maxwell–Garnett. For porous Si-SiO_2, we used a three-component model developed from the J.E. Lugo model. Figure 5a,b shows the refractive index values, and Figure 5c,d depicts the extinction coefficient values of PS and Porous SiO_2, respectively, obtained from both models. The continuous lines in Figure 5a correspond to a high refractive index (red line) and a low refractive index (blue line) for PS. Their extinction coefficients are displayed with dotted lines (red and blue) in Figure 5c, where these values were used to fabricate PS microcavities filters in the blue range. The continuous lines of different colors in Figure 5b between 1.45 and 1.6, and the green line with values between 1.35 and 1.5, represent high and low refractive indexes for the porous Si-SiO_2 layers (see Figure 5b). Their corresponding extinction coefficients are shown below the value of 0.1 (dotted lines) in Figure 5d. The vertical black dotted lines that intersect the complex refractive index components (Figure 5) indicate the exact values utilized to design MF in the blue and UV range.

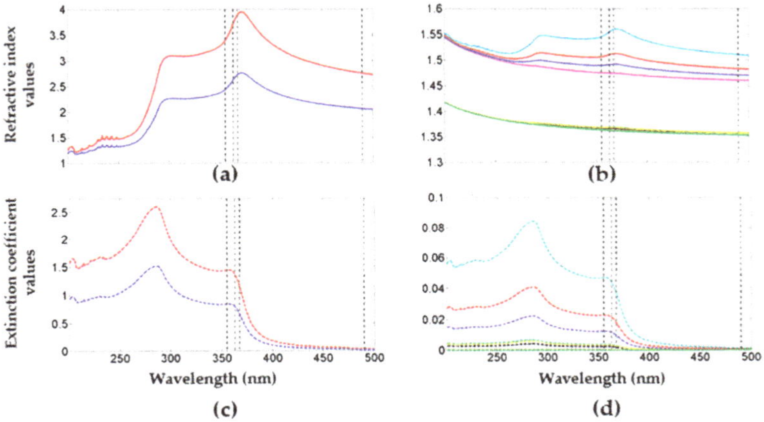

Figure 5. Theoretical values of the complex refractive index for PS layers and porous Si-SiO_2 layers. (**a**,**b**) shows the refractive index values for PS and porous Si-SiO_2, and (**c**,**d**) depicts the extinction coefficient values of PS and Porous Si-SiO_2, respectively, obtained from both models.

Figure 5 shows the refractive index and extinction coefficient are high when the PS is not oxidized. Meanwhile, the complex refractive index component values for porous Si-SiO_2 are lower than the PS values. The growth of SiO_2 inside PS layers decreased the refractive index and extinction coefficient. Moreover, during the PS layers' oxidation process, some air and Si fractions were replaced by SiO_2, causing a lattice expansion and a decrement on the layer's porosity, as shown in Table 1.

Table 1. Theoretical values of porosity before and after dry oxidation, thickness, Si fraction, and oxide fraction of five microcavities.

Filter	Porosity (%)	f_{Si} (%)	f_{ox} (%)	p_{ox} (%)	Average Thickness (nm)
MF1	36 57	64 43	——	——	$d_H = 35$ $d_L = 68.7$
MF2	——	0.23 0.25	98.51 78.86	1.27 20.89	$d_H = 46.5$ $d_L = 78.0$
MF3	——	0.78 0.15	97.88 79.02	1.34 20.83	$d_H = 47.9$ $d_L = 78.8$
MF4	——	1.42 0.22	97.14 78.91	1.44 20.87	$d_H = 47.7$ $d_L = 78.8$
MF5	——	2.92 0.01	95.41 79.23	1.67 20.76	$d_H = 47.5$ $d_H = 78.1$

Knowing the refractive index of each layer that made up the microcavity allowed estimation of the theoretical thickness of each layer by fitting the theoretical transmission and the reflection spectrum (applying the matrix method) with its experimental result. The thickness values displayed in Table 1 correspond to theoretical thickness for different microcavities. On the other hand, an increase in the thickness for the oxidized microcavities (MF2, MF3, MF4, and MF5) was shown after dry oxidation compared to the thickness of the unoxidized microcavity (MF2).

Additionally, the thickness of an oxidized microcavity was obtained using SEM measurements. Figure 1 (down panel) shows where the porous Si-SiO_2 UV filter has a total thickness of 1.92 μm and it can observe a defect between two BRF. The light gray thickness (d_H = 48 nm) corresponds to layers with high refractive index (low porosity) and the dark gray thickness (d_H = 75 nm) depicts the layers with low refractive index (high porosity). Meanwhile, the defect in the microcavity had twice the thickness of the high porosity layer. The thickness found using both methods showed a difference of a few nanometers, and both results give an approximate thickness value for each layer of an oxidized microcavity filter.

Furthermore, the optical path was modified by decreasing the refractive index and increasing the physical thickness. Finally, the vertical black dotted lines that intersected with the complex refractive index components (Figure 5) indicated the exact values utilized to design MF in the blue and UV range.

3.4. Optical Losses Due to Light Absorption

Optical losses in the VIS and UV range due to light absorption are found frequently in PS. This absorption loss impedes the fabrication of PS filters in the UV range, but these losses can be decreased if the PS structure is thermally oxidized in an oxygen environment [35,36]. Figure 6 (MF1) shows the theoretical (a) and experimental (b) absorbance spectrum of five microcavities. As observed, the absorption was dominant in the VIS and UV range, while in the infrared range, the absorption did not play an important role. It has been reported that optical losses are mainly due to dispersion in the infrared range [41,42]. Moreover, Figure 6a (MF2-MF5) shows the theoretical absorption spectra of four oxidized PS MCs. Figure 6b displays the experimental absorption spectra of the same four oxidized PS MCs filters using different oxygen flows. These plots showed an absorption increase when the oxygen flow builds up. The average absorbance spectra of the oxidized microcavities showed less optical losses due to absorption. Their amplitude decreased more than 70% in the UV range and almost disappeared within the VIS range after dry oxidation. We obtained a good fit between the theoretical and experimental absorption spectra in the VIS and UV range.

These results showed that the defect mode of the PS microcavities filters had a wavelength shift of 134 nm to lower wavelengths when the PS microcavities were oxidized. It also had a wavelength shift of 13 nm to lower wavelengths when a maximum oxygen flow of 4.52 SLPM was applied. Increasing the oxygen flow made the porous Si-SiO_2 microcavities exhibit less amplitude in the transmission and reflection spectrum inside the UV range. High oxygen flow did not allow oxygen particles to incorporate into the PS structure. Thus, there was less SiO_2 formation and consequently, optical absorption increased. Besides, a decrease of the PBG bandwidth was achieved by incorporating SiO_2 within the PS microcavities. This bandwidth decrease happened because there was less contrast between the high refractive index and the low refractive index of the porous Si-SiO_2 layers.

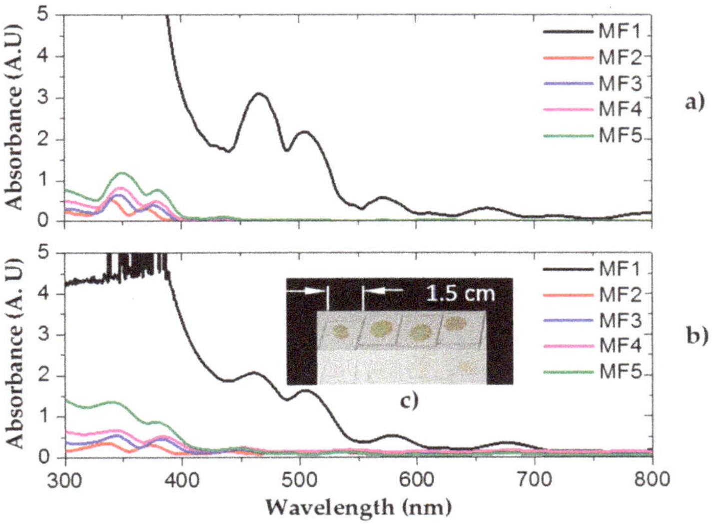

Figure 6. Theoretical (**a**) and experimental (**b**) absorption spectra of MFs in the blue and UV range. (**c-up**) Four unoxidized PS microcavities on a quartz substrate. (**c-bottom**) Four oxidized porous SiO_2 microcavities.

3.5. Photocurrent and Responsivity Measurements of a Commercial Photodetector without and with Filters

Porous Si-SiO_2 UV microcavities filters were used to modulate the photocurrent of a broad photodetector (GVGR-T10GD) with a detection range from 300 to 510 nm. Figure 7 depicts the measured photocurrent of a commercial photodetector (gray line) without and with filters (black, red, blue, pink lines). The MF produced a photocurrent maximum in the UV range, which corresponded to the localized mode of the porous Si-SiO_2 microcavity filters positioned between 356 and 364 nm. This meant that porous Si-SiO_2 filters cut off the photocurrent range from 300 to 350 nm. Therefore, in the short-wavelength range, the photons were absorbed and did not contribute to the photocurrent, whereas in the VIS range, photons with wavelength light from 390 to 510 nm were allowed to pass and they contributed to increasing the photocurrent. Besides, all photocurrent spectra modulated with porous Si-SiO_2 microcavities showed a decline in the UV range, which was attributed to the transmission amplitude of the microcavity as it decreased when the oxygen flow increased.

Figure 7b shows the normalized photocurrents of a commercial photodetector modulated with porous Si-SiO_2 microcavities filters. We observed that the photocurrent peak at the localized mode wavelength decreased when the filters with less transmission amplitude were used.

Spectra responsivities were obtained using a 150 W Xenon arc lamp as a light source and a monochromator with a light wavelength from 300 to 510 nm at 5 V applied bias. Thus, we calculated the responsivity from:

$$R = \frac{I_{ph}}{P_{inc}} \quad (1)$$

where R is the responsivity of the modulate photodetector with porous Si-SiO_2 microcavities filters, I_{ph} is the photocurrent in Ampers, and P_{inc} is the power of the Xenon lamp in watts. The photodetector responsivities without and with porous Si-SiO_2 microcavities filters were found in the range from 300 to 510 nm. The result is shown in Figure 8.

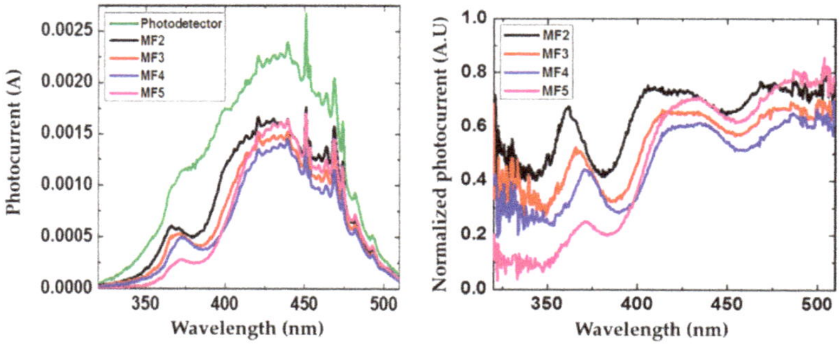

(a) (b)

Figure 7. Photocurrents spectra of a commercial photodetector with and without porous Si-SiO_2 microcavity filters. (**a**) The photocurrent spectra of a commercial photodetector (green line) and its modulated photocurrent with porous Si-SiO_2 microcavities filters (black, red, blue, and pink lines) and (**b**) the normalized photocurrent.

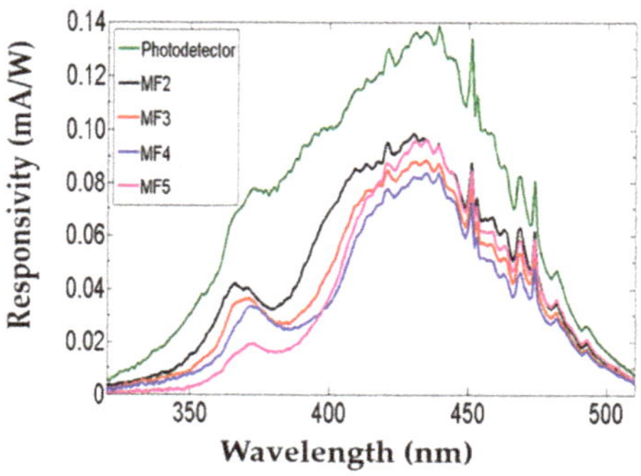

Figure 8. Spectral responsivity comparison of a commercial photodetector without (line green) and with porous Si-SiO_2 filters (black, red, blue, and pink lines).

The spectral responsivity of a commercial photodetector was modified using different porous Si-SiO_2 filters. It decreased when porous Si-SiO_2 filters with low transmission in the UV range were placed on the commercial photodetector. However, a maximum peak of detection corresponding to the localized mode of the microcavity was observed in the UV-A range. In the VIS range, an increase of the responsivity was displayed using the MF5 filter. It was due to an increase in its transmission. The spectral responsivity shape was the same as the photocurrent. We also observed that the sensitivity of the photodetector with a filter was lower than the photodetector without a filter.

PS Bragg reflectors and microcavities were used to filter incident light reaching the Si photosensitive wafer. To tailor its spectral response, they were designed to different wavelengths and integrated above the p-n junction of a silicon photodiode. In this way, they were converted in an array of PS detectors sensitive to the color [16,31], where the sensitivity peak could be tuned along from green to near-infrared, and where sharp peaks in the spectral responsivity were achieved using microcavities and color-sensitivity by Bragg-reflectors [16,43]. The transmittance of the PS filters mainly modulates the spectral responsivity of the silicon photodiode. Therefore, when a Bragg reflector is used to modulate the spectral response of the Si photodiode, its reflectance maximum is nearly insensitive [16,37].

The study on the responsivity has focused on the range visible and infrared placing different kinds of filters on Si photodetectors. A narrowing in the responsivity spectrum of the Si photodetector was observed from the green to infrared range. Moreover, a decrease in the responsivity spectrum of the PS detectors sensitive to the color was displayed, which was compared to the responsivity spectrum of the Si photodetector without filters. The difference between both results was approximately a one magnitude order [2,37]. On the other hand, MF and BRF are used to modify the responsivity of a Si photodetector in the orange range, where the photodetector's responsivity with a filter was stronger. It reduced the undesired effects in the responsivity spectrum [16,43].

Furthermore, it has been reported that below 500 nm, no photocurrent is detected due to the low responsivity of a silicon photodetector. Its modulation using UV filters cannot be achieved because the absorption on PS is dominant from the blue to UV range [37]. Thus, it is not possible to modulate the responsivity of broadband photodetectors in that part of the electromagnetic spectrum. To address this limitation, we employed dry oxidation to decrease the absorption losses in the UV-A range. The transmission improvement in the porous Si-SiO_2 UV filters allowed us to modulate the responsivity of a commercial broadband photodetector.

4. Conclusions

In this work, porous Si-SiO_2 UV filters were manufactured based on Si and SiO_2. We fabricated both symmetric and antisymmetric filters in the blue region, where the antisymmetric structures gave us better results. This result was because we tried to reduce the number of layers to obtain porous silicon MFs as close as possible to the short wavelength region. Furthermore, we noticed during the dry oxidation that the number of layers had an enormous influence on the growth of SiO_2 into the porous silicon structure. This is because large structures can be transformed entirely in porous SiO_2. Besides, the presence of a higher quantity of silicon would cause absorption losses in the porous silicon structure; therefore, it could not be possible to obtain porous SiO_2 MF in the UV range.

The filters were placed on quartz substrates to obtain their transmission and reflection spectra. First, we observed the onset of a localized mode in the blue region in unoxidized MFs, and second, we observed an optical shift of the localized mode location towards small wavelengths (UV region). The shift was due to the oxidation process. The ultimate consequence of the oxidation procedure was the phase change of the PS skeleton, which was converted into SiO_2. Consequently, the refractive index decreased and the extinction values also decreased, although more drastically. The refractive index decrease is the origin of the localized mode location shift, whereas the extinction value decrease is related to a severe light absorption decrement, which means less optical losses.

The filters on quartz substrates were placed on a commercial photodetector. The light coming from a Xenon lamp was filtered out by the filters and changed the reflection at the top photodetector surface, achieving modulation of the photodetector responsivity. As a result, the spectral responsivity obtained matched the UV filter transmission spectrum. The filters' spectral responses were modified by applying different oxidation flows in the oxidation process. When a high flow was used, the transmission in the localized mode wavelength decayed. We have shown that the photodetector became more selective in the UV range using porous UV Si-SiO_2 filters, where an average responsivity peak arose due to the localized mode of the microcavity. The position and amplitude of the localized mode microcavity can be manipulated and tuned in the UV range, by applying different oxygen flows during dry oxidation. In this work, we demonstrated that the responsivity of a commercial photodetector could be modulated using porous Si-SiO_2 UV filters. Moreover, the responsivity peak may be fully tunable, depending on the filter design from UV-A to near-infrared. In the future, the modulated responsivity could be enhanced using Rugate filters, and these filters could be implemented or integrated into Si-based photodetectors, thereby achieving better sensitivity to UV light. Rugate filters could exhibit a unique maximum peak in all the detection ranges of the UV photodetector. It has been reported that Rugate filters have a narrow PBG and the sidelobes on each side of these filters are smaller than other kinds of filters [29].

Other possible applications of these MFs are to embed quantum dots and liquid crystals inside porous Si-SiO_2 UV filters to match their emission spectra with the microcavity localized mode. In this way, we would enhance and modulate the photoluminescence in the UV range by experimentally tuning the localized mode of the microcavity. Porous Si-SiO_2 UV filters may be used as antireflection coatings [9,44] to enhance the efficiency and photocurrent in solar cells. This is because the localized mode of the microcavity shows a maximum amplitude of 70% in the UV range, whereas, in the VIS range, more than 80% of the light is transmitted. Decreasing reflection and increasing transmission in the UV filters could be raised by the efficiency of solar cells in the high-energy range. It could be achieved if a UV filter attains high transmission close to 100%.

One significant contribution of this work is that our microcavities are based on silicon solely. This contribution is important because up to now, silicon photodiodes and solar cells can only work in the UV region. However, silicon photodetectors have a broad response, and our UV microcavities can make them more selective in that range. In the future, the possibility of integrating these UV filters along with porous silicon photodetectors may be a reality, as well as UV light-emitting devices based only on silicon and silicon dioxide technology. Moreover, as shown above, similar approaches to modulate the responsivity of a photodetector have been proposed. Nevertheless, to the best of our knowledge, this is the first time that it has been done within the UV range.

What is the advantage of using porous SiO_2 compared to other materials with similar optical properties? Some authors have reported that several fluorides with bandgap energies around 10 eV are suitable for filtering applications in the UV wavelength range, even below 200 nm. An example is a mixture of lanthanum fluoride (LaF_3) and magnesium fluoride (MgF_2), which have been used as high reflectance coatings. They showed strong absorption in the VIS and UV range, as well as a maximum reflectance higher than 90% at 180 nm, while their transmittance was lower than 2%. Their refractive indexes were found in the range from 1.41 to 1.80. High reflectance coatings have been produced by ion-assisted deposition, ion-beam sputtering, and electron-beam evaporation. However, several difficulties exist in these materials due to optical losses, mechanical stress of the thin films, small refractive index difference between both materials, and the optical inhomogeneity of LaF_3 films [45,46].

On the other hand, many oxides are an essential class of coating materials, which react at high temperatures. HfO_2 (hafnium oxide) and ZrO_2 (zirconium oxide) are examples of transparent oxides, where HfO_2 films are produced by electron-beam evaporation, reactive sputtering, and ion-assisted deposition. ZrO_2 films are prepared by electron-beam evaporation, which showed strong inhomogeneity of the refractive index. HfO_2 and ZrO_2 are common materials used to manufacture UV filters due to their transparency, low-absorption, and high refractive index in that optical range. A long pass edge filter composed of 91 layers of HfO_2/SiO_2 deposited on a fused silica substrate by ion-assisted deposition (IAD) has been reported. The total thickness of the coating was around 3 μm [47–49]. The 50% transmission cut-on edge of this particular filter was at 256 nm, and the transmission in the passband average above 90% was from 260 nm to 1200 nm. Moisture stable HfO_2 films and stacks containing HfO_2 and SiO_2 can be deposited at lower substrate temperatures (100 °C to 130 °C) using an argon oxygen gas mixture IAD process. Most thin-films deposited without ion-assistance are porous and sensitive to moisture going in and out of the voids, thereby causing an apparent shift in the refractive index depending on the relative humidity [50]. HfO_2 films obtained by ion-assisted electron-beam evaporation showed low-absorption, with a refractive index of 2.19. In non-ion-assisted, the refractive index was 2.06, and the absorption coefficient was 0.006 and 0.003 for ion-assisted and non-ion-assisted [51]. ZrO_2 films have been deposited by magnetron sputtering at different argon partial pressure values, where the films of ZrO_2 showed a maximum transmission of 80% from 300 to 800 nm. This material has a high refractive index and low absorption above 240 nm to the IR range (below 8 mm) [52]. Oxide films are also widely used in multilayers systems, such as cold light mirrors, heat-reflecting filters, color separators, narrow band interference filters, and laser coatings. ZrO_2 and HfO_2 have disadvantages; for instance, they form uneven surfaces during evaporation, which often cause inhomogeneous thickness distributions and inferior thickness reproducibility. It has been investigated that no pure oxides exist with refractive indices of about 2.2. Available mixtures of silver (Ag)–silicon dioxide (SiO_2) can be found, which have been used as thin-film bandpass filters for the UV range obtained by radiofrequency sputtering. In this filter, the suppression of undesired visible and infrared parts of the spectrum was achieved. Some structures applicable for bandpass filtering in the UV have high transmission in the passband but a limited range of out-band blocking. An additional blocking component such as monolithic deposited silver has been used to remove unwanted out-of-band radiation. However, this also reduces the overall transmission through the filter. Some pairs of Ag/SiO_2 layers and a layer of silica were added to serve as an antireflection coating, where the band-pass filter was designed with the maximum transmission in the UV-A range and with a resonance at 320 nm. The use of transparent metals can ensure a transmission decrease of several orders of magnitude in the visible and infrared wavelength range, at the same time, fully preserving transparency in the ultraviolet spectrum [53].

Solar-blind deep-UV band-pass filters based on a mixture of aluminum (Al) and SiO_2 have been developed. The filters showed a 27% transmission peak at 290 nm, a band-pass from 250 to 350 nm, and a rejection ratio to visible light of 20 dB, where the peak of transmission could be tuned by adjusting the metal nano-grind dimensions [54]. An alternative approach to solar-blind UV detection is to integrate Si-based photodetectors with solar-blind UV-pass filters to reject visible and IR light in the solar spectrum. Moreover, the designed and fabricated filters are fully applicable for the enhancement of UV silicon detectors since one of the problems with the use of silicon photodetectors is the avoidance of the visible and infrared components of radiation. Under standard conditions, it often exceeds the UV components by three to four orders of magnitude, thereby completely masking the useful signal. A straightforward way to avoid the problem is to use bandpass filters for the UV strong rejection of unwanted visible and infrared radiation in the range of the detector sensitivity. However, a disadvantage of Ag/SiO_2 and Al/SiO_2 structures is that they present reduced transmission (less than 50%) in the UV range due to the absorption of the Ag and Al.

Moreover, dielectric interference filters have also been demonstrated for solar-blind UV applications. However, this type of filter requires a thick stack of multiple dielectric materials layers that must be deposited with precise thickness and uniformity control, with high quality over a very long deposition time. This inevitably increases production costs and reduces the uniform area. The filters also need accurate control of pressure and temperature.

In our work, dielectric interference filters in the UV were achieved by electrochemical etching, followed by two-stage dry oxidation. Our method was easy, cheap, and fast to fabricate compared to the methods mentioned before. The thickness of each layer that made up the UV dielectric microcavity was precise and uniform, and the fabrication time of these dielectric microcavities was shorter than other methods. Besides, physical vapor deposition infrastructure is expensive, and ion-assisted deposition is a costly and slow technique. Additionally, dielectric oxide films do not show sufficient thermal stability because the structure of the oxide films is easily converted from amorphous to polycrystalline and it reacts with the Si substrate.

To conclude, the porous Si-SiO_2 UV filters (MF2, MF3, and MF4) can be applied as UV hot mirrors or UV bandpass filters since they display more than 70% of the transmitted light in the UV-VIS range. Meanwhile, the MF5 filter can be used as a UV blocking filter since it cuts off more UV light than other filters, and it shows a maximum transmission amplitude of 80% in the VIS-IR.

Supplementary Materials: The following are available online at http://www.mdpi.com/2079-4991/10/2/222/s1.

Author Contributions: Conceptualization, M.R.J.-V., G.G., and J.E.L.; Formal analysis, M.R.J.-V. and J.E.L.; Funding acquisition, G.G., J.C., M.G., and J.F.; Investigation, M.R.J.-V.; Methodology, M.R.J.-V., G.G., and J.E.L.; Writing—original draft, M.R.J.-V. and J.E.L.; Writing—review & editing, G.G., J.C., F.M.-M., A.C., M.G., R.D., J.F., and J.E.L. All authors have read and agreed to the published version of the manuscript.

Funding: This work was supported by a National Sciences and Engineering Research Council of Canada Discovery operating grant.

Acknowledgments: R.J. wants to give thanks to CONACYT for the scholarship granted during her doctorate studies. The authors acknowledge D.L. Serrano for the technical assistance provided in the UV-VIS-NIR measurements.

Conflicts of Interest: The authors declare no conflict of interest. The founding sponsors had no role in the design of the study; in the collection, analyses, or interpretation of data; in the writing of the manuscript, and in the decision to publish the results.

References

1. Ramizy, A.; Aziz, W.J.; Hassan, Z.; Omar, K.; Ibrahim, K. Improved performance of solar cell based on porous silicon surfaces. *Opt. Int. J. Light Electron Opt.* **2011**, *122*, 2075–2077. [CrossRef]
2. Torres-Costa, V.; Martin-Palma, R. Application of nanostructured porous silicon in the field of optics. A review. *J. Mater. Sci.* **2010**, *45*, 2823–2838. [CrossRef]
3. Pham, V.H.; Van Nguyen, T.; Bui, H. Nano porous silicon microcavity sensor for determination organic solvents and pesticide in water. *Adv. Nat. Sci. Nanosci. Nanotechnol.* **2014**, *5*, 045003. [CrossRef]
4. Li, W.; Liu, Z.; Fontana, F.; Ding, Y.; Liu, D.; Hirvonen, J.T.; Santos, H.A. Tailoring porous silicon for biomedical applications: From drug delivery to cancer immunotherapy. *Adv. Mater.* **2018**, *30*, 1703740. [CrossRef] [PubMed]
5. Acquaroli, L.N.; Kuchel, T.; Voelcker, N.H. Towards implantable porous silicon biosensors. *RSC Adv.* **2014**, *4*, 34768–34773. [CrossRef]
6. Rodriguez, G.A.; Hu, S.; Weiss, S.M. Porous silicon ring resonator for compact, high sensitivity biosensing applications. *Opt. Express* **2015**, *23*, 7111–7119. [CrossRef]
7. Ismail, R.A.; Alwan, A.M.; Ahmed, A.S. Preparation and characteristics study of nano-porous silicon UV photodetector. *Appl. Nanosci.* **2017**, *7*, 9–15. [CrossRef]
8. Maier-Flaig, F.; Rinck, J.; Stephan, M.; Bocksrocker, T.; Bruns, M.; Kübel, C.; Powell, A.K.; Ozin, G.A.; Lemmer, U. Multicolor silicon light-emitting diodes (SiLEDs). *Nano Lett.* **2013**, *13*, 475–480. [CrossRef]
9. Osorio, E.; Urteaga, R.; Acquaroli, L.N.; García-Salgado, G.; Juaréz, H.; Koropecki, R.R. Optimization of porous silicon multilayer as antireflection coatings for solar cells. *Sol. Energy Mater. Sol. Cells* **2011**, *95*, 3069–3073. [CrossRef]
10. Aziz, W.J.; Ramizy, A.; Ibrahim, K.; Hassan, Z.; Omar, K. The effect of anti-reflection coating of porous silicon on solar cells efficiency. *Opt. Int. J. Light Electron Opt.* **2011**, *122*, 1462–1465. [CrossRef]
11. Raut, H.K.; Nair, A.S.; Dinachali, S.S.; Ganesh, V.A.; Walsh, T.M.; Ramakrishna, S. Porous SiO_2 anti-reflective coatings on large-area substrates by electrospinning and their application to solar modules. *Sol. Energy Mater. Sol. Cells* **2013**, *111*, 9–15. [CrossRef]
12. Welser, R.E.; Sood, A.W.; Pethuraja, G.G.; Sood, A.K.; Yan, X.; Poxson, D.J.; Cho, J.; Schubert, E.F.; Harvey, J.L. Broadband nanostructured antireflection coating on glass for photovoltaic applications. In Proceedings of the 2012 38th IEEE Photovoltaic Specialists Conference, Austin, TX, USA, 3–8 June 2012; pp. 3339–3342.
13. Zhong, F.; Wu, Z.; Guo, J.; Jia, D. Porous silicon photonic crystals coated with Ag nanoparticles as efficient substrates for detecting trace explosives using SERS. *Nanomaterials* **2018**, *8*, 872. [CrossRef] [PubMed]
14. McInnes, S.; Macdonald, T.; Parkin, I.; Nann, T.; Voelcker, N. Electrospun composites of polycaprolactone and porous silicon nanoparticles for the tunable delivery of small therapeutic molecules. *Nanomaterials* **2018**, *8*, 205. [CrossRef] [PubMed]
15. Ksenofontova, O.; Vasin, A.; Egorov, V.; Soldatenkov, F.Y.; Terukov, E.; Ulin, V.; Ulin, N.; Kiselev, O. Porous silicon and its applications in biology and medicine. *Tech. Phys.* **2014**, *59*, 66–77. [CrossRef]
16. Krüger, M.; Marso, M.; Berger, M.; Thönissen, M.; Billat, S.; Loo, R.; Reetz, W.; Lüth, H.; Hilbrich, S.; Arens-Fischer, R. Color-sensitive photodetector based on porous silicon superlattices. *Thin Solid Films* **1997**, *297*, 241–244. [CrossRef]
17. Hunkel, D.; Marso, M.; Butz, R.; Arens-Fischer, R.; Lüth, H. Integrated photometer with porous silicon interference filters. *Mater. Sci. Eng. B* **2000**, *69*, 100–103. [CrossRef]
18. Kochergin, V.; Foell, H. Commercial applications of porous Si: Optical filters and components. *Phys. Status Solidi C* **2007**, *4*, 1933–1940. [CrossRef]
19. Jiménez Vivanco, M.d.R.; García, G.; Doti, R.; Faubert, J.; Lugo Arce, J.E. Time-Resolved Spectroscopy of Ethanol Evaporation on Free-Standing Porous Silicon Photonic Microcavities. *Materials* **2018**, *11*, 894. [CrossRef]

20. Kochergin, V.; Sanghavi, M.; Swinehart, P.R. Porous silicon filters for low-temperature far IR applications. In Proceedings of the Infrared Spaceborne Remote Sensing 2005, San Diego, CA, USA, 29 August 2005; p. 58830T.
21. Morales-Morales, F.; Palacios-Huerta, L.; Cabañas-Tay, S.; Coyopol, A.; Morales-Sánchez, A. Luminescent Si quantum dots in flexible and semitransparent membranes for photon down converting material. *Opt. Mater.* **2019**, *90*, 220–226. [CrossRef]
22. Dovzhenko, D.; Osipov, E.; Martynov, I.; Samokhvalov, P.; Eremin, I.; Kotkovskii, G.; Chistyakov, A. Porous silicon microcavity modulates the photoluminescence spectra of organic polymers and quantum dots. *Mater. Today. Proc.* **2016**, *3*, 485–490. [CrossRef]
23. Zhao, Y.; Gaur, G.; Retterer, S.T.; Laibinis, P.E.; Weiss, S.M. Flow-through porous silicon membranes for real-time label-free biosensing. *Anal. Chem.* **2016**, *88*, 10940–10948. [CrossRef] [PubMed]
24. Weiss, S.M.; Ouyang, H.; Zhang, J.; Fauchet, P.M. Electrical and thermal modulation of silicon photonic bandgap microcavities containing liquid crystals. *Opt. Express* **2005**, *13*, 1090–1097. [CrossRef] [PubMed]
25. Zhao, C.; Wada, T.; De Andrade, V.; Gürsoy, D.; Kato, H.; Chen-Wiegart, Y.-c.K. Imaging of 3D morphological evolution of nanoporous silicon anode in lithium ion battery by X-ray nano-tomography. *Nano Energy* **2018**, *52*, 381–390. [CrossRef]
26. Chen, X.; Bi, Q.; Sajjad, M.; Wang, X.; Ren, Y.; Zhou, X.; Xu, W.; Liu, Z. One-dimensional porous silicon nanowires with large surface area for fast charge–discharge lithium-ion batteries. *Nanomaterials* **2018**, *8*, 285. [CrossRef] [PubMed]
27. Torres-Costa, V.; Mäkilä, E.; Granroth, S.; Kukk, E.; Salonen, J. Synaptic and Fast Switching Memristance in Porous Silicon-Based Structures. *Nanomaterials* **2019**, *9*, 825. [CrossRef] [PubMed]
28. Ishikura, N.; Fujii, M.; Nishida, K.; Hayashi, S.; Diener, J. Porous silicon based extended-bandwidth rugate filters for mid-infrared application. *Infrared Phys. Technol.* **2010**, *53*, 292–294. [CrossRef]
29. Ilyas, S.; Böcking, T.; Kilian, K.; Reece, P.; Gooding, J.; Gaus, K.; Gal, M. Porous silicon based narrow line-width rugate filters. *Opt. Mater.* **2007**, *29*, 619–622. [CrossRef]
30. Pérez, K.S.; Estevez, J.O.; Méndez-Blas, A.; Arriaga, J.; Palestino, G.; Mora-Ramos, M.E. Tunable resonance transmission modes in hybrid heterostructures based on porous silicon. *Nanoscale Res. Lett.* **2012**, *7*, 392. [CrossRef]
31. Thanissen, M.; Berger, M.; Kruger, M.; Billat, S.; Loni, A.; Arens-Fischer, R.; Luth, H.; Theib, W. Microoptical applications of porous silicon superlattices. In Proceedings of the Digest IEEE/Leos 1996 Summer Topical Meeting, Advanced Applications of Lasers in Materials and Processing, Keystone, CO, USA, 5–9 August 1996; pp. 47–48.
32. Massad-Ivanir, N.; Bhunia, S.K.; Jelinek, R.; Segal, E. Porous Silicon Bragg Reflector/Carbon Dot Hybrids: Synthesis, Nanostructure and Optical Properties. *Front. Chem.* **2018**, *6*, 574. [CrossRef]
33. Huang, J.; Li, S.; Chen, Q.; Cai, L. Optical characteristics and environmental pollutants detection of porous silicon microcavities. *Sci. China Chem.* **2011**, *54*, 1348. [CrossRef]
34. Gelloz, B.; Koshida, N. Stabilization and operation of porous silicon photonic structures from near-ultraviolet to near-infrared using high-pressure water vapor annealing. *Thin Solid Films* **2010**, *518*, 3276–3279. [CrossRef]
35. Morales, F.; García, G.; Luna, A.; López, R.; Rosendo, E.; Diaz, T.; Juárez, H. UV distributed Bragg reflectors build from porous silicon multilayers. *J. Eur. Opt. Soc. Rapid Publ.* **2015**, *10*. [CrossRef]
36. Vivanco, M.d.R.J.; Salgado, G.G.; López, J.C.; Agarwal, V.; Caballero, G.N.; Becerril, T.F.D.; Andrés, E.R.; Solís, A.C.; Isasmendi, R.G.; Trujillo, R.R. Porous silicon-based UV microcavities (Conference Presentation). In Proceedings of the Nanophotonic Materials XV 2018, San Diego, CA, USA, 17 September 2018; p. 1072005.
37. Torres-Costa, V.; Martín-Palma, R.; Martínez-Duart, J. All-silicon color-sensitive photodetectors in the visible. *Mater. Sci. Eng. C* **2007**, *27*, 954–956. [CrossRef]
38. Xu, Z.; Chen, G.; Abou-Galala, F.; Leonardi, M. Experimental performance evaluation of non-line-of-sight ultraviolet communication systems. In Proceedings of the Free-Space Laser Communications VII 2007, San Diego, CA, USA, 25 September 2007; p. 67090Y.
39. Van Groesen, E.; Sopaheluwakan, A.; Andonowati, A. Direct characterization of states and modes in defect grating structures. *J. Nonlinear Opt. Phys. Mater.* **2004**, *13*, 155–173. [CrossRef]
40. Van Groesen, E.W.; Sopaheluwakan, A.; Andonowati, A. Defect grating modes as superimposed grating states. In Proceedings of the 8th Annual Symposium IEEE/LEOS Benelux Chapter 2003, Enschede, The Netherlands, 20–21 November 2003; pp. 273–276.

41. Lérondel, G.; Romestain, R.; Madéore, F.; Muller, F. Light scattering from porous silicon. *Thin Solid Films* **1996**, *276*, 80–83. [CrossRef]
42. Ghulinyan, M.; Oton, C.; Bonetti, G.; Gaburro, Z.; Pavesi, L. Free-standing porous silicon single and multiple optical cavities. *J. Appl. Phys.* **2003**, *93*, 9724–9729. [CrossRef]
43. Kruiger, M.; Berger, M.; Marso, M.; Thonissen, M.; Hilbrich, S.; Theib, W.; Loo, R.; Eickhoff, T.; Reetz, W.; Grosse, P. Integration of Porous Silicon Interference Filters in Si-Photodiodes. In Proceedings of the ESSDERC'96: Proceedings of the 26th European Solid State Device Research Conference, Bologna, Italy, 9–11 September 1996; pp. 891–894.
44. Yao, Y.; Lee, K.T.; Sheng, X.; Batara, N.A.; Hong, N.; He, J.; Xu, L.; Hussain, M.M.; Atwater, H.A.; Lewis, N.S. Porous Nanomaterials for Ultrabroadband Omnidirectional Anti-Reflection Surfaces with Applications in High Concentration Photovoltaics. *Adv. Energy Mater.* **2017**, *7*, 1601992. [CrossRef]
45. Sarto, F.; Nichelatti, E.; Flori, D.; Vadrucci, M.; Santoni, A.; Pietrantoni, S.; Guenster, S.; Ristau, D.; Gatto, A.; Trovò, M. Vacuum-ultraviolet optical properties of ion beam assisted fluoride coatings for free electron laser applications. *Thin Solid Films* **2007**, *515*, 3858–3866. [CrossRef]
46. Ristau, D.; Günster, S.; Bosch, S.; Duparré, A.; Masetti, E.; Ferré-Borrull, J.; Kiriakidis, G.; Peiró, F.; Quesnel, E.; Tikhonravov, A. Ultraviolet optical and microstructural properties of MgF_2 and LaF_3 coatings deposited by ion-beam sputtering and boat and electron-beam evaporation. *Appl. Opt.* **2002**, *41*, 3196–3204. [CrossRef]
47. Jerman, M.; Qiao, Z.; Mergel, D. Refractive index of thin films of SiO_2, ZrO_2, and HfO_2 as a function of the films' mass density. *Appl. Opt.* **2005**, *44*, 3006–3012. [CrossRef]
48. Al-Kuhaili, M.; Durrani, S.; Khawaja, E. Characterization of hafnium oxide thin films prepared by electron beam evaporation. *J. Phys. D Appl. Phys.* **2004**, *37*, 1254. [CrossRef]
49. Al-Kuhaili, M. Optical properties of hafnium oxide thin films and their application in energy-efficient windows. *Opt. Mater.* **2004**, *27*, 383–387. [CrossRef]
50. Morton, D.E.; Jensen, T.R. *Ion-Assisted Deposition of Moisture Stable HfO_2 Thin-Films*; Optical Society of America, OIC: Santa Ana Pueblo, NM, USA, 2001. [CrossRef]
51. Jensen, T.R.; Warren, J.; Johnson, R.L. Ion-assisted deposition of moisture-stable hafnium oxide films for ultraviolet applications. *Appl. Opt.* **2002**, *41*, 3205–3210. [CrossRef]
52. Patel, U.S.; Patel, K.H.; Chauhan, K.V.; Chawla, A.K.; Rawal, S.K. Investigation of various properties for zirconium oxide films synthesized by sputtering. *Procedia Technol.* **2016**, *23*, 336–343. [CrossRef]
53. Jakšić, Z.; Maksimović, M.; Sarajlić, M. Silver–silica transparent metal structures as bandpass filters for the ultraviolet range. *J. Opt. A. Pure Appl. Opt.* **2004**, *7*, 51. [CrossRef]
54. Li, W.-D.; Chou, S.Y. Solar-blind deep-UV band-pass filter (250–350 nm) consisting of a metal nano-grid fabricated by nanoimprint lithography. *Opt. Express* **2010**, *18*, 931–937. [CrossRef] [PubMed]

nanomaterials

MDPI

Article

Deep Subwavelength-Scale Light Focusing and Confinement in Nanohole-Structured Mesoscale Dielectric Spheres

Yinghui Cao [1], Zhenyu Liu [2], Oleg V. Minin [3,4] and Igor V. Minin [4,*]

1 College of Computer Science and Technology, Jilin University, 2699 Qianjin Street, Changchun 130012, China; caoyh@jlu.edu.cn

2 Changchun Institute of Optics, Fine Mechanics and Physics, 3888 East Nanhu Road, Changchun 130033, China; liuzy@ciomp.ac.cn

3 Radiohpysical Department, Tomsk State University, 30 Lenin Avenue, Tomsk 634050, Russia; oleg.minin@ngs.ru

4 Nondestructive Testing School, Tomsk Polytechnic University, 36 Lenin Avenue, Tomsk 634050, Russia

* Correspondence: prof.minin@gmail.com; Tel.: +7-383-361-0731

Received: 28 December 2018; Accepted: 30 January 2019; Published: 1 February 2019

Abstract: One of the most captivating properties of dielectric mesoscale particles is their ability to form a sub-diffraction limited-field localization region, near their shadow surfaces. However, the transverse size of the field localization region of a dielectric mesoscale particle is usually larger than $\lambda/3$. In this present paper, for the first time, we present numerical simulations to demonstrate that the size of the electromagnetic field that forms in the localized region of the dielectric mesoscale sphere can be significantly reduced by introducing a nanohole structure at its shadow surface, which improves the spatial resolution up to $\lambda/40$ and beyond the solid immersion diffraction limit of $\lambda/2n$. The proposed nanohole-structured microparticles can be made from common natural optical materials, such as glass, and are important for advancing the particle-lens-based super-resolution technologies, including sub-diffraction imaging, interferometry, surface fabrication, enhanced Raman scattering, nanoparticles synthesis, optical tweezer, etc.

Keywords: nanohole; microsphere; subwavelength-scale light focusing

1. Introduction

Photonic Nanojet (PNJ) is the phenomenon of subwavelength-scale light focusing that is generated by dielectric microparticles. It has been widely applied in different areas, including laser cleaning, nanolithography, super-resolution imaging, enhanced Raman scattering, non-linear fluorescence enhancement, etc. [1–4]. However, one limitation for PNJ is its minimum beam width of $\lambda/3$ [1] (where λ is the wavelength of the incident light) and, thus, more efforts are needed to further reduce the focal spot sizes of the PNJ.

Over the past decade, many studies have been focused on decreasing the width of the PNJ to be as small as possible. If this could be achieved, PNJs would be able to provide new pathways for the trapping of nanoparticles and cells [5,6], subwavelength optical imaging [7], ultrafast all-optical switching [8], etc. Multilayer [9] and a two-layer [10] graded refractive index dielectric particles have also been considered. By using the anomalously intensity-enhanced apodization effect, PNJ with a Full Width at Half Maximum (FWHM) focal spot size that is less than 0.3λ have been attained in previous studies [11,12]. PNJ with a FWHM focal spot size of 0.29λ, which was generated by overstepping the upper refractive index limit has also been reported in the literature [13].

Nanohole-structured dielectric objects have also been considered in the literature. For example, as shown in Reference [14], the effects of the total light far-field scattering of deep holes in the spherical particles with a refractive index near 1.05, were investigated, which showed that these structural details had a negligible influence. In Reference [15], it was shown that the PNJ beam sizes shrank by nearly 28%, due to the introducing of concentric rings on the illumination side of microspheres. To control the unidirectional scattering by the spectral overlapping of the Mie-type resonant modes, a dielectric high-index nanocylinder (n ~ 3.4), with an axial nanohole, has been considered in some studies [16]. Nevertheless, although several methods of light localization have been investigated, deep subwavelength localization, based on a single dielectric mesoscale particle, has not yet been realized.

In this work, a nanohole-structured dielectric microsphere is proposed for deep subwavelength-scale light focusing and strong light confinement, well below the diffraction limit. The field enhancement from the nanostructured mesoscale dielectric microsphere is due to the permittivity contrast between the material of the microsphere and the nanohole. The proposed nanohole-structured microparticles can be made from common dielectric materials, such as glass or latex, and has the advantage of the ability to tailor the spatial region of light confinement and enhancement by choosing the proper geometry, shape, and size of the nanohole. The approach provides a wide platform for deep subwavelength focusing and imaging, which offers the capability of sub-diffraction techniques for microscopy systems, light–matter interactions, interferometry, surface fabrication, enhanced Raman scattering, optical tweezer, and information processing.

This paper has been organized as follows. In Section 2, microspheres without a nanohole have been numerically modeled and analyzed. In Section 3, nanohole-structured microspheres have been simulated and analyzed. Finally, in Section 4, conclusions have been drawn.

2. Dielectric Microspheres without Hole

This section describes the dielectric microspheres with a fixed refractive index of n = 1.5 and different sphere diameters of D_s = 1.5λ, 2.5λ, 3.5λ and 4.5λ, which were numerically modeled and simulated. For λ = 600 nm, the simulated dielectric spheres had a diameter of D_s = 0.9, 1.5, 2.1, and 2.7 μm, respectively. We selected a refractive index of n =1.5, because in an optical band, many commonly used dielectric materials have a refractive index nearly equal to 1.5, such as glass, PMMA, fused silica, etc. [17]. We chose the popular spherical shape of a microparticle for the photonic nanojet formation [4]. The dielectric microspheres were modeled and simulated by using the commercial software COMSOL Multiphysics, which is based on the finite elements method (FEM). In this simulation, a non-uniform mesh was employed to reduce the computational cost and the Perfect Matched Layer (PML) was applied as the boundary condition. The incident light was assumed to be a plane wave that propagates along the z-axis, with a linear polarization along the y-axis. A schematic diagram for the simulated dielectric microsphere is given in Figure 1a.

(a) Microsphere without hole (b) Microsphere with a through hole (c) Microsphere with a blind hole

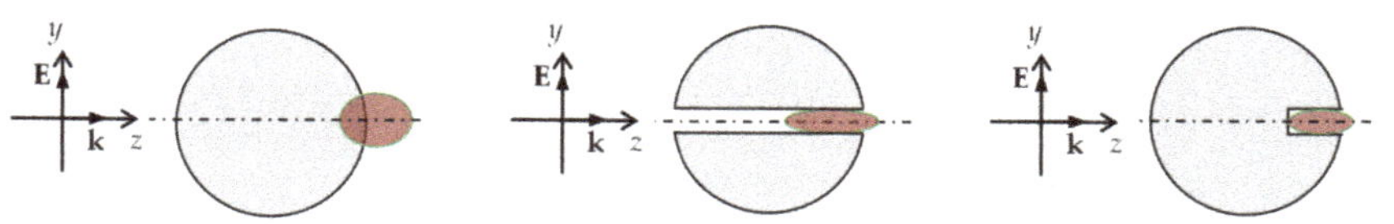

Figure 1. Schematic diagrams for the simulated dielectric sphere (a) without a nanohole, (b) with a through hole, and (c) with a blind hole. The incident light is a plan wave that propagates along the z-axis, **k** is the wave vector. The incident field **E** is polarized along the y-axis. The oval-shaped zones in dark red color indicate the focal spots, and dash-dotted lines indicate the symmetrical axes for the spheres.

Figure 2 shows the light intensity distribution around the simulated dielectric spheres. The light focusing properties of the simulated microspheres are shown in Table 1, including the FWHM focal spot sizes along the x, y, and z-axis (S_x, S_y, S_z), the focal volumes V (which were obtained by volume integration inside the FWHM focal spot) and the maximum light intensity I_{max}, in the focal spot. After combining Figure 2 and Table 1, we found that, with an increase in the sphere diameter from $D_s = 1.5\lambda$ to 3.5λ, the focal volume V decreased from $0.057\lambda^3$ to $0.051\lambda^3$, while the maximum light intensity I_{max} increased from $16.1I_0$ to $61.1I_0$ (I_0 is the light intensity of the incident light). Furthermore, when the sphere diameter was increased to $D_s = 4.5\lambda$, the focal spot split into two parts, with the major part of the focal spot expanding into the surrounding medium, and a smaller part staying inside the microsphere [2,4]. Thus, for the microspheres with a refractive index of $n = 1.5$, when the sphere had a smaller diameter ($D_s < 4.5\lambda$), the light incident on the sphere was focused on the shadow surface (the surface of the particle that is opposite to the irradiated side [4]) of the sphere. When the sphere was large enough ($D_s \geq 4.5\lambda$), PNJ formed beyond the shadow surface of the sphere [2,4], and the second focal spot formed inside the sphere and close to its shadow surface, due to light reflection at the inner side of the shadow surface [4]. It could be noted that, in this case, the maximal field intensity was slightly reduced (Table 1).

Table 1. Light focusing properties of the microspheres with a fixed refractive index of $n = 1.5$ and different sphere diameters (D_s).

D_s	S_x	S_y	S_z	V	I_{max}
1.5λ	0.33λ	0.77λ	0.50λ	$0.057\lambda^3$	$16.1I_0$
2.5λ	0.32λ	0.78λ	0.43λ	$0.055\lambda^3$	$35.5I_0$
3.5λ	0.35λ	0.74λ	0.49λ	$0.051\lambda^3$	$61.1I_0$
4.5λ	0.50λ	0.81λ	0.81λ	$0.173\lambda^3$	$57.4I_0$

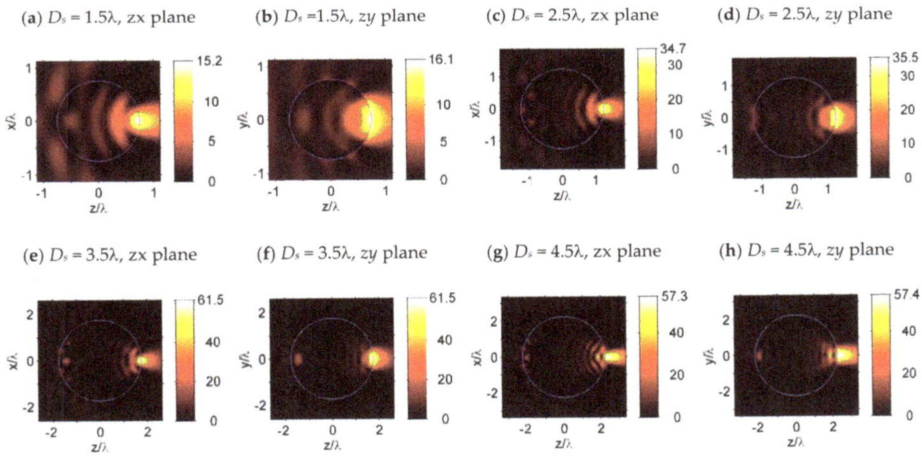

Figure 2. Light intensity distribution of a simulated microsphere with a sphere diameter of (**a**,**b**) $D_s = 1.5\lambda$, (**c**,**d**) $D_s = 2.5\lambda$, (**e**,**f**) $D_s = 3.5\lambda$, and (**g**,**h**) $D_s = 4.5\lambda$. Subfigures (**a**,**c**,**e**,**g**) are plotted in the zx-plane, which is perpendicular to the plane of polarization; subfigures (**b**,**d**,**g**), and (**h**) are plotted in the zx-plane, which is the plane of polarization. The Full Width at Half Maximum (FWHM) focal spot of the simulated microspheres are indicated by the contour lines at the value of half maximum light intensity $0.5I_{max}$, which are plotted by the solid green lines.

3. Nanohole-Structured Dielectric Microspheres

This section describes the nanohole-structured microspheres with a sphere diameter of $D_s = 3.5\lambda$ and a refractive index of $n = 1.5$, which were simulated and analyzed. To make the structure clear, schematic diagrams for the simulated microspheres with a through hole and a blind hole are depicted in Figure 1b,c, respectively.

3.1. Microspheres with a through Hole

First, the nanohole-structured microsphere with a through hole of diameter $d_h = \lambda/5$, $\lambda/10$, and $\lambda/15$ were simulated; light intensity around the simulated microspheres are shown in Figure 3. The corresponding focal spot properties, including the FWHM focal spot sizes (S_x, S_y, S_z), the focal volumes V, and the maximum light intensity I_{max}, in the focal spot, are shown in Table 2. After comparing Figure 3a,b with Figure 2e,f, we found that the focal spot sizes ($S_{x,y,z}$) and focal volume V of the $\lambda/5$-nanohole-structured microsphere were even larger than that of the microspheres without a nanohole, which could be explained as the weakening of the light focusing capability of the dielectric microsphere, due to the comparatively large $\lambda/5$-sized hole. When the hole size was reduced to be smaller than $\lambda/10$, the focal spot sizes and focal volume were reduced considerably, as shown in Figure 3c–h and Table 2. Finally, the features of the FWHM focal spot of the microspheres with $\lambda/15$-sized hole are plotted in Figure 3g, where the green solid lines, the green dashed lines, and the gray solid lines represent the contour lines with a value of $0.5I_{max}$, $0.8I_{max}$, and $0.9I_{max}$, respectively. A comparison of the data in Tables 1 and 2 shows that, with a hole of $\lambda/15$ diameter, the maximum field intensity near the shadow surface of the particle is increased by nearly two times, with a significant decrease in the field localization volume, compared to the unstructured particle.

Table 2. Light focusing properties of the simulated nanohole-structured microspheres with a sphere diameter of $D_s = 3.5\lambda$, refractive index of $n = 1.5$, and a through hole of $d_h = \lambda/15$ in diameter.

D_s	S_x	S_y	S_z	V	I_{max}
$\lambda/5$	0.39λ	0.81λ	0.63λ	$0.064\lambda^3$	$48.6I_0$
$\lambda/10$	0.25λ	0.42λ	0.27λ	$0.0033\lambda^3$	$93.5I_0$
$\lambda/15$	0.20λ	0.14λ	0.25λ	$0.0012\lambda^3$	$112I_0$

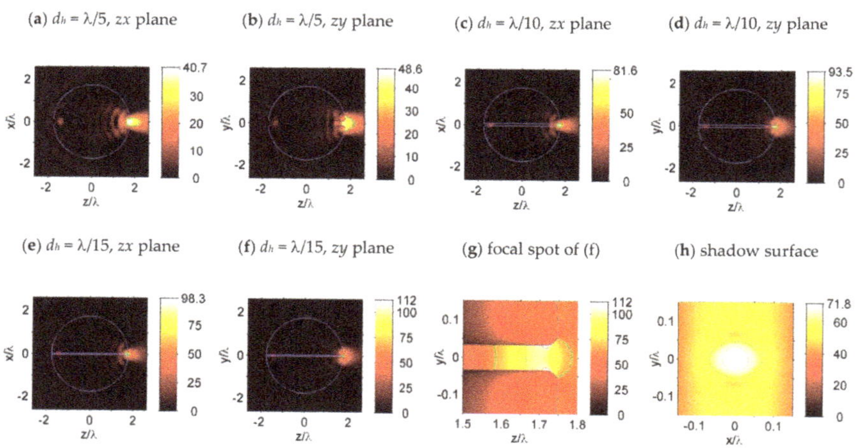

Figure 3. Light intensity ($|E|^2$) of the simulated dielectric microspheres with a nanohole of size (**a**,**b**) $d_h = \lambda/5$, (**c**,**d**) $d_h = \lambda/10$, and (**e**–**h**) $d_h = \lambda/15$. The sphere diameter and refractive index are set as $D_s = 3.5\lambda$ and $n = 1.5$. Features of the focal spot in (**f**) are plotted in (**g**), where the green solid lines, the gray dashed lines, and the gray solid lines indicate the contour lines at values $0.5I_{max}$, $0.8I_{max}$, and $0.9I_{max}$, respectively. For a sphere with a hole size of $d_h = \lambda/15$, light intensity along a plane that is 0.001λ from the shadow surface is plotted in (**h**), with a circle plotted in gray dashed lines, indicating the projection of the hole interface.

3.2. Microspheres with a Blind Hole

After this, dielectric microspheres with a blind hole at the shadow surface were also simulated and the results are shown in Figure 4. In the simulation, the hole diameters were set to be $d_h = \lambda/5$, $\lambda/10$, and $\lambda/40$, respectively, with a hole depth of $3d_h$. The corner radius at the opening of the hole, as well as the fillet radius at the blind end of the hole, were both set to be $r = d_h/2$. From Figure 4a,b, we can see that for a sphere with a hole size of $d_h = \lambda/5$, the focusing capability of the sphere is weakened by the hole, compared to a sphere without a hole in Figure 1e,f. When the hole sizes are decreased below $\lambda/10$, the focused light spot was mainly confined to the blind hole, as shown in Figure 4c–h. "Hot spots" can be observed in Figure 4e,h, which were located inside the hole and near the opening of the hole in the polarization plane (z–y plane), as indicated by the contour lines at the value of $0.85I_{max}$ and $0.9I_{max}$.

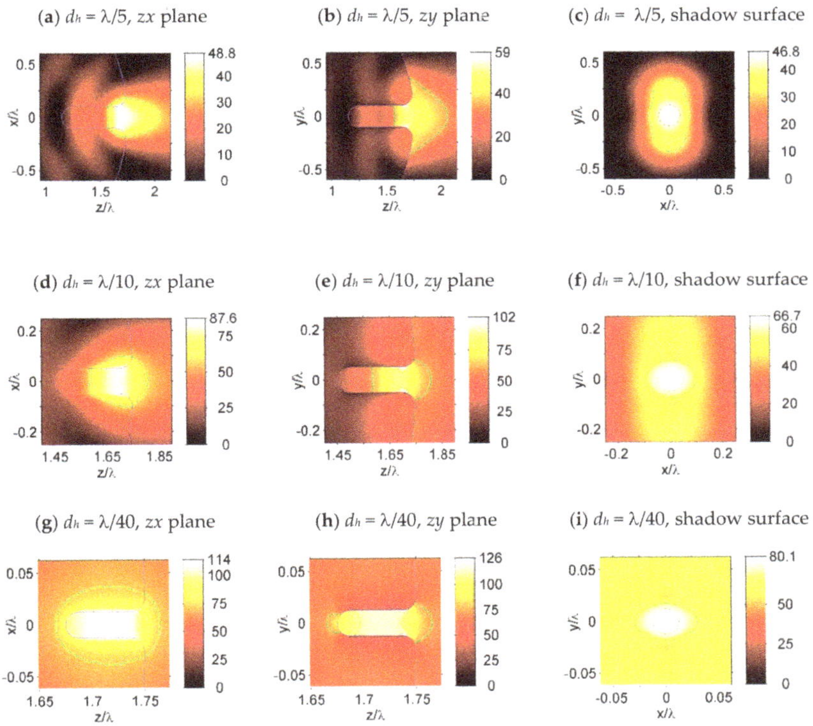

Figure 4. Light intensity ($|E|^2$) of the simulated dielectric microspheres with a blind nanohole of the diameter (**a–c**) $d_h = \lambda/10$, (**d–f**) $d_h = \lambda/10$, and (**g–i**) $d_h = \lambda/40$. The sphere diameter and refractive index are set as $D_s = 3.5\lambda$ and $n = 1.5$, with a hole depth of $3d_h$. Subfigures (**a**,**d**,**g**) are plotted in the zx-plane, which is perpendicular to the polarization plane. Subfigures (**b**,**e**,**h**) are plotted in the zy-plane, which is the polarization plane. To indicate the "hot spots" in subfigures (**e**,**h**,**i**), contour lines at the value of $0.5I_{max}$, $0.85I_{max}$, and $0.9I_{max}$ are plotted by the green solid lines, the green dashed lines, and the gray solid lines, respectively. Subfigures (**c**,**f**,**i**) show the light intensity along a plane that is 0.001λ from the shadow surface of the sphere and the circles, which are plotted by gray dashed lines, indicating the projection of the hole interface in the shadow surface.

4. Discussion

The effect of 3D-field localization inside and near the open boundary of a hole, which is clearly visible in Figures 3 and 4, is similar to the horizontal slot wave-guide in 2D dimensions [18], as a result of this slot concentrating on the optical energy in the low refractive index region of the wave-guide [19]. It should be noticed that in Reference [18], the refractive index contrast was found to be more than 2. Moreover, in our case, the hole was surrounded by dielectric material with axial symmetry, and the boundaries of such a three-dimensional hole structure were essential for the field localization and enhancement, as is shown in Figures 3 and 4. Furthermore, the high E-field confinement, and the large discontinuity in the air hole, is clearly visible in Figure 5. The E-field confinement in the air hole is dictated by the dielectric refractive index contrast [18,19]. It must be mentioned that, in our case, unlike the slot-wave-guides, the field became localized only near the shadow surface of the particle and not along the entire hole. The observed field enhancement effect was analogous to the lens effect. This allowed us to choose the length of the hole, which was not necessary to be equal to the diameter of the particle, but only a part of it (not the through hole). For example, this could be the length of the field localization region inside the particle, near the shadow surface.

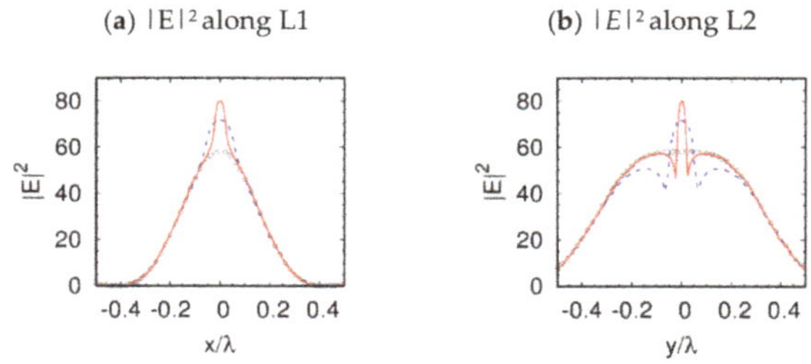

Figure 5. Light intensity along the two imaginary cutting lines L1 and L2, which lie on the shadow surface of the simulated microspheres. L1 is defined as $y = 0$ and $z = D_S/2 + 0.001\lambda$, which is parallel to the x-axis and perpendicular to the polarization direction. L2 is defined as $x = 0$ and $z = D_S/2 + 0.001\lambda$, which is parallel to the y-axis, as well as the polarization direction. The gray circles, the blue dashed lines, and the red solid lines indicate the light intensity of the 3.5λ-diameter microsphere without hole, the microsphere with a through hole of $\lambda/15$ (see Figure 3e–h), and the microsphere with a blind hole of $\lambda/40$ (see Figure 4g–i), respectively.

In addition, the proposed nanohole-structured sphere had several unique properties. For example, it could produce a high-field intensity and optical power in relatively low-index materials (air), at levels that cannot be achieved through a conventional PNJ produced by spheres with the same diameter [2–4]. This property allowed for highly efficient interactions between materials and the fields, in the hole. Moreover, it could produce a strong field confinement which is localized in a nanometer-sized hole, obtaining a resolution that is comparable to the nanohole size, beyond the diffraction limit. Thus, such nanohole-structured mesoscale particles can be used to significantly enhance the efficiency of the near-field probes, and to increase the sensitivity of the compact optical sensing devices.

5. Conclusions

In this work, a nanohole-structured dielectric mesoscale microsphere for a subwavelength-scale light focusing and confinement has been proposed. Numerical simulations showed that light can be focused and confined, considerably, in the nanohole that is located at the shadow surface of the dielectric microparticles, achieving a resolution beyond the solid immersion diffraction limit ($\lambda/2n$) [20], with a plane wave illumination. The introduction of a nanohole on the shadow surface of a dielectric particle, allows us to "compress" the field localization characteristic of a photonic nanojet, to the size of this nanohole. The proposed nanohole-structured microparticles are important for advancing the particle-lens-based super-resolution technologies, including sub-diffraction imaging, interferometry, surface fabrication, enhanced Raman scattering, optical tweezer, and so on.

Author Contributions: I.V.M. and O.V.M., initiation and supervision; Y.C., simulation and visualization of the results; Z.L., methodology and verification of simulation; all authors wrote the paper and discussed the results.

Funding: This research was funded by the National Natural Science Foundation of China, Grant Number 51675506.

Acknowledgments: The work by Igor V. Minin was carried out within the framework of the Tomsk Polytechnic University Competitiveness Enhancement Program.

Conflicts of Interest: The authors declare no conflict of interest.

References

1. Heifetz, A.; Kong, S.C.; Sahakian, A.V.; Taflove, A.; Backman, V. Photonic Nanojets. *J. Comput. Theor. Nanosci.* **2009**, *6*, 1979–1992. [CrossRef] [PubMed]
2. Minin, I.V.; Minin, O.V. *Diffractive Optics and Nanophotonics: Resolution below the Diffraction Limit*; Springer International Publishing: New York, NY, USA, 2016.
3. Ferrand, P.; Wenger, J.; Devilez, A.; Pianta, M.; Stout, B.; Bonod, N.; Popov, E.; Rigneault, H. Direct imaging of photonic nanojets. *Opt. Express* **2008**, *16*, 6930–6940. [CrossRef] [PubMed]
4. Lukyanchuk, B.S.; Dominguez, R.P.; Minin, I.V.; Minin, O.V.; Wang, Z. Refractive index less than two: Photonic nanojets yesterday, today and tomorrow. *Opt. Mater. Express* **2017**, *7*, 1820–1847. [CrossRef]
5. Li, Y.; Xin, H.; Liu, X.; Zhang, Y.; Lei, H.; Li, B. Trapping and detection of nanoparticles and cells using a parallel photonic nanojet array. *ACS Nano* **2016**, *10*, 5800–5808. [CrossRef] [PubMed]
6. Minin, I.V.; Minin, O.V.; Pacheco-Peña, V.; Beruete, M. Subwavelength, standing-wave optical trap based on photonic jets. *Quantum Electron.* **2016**, *46*, 555–557. [CrossRef]
7. Yang, H.; Trouillon, R.; Huszka, G.; Gijs, M. Super-resolution imaging of a dielectric microsphere is governed by the waist of its photonic nanojet. *Nano Lett.* **2016**, *16*, 4862–4870. [CrossRef] [PubMed]
8. Born, B.; Krupa, J.; Gagnon, S.; Holzman, J. Integration of photonic nanojets and semiconductor nanoparticles for enhanced all-optical switching. *Nat. Commun.* **2015**, *6*, 8097. [CrossRef] [PubMed]
9. Liu, C.; Yen, T.; Minin, O.V.; Minin, I.V. Engineering photonic nanojet by a graded-index micro-cuboid. *Physica E* **2018**, *98*, 105–110. [CrossRef]
10. Shen, Y.; Wang, L.; Shen, J. Ultralong photonic nanojet formed by a two-layer dielectric microsphere. *Opt. Lett.* **2014**, *39*, 4120–4123. [CrossRef] [PubMed]
11. Yue, L.; Yan, B.; Monks, J.; Dhama, R.; Wang, Z.; Minin, O.V.; Minin, I.V. Intensity-enhanced apodization effect on an axially illuminated circular-column particle-lens. *Ann. Phys.* **2018**, *530*, 1700384. [CrossRef]
12. Yue, L.; Yan, B.; Monks, J.; Wang, Z.; Tung, N.T.; Lam, V.D.; Minin, O.V.; Minin, I.V. Production of photonic nanojets by using pupil-masked 3D dielectric cuboid. *J. Phys. D Appl. Phys.* **2017**, *50*, 175102. [CrossRef]
13. Gu, G.; Song, J.; Liang, H.; Zhao, M.; Chen, Y.; Qu, J. Overstepping the upper refractive index limit to form ultra-narrow photonic nanojets. *Sci. Rep.* **2017**, *7*, 5635. [CrossRef] [PubMed]
14. Latimer, P. Light scattering by a structured particle: The homogeneous sphere with holes. *Appl. Opt.* **1984**, *23*, 1844. [CrossRef] [PubMed]
15. Wu, M.X.; Huang, B.J.; Chen, R.; Yang, Y.; Wu, J.F.; Ji, R.; Chen, X.D.; Hong, M.H. Modulation of photonic nanojets generated by microspheres decorated with concentric rings. *Opt. Express* **2015**, *23*, 20096–20103. [CrossRef] [PubMed]
16. Isro, S.D.; Iskandar, A.A.; Kivshar, Y.S.; Shadrilov, I.V. Engineering scattering patterns with asymmetric dielectric nanorods. *Opt. Express* **2018**, *26*, 32624. [CrossRef] [PubMed]
17. Refractive Index Database. Available online: https://refractiveindex.info (accessed on 21 January 2019).
18. Almeida, V.R.; Xu, Q.; Barrios, C.A.; Lipson, M. Guiding and confining light in void nanostructure. *Opt. Lett.* **2004**, *29*, 1209–1211. [CrossRef] [PubMed]
19. Barrios, C.A.; Lipson, M. Electrically driven silicon resonant light emitting device based on slot-waveguide. *Opt. Express* **2005**, *13*, 10092. [CrossRef] [PubMed]
20. Mason, D.R.; Jouravlev, M.V.; Kim, K.S. Enhanced resolution beyond the Abbe diffraction limit with wavelength-scale solid immersion lenses. *Opt. Lett.* **2010**, *35*, 2007–2009. [CrossRef] [PubMed]

nanomaterials

MDPI

Article

Nanopillar Diffraction Gratings by Two-Photon Lithography

Julia Purtov [1,2,*], Peter Rogin [1], Andreas Verch [1], Villads Egede Johansen [3] and René Hensel [1,*]

1 INM—Leibniz Institute for New Materials, Campus D2 2, 66123 Saarbrücken, Germany; Peter.Rogin@leibniz-inm.de (P.R.); Andreas.Verch@leibniz-inm.de (A.V.)
2 Department of Materials Science and Engineering, Saarland University, 66123 Saarbrücken, Germany
3 Department of Chemistry, University of Cambridge, Lensfield Road, Cambridge CB2 1EW, UK; villads.egede.johansen@gmail.com
* Correspondence: julia-purtov@web.de (J.P.); rene.hensel@leibniz-inm.de (R.H.)

Received: 7 September 2019; Accepted: 15 October 2019; Published: 19 October 2019

Abstract: Two-dimensional photonic structures such as nanostructured pillar gratings are useful for various applications including wave coupling, diffractive optics, and security features. Two-photon lithography facilitates the generation of such nanostructured surfaces with high precision and reproducibility. In this work, we report on nanopillar diffraction gratings fabricated by two-photon lithography with various laser powers close to the polymerization threshold of the photoresist. As a result, defect-free arrays of pillars with diameters down to 184 nm were fabricated. The structure sizes were analyzed by scanning electron microscopy and compared to theoretical predictions obtained from Monte Carlo simulations. The optical reflectivities of the nanopillar gratings were analyzed by optical microscopy and verified by rigorous coupled-wave simulations.

Keywords: nanostructures; optical pillar gratings; photonic crystals; two-photon lithography; direct laser writing

1. Introduction

Two-dimensional photonic structures such as periodical pillar gratings are applicable for light coupling devices [1], solar devices [2], sensors [3], encoders [4], holographic structures [5], or security features [6]. The optical characteristics of such gratings are very sensitive to the shape, diameter, and height of the nanostructures, as well as the pitch and periodicity of the array [7–11]. Therefore, these features require strict tolerances to ensure uniformity with virtually no defects within the grating. Simultaneously, there is a desire to enable flexible manufacturing of optical elements, as each application has its own requirements regarding design and feature size. In this context, two-photon lithography (TPL) is a promising candidate for the fabrication of nanostructured gratings with tunable optical properties.

In recent years, TPL has been established as a suitable technique for the fabrication of complex three-dimensional structures with submicron resolution [12–15]. Its versatility has been demonstrated by various applications ranging from microfluidic devices [16], micropatterned adhesives [17,18], biological and mechano-sensitive scaffolds [19–22], and optical devices, such as lenses [23] and photonic crystals [24,25]. In a typical TPL process, a focused, femtosecond-pulsed, near-infrared laser (λ = 780 nm) exposes a photoresist, that is composed of reactive oligomers and a photoinitiator. However, the photoreaction is only initiated when two photons excite the initiator concertedly. As a result, the initiator molecule decomposes into radicals, which induce a cross-linking reaction of the oligomers. Such a two-photon event is very rare, so that the probability of a two-photon excitation, and thus the start of polymerization, is only given in the focal region where the photon density is highest [15]. The polymerization reaction is in most cases terminated by oxygen quenching of radicals [26]. The size of the voxel, i.e., the volume element at which the polymerization occurs, is a function of the beam [15,27,28], the exposure parameters [29,30], and the chemical and physical properties of the photoresist [26,29,31,32]. In order to estimate the lateral voxel size S, Sun et al. [33] empirically derived an equation based on the diffraction limit l_{diff}, a material constant α, the applied laser power E_{re}, and the threshold laser power E_{th} required for the cross-linking reaction:

$$S = l_{diff}\alpha\sqrt{\frac{\ln(E_{re}/E_{th})}{4k\ln 2}} \quad (1)$$

where k equals 1 or 2 for a single or two-photon excitation, respectively. According to Equation (1), S decreases with E_{re} and reaches 0 for $E_{re} = E_{th}$. From this it can be concluded that the smallest lateral feature sizes are achieved for laser powers close to the polymerization threshold of the photoresist. Small feature sizes, however, are highly susceptible to deformations and collapse due to their low mechanical resistance [34–36]. In particular, nanostructures with high aspect ratios and low mechanical stiffness are prone to such defects [37,38].

In the present study, we report on the fabrication of nanopillar diffraction gratings. These are fabricated via TPL and laser energies close to the polymerization threshold of the photoresist (10–23 mW) combined with a recently reported, improved development routine [36]. The obtained structures are analyzed by scanning electron microscopy. To enable a prediction of the size of the nanopillars for further investigations, numerical simulations based on a Monte Carlo algorithm were performed. The optical properties of the gratings with pillar diameters between 120 and 430 nm and heights ranging from 330 to 1315 nm were corroborated by rigorous coupled-wave analysis (RCWA) simulations.

2. Experimental

2.1. Two-Photon Lithography (TPL)

Diffraction gratings were fabricated from a negative tone photoresist IP-Dip (Nanoscribe, Eggenstein-Leopoldshafen, Germany) on fused silica substrates using the Professional GT two-photon lithography system (Nanoscribe, Eggenstein-Leopoldshafen, Germany). The system consisted of a 63× objective (NA = 1.4, Carl Zeiss, Oberkochen, Germany) and a femtosecond pulsed IR-laser (λ = 780 nm, 80 MHz laser repetition rate, and 100–200 fs pulse length, TOPTICA Photonics AG, Graefelfing, Germany). The system was operated in 'dip-in' mode, where the objective is immersed into the photoresist. The nanopillars were arranged with a center-to-center distance of 1 µm in a square lattice of 50 × 50 µm. Each pillar consisted of four vertically stacked voxels, whose focal points were separated by 300 nm, whereby the lowest voxel was placed 200 nm below the substrate-resist interface to ensure appropriate attachment to the fused silica substrate. The exposure and settling times were set to 0.1 and 2 ms, respectively. All diffraction gratings were fabricated on the same substrate with altered laser powers ranging from 10 to 23 mW for different gratings. To improve the mechanical stability of the nanopillars, the development was performed according to Purtov et al. [36]. Structures were developed for 20 min in PGMEA (Sigma-Aldrich, Steinheim, Germany), after which 70% of the solution was carefully replaced with isopropanol (Sigma-Aldrich, Steinheim, Germany) without exposing the structures to air. Such a solvent exchange was repeated three times, separated by a residence time of 10 min. Subsequently, a UV-post-crosslinking was applied (t = 300 s, λ_{UV} = 365 nm, 350 mW, OmniCure S1500A, igb-tech, Friedelsheim, Germany) before structures were removed from the liquid and air-dried.

2.2. Scanning Electron Microscopy (SEM)

Samples were fixed on a metallic sample holder and investigated at tilt angles of 0° and 40° using a Quanta 250 FEG (FEI, Eindhoven, The Netherlands) equipped with an Everhart-Thornley-Detector (ETD) in high-vacuum mode. Copper tape was placed close to the nanostructures to avoid charging, as no conductive coating was applied to preserve the optical properties of the arrays. The spot size and the acceleration voltage were set to 2.0 and 2 kV, respectively. The measured pillar heights in micrographs were corrected for the sample tilt.

2.3. Optical Microscopy

The optical reflection characteristics of the pillar arrays were investigated using an optical microscope (Eclipse LV100ND, Nikon, Tokyo, Japan) equipped with a 20× color-corrected objective (NA = 0.45). The microscope was operated in bright field mode with a fully opened illumination aperture upon a white balance using a white sheet of paper.

3. Numerical Simulations

3.1. Voxel Sizes

In order to gain a better understanding of the size- and shape-changing effects during the TPL-process, the obtained nanostructures were analyzed and compared to theoretical pillar sizes derived from simulations considering the different laser powers applied. Since the Gaussian beam formalism based on the paraxial approximation is not appropriate to describe the tightly focused beam used in the experiment, the electric field distribution around the focal spot had to be determined by numerical integration. This integration was performed using the Huygens' principle, i.e., by assuming the field distribution to be the result of superimposing fields originating from an ensemble of emitting elementary sources. This approach was implemented in a self-written software executing a Monte Carlo algorithm.

The coordinate system of the simulation was defined with the light incident from the +z-direction. The origin of the coordinate system was set to the focal spot of the focussed beam. The elementary emitters were assumed to be dipole oscillators distributed in a planar arrangement parallel to the x-y focal plane with a normal distance z_0. The plane of the dipole oscillators can be regarded as the exit pupil of the focusing objective with z_0 being the working distance. The phase of the dipoles $\varphi(r)$ as a function of distance r from the z-axis was adjusted to result in a constant phase at the origin of the coordinate system (i.e., the center of the focal spot), which gives focusing:

$$\varphi = -\frac{2\pi}{\lambda}\cdot\sqrt{z_0^2 + r^2} \tag{2}$$

The amplitude of the dipole strength per unit area $\vec{P}$ followed a Gaussian radial profile, while the polarization was assumed to be circular in order to result in a rotationally symmetric field distribution around the focal spot. By arbitrarily setting all constant factors to 1, the full description of the radial distribution of the dipole strength is given by:

$$\vec{P}(r) = e^{-(r/w)^2}\cdot\begin{pmatrix} 1 \\ i \\ 0 \end{pmatrix}\cdot e^{i\varphi(r)} \tag{3}$$

The width w of the Gaussian function was calculated from the half divergence angle α, (experimentally determined to be 31.4 degrees by analyzing the beam profile as a function of the z-coordinate) and the working distance z_0, as

$$w = z_0\cdot tan(\alpha) \tag{4}$$

The distribution described above was then cut-off at a finite maximum radius r_{max} representing the finite opening of the focusing lens, which is defined by

$$arctan(r_{max}/z_0) = arcsin\left(\frac{NA}{n}\right) \tag{5}$$

where NA is the numerical aperture of the lens and n the refractive index of the medium.

The field distribution near the focal spot was calculated as the superposition of elementary waves emerging from the emitter distribution (Equation (3)). To do this, the field in the volume surrounding the focal spot was mapped to a two-dimensional array of pixels addressed by axial and radial coordinates in the x-z-plane, taking advantage of the deliberately introduced rotational symmetry of the source. The resolution of this field map was chosen to be $\lambda/50$, where λ was set to 780/1.52 = 513 nm (the wavelength of the laser divided by the refractive index of the photoresist). The Monte Carlo algorithm repeatedly picked a randomly selected pair of a pixel $\vec{x}_f$ in this map and a point $\vec{x}_s$ in the source distribution in order to calculate the contribution of the source emitter to the selected pixel. The latter was taken to be the field of an elementary dipole equivalent to the emitter strength $\vec{P} = \vec{P}(\vec{x}_s)$. Omitting constant factors, the elementary dipole field is given by:

$$\vec{E}(\vec{x}_f) = \left\{\frac{\left(\vec{n}\times\vec{P}\right)\times\vec{n}}{R} + \left[3\cdot\vec{n}\cdot\left(\vec{n}\cdot\vec{P}\right) - \vec{P}\right]\cdot\left(R^{-3} + i\cdot R^{-2}\right)\right\}\cdot e^{iR} \tag{6}$$

where $R = \frac{2\pi}{\lambda}\left|\vec{x}_f - \vec{x}_s\right|$ is the distance between the source and the pixel in the field map scaled by the wavelength, and $\vec{n} = \frac{\vec{x}_f - \vec{x}_s}{\left|\vec{x}_f - \vec{x}_s\right|}$ is the normalized vector pointing from the source point to the pixel. All contributions to one pixel originating from different source points (typically 10^5 contributions per pixel) were averaged into one field vector approximating the electrical field with full phase information at the center of each pixel. The square of the above field vector, $I = \vec{E}^2$, is a relative intensity proportional to the physical intensity; the factor of proportionality arising from all the constants explicitly or implicitly omitted above. To calculate voxel sizes as a function of the laser power, this intensity distribution needs to be correlated to the laser power. We start by expressing the latter as a factor f times the threshold laser power (determined to be 9.3 mW by the analysis of the observed pillar diameters according to Equation (1); see Figure 1b). On the other hand, the above intensity distribution has a maximum I_{max} at the focal spot. At the threshold laser power ($f = 1$), this maximum is equivalent to the polymerization threshold I_{th}, $I_{max} = I_{th}$. At a higher laser power, the whole intensity distribution is multiplied by f, and polymerization is initiated wherever the scaled intensity exceeds the threshold. The theoretical height of a voxel is thus determined as the size of the interval in a longitudinal section through the intensity map where $f\cdot I > I_{th}$. To derive the width of the voxel, the same analysis is applied to the cross-section.

The resulting theoretical voxel sizes were used to calculate the theoretical height h of the pillars as follows:

$$h = (a-1)b + c + d/2 \tag{7}$$

where $a = 4$ is the number of stacked voxels, $b = 300$ nm is the vertical center-to-center distance between voxels, $c = -200$ nm is the centre distance of the first voxel from the substrate interface, and d is the height of an individual voxel obtained from the numerical simulations. The resulting theoretical pillar heights and diameters were further used to calculate the initial aspect ratios of pillars and to estimate the shrinkage by comparing the theoretical with the experimental values (see Table S1 in the Supplementary Materials).

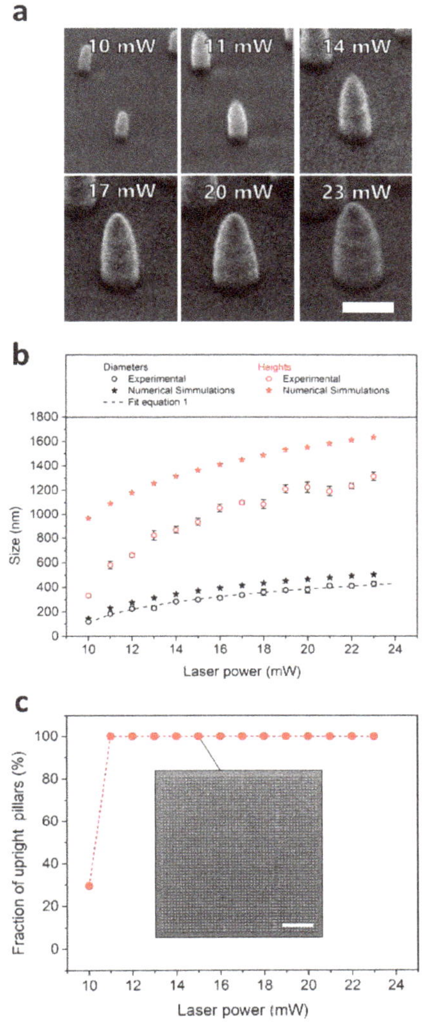

Figure 1. Sizes of nanopillars as a function of the applied laser power. (**a**) Scanning electron micrographs of nanopillars fabricated with different laser powers. The scale bar is 500 nm. (**b**) Diameters (black symbols) and heights (red symbols) of the nanopillars obtained from two-photon lithography (open circles) compared to numerical simulations (filled stars). The dashed line shows the fit of the pillar diameters using Equation (1) to estimate the threshold laser power of the photo resist. (**c**) Defect rates of optical gratings expressed as fractions of upright pillars in dependence on the applied laser power. The values were obtained from Scanning Electron Microscopy (SEM)-images as shown for 15 mW in the insert. The scale bar is 10 μm.

3.2. Optical Spectra

Although no reflection spectra were recorded due to instrumental limitations, an attempt was made to correlate the observed colors with simulated optical spectra. These were calculated from simulated diffraction efficiencies taking into account the different diffraction orders and the finite aperture angle of the microscope objective used. Diffraction was simulated by rigorous coupled-wave analysis (RCWA) using the electromagnetic solver, S4, developed at the Stanford University [39,40]. In these simulations, the pillar shape was approximated as a cylinder with an ellipsoidal tip. The radii of the cylinders r_c were set to the experimentally determined radii of pillars fabricated at different laser powers. The cylinder heights h_c were calculated with $h_c = h_p - h_e$, where h_p was the measured pillar height. The height of the ellipsoidal tip h_e was assigned to half the voxel height evaluated by Monte Carlo simulations described above, as this parameter was hard to determine experimentally. Using this set of parameters, the pillar envelope function can be described as follows:

$$y(z) = \begin{cases} r_c & \text{for } z \leq h_c \\ r_c\sqrt{1-\left(\frac{z-h_c}{h_e}\right)} & \text{for } h_c < z \leq h_p \end{cases} \tag{8}$$

where the z-axis is assumed to be normal to the substrate surface, and thus corresponds to the structure expansion in the vertical direction. The pillars were discretized in slices of 10 nm. A periodic boundary condition was applied with a box size of 1 × 1 μm (in accordance with the pillar center-to-center distance) as well as an incidence angle of 0°. The refractive index and extinction coefficient of the cured photoresist and the fused silica substrate were taken from ellipsometry measurements (see Figure S1 in the Supplementary Materials).

In experiments, the observed colors originate from reflected light that is diffracted by the grating. Reflection occurred at two different interfaces, i.e., the pillar-substrate interface and the backside of the 1 mm fused silica substrate. Due to the opening angle of the objective with NA = 0.45, only diffraction angles of equal to or below 27° were collected. To quantify the contribution of different diffraction orders to the coloration, an overlap factor *OF* between the incident light and the light cones of different diffraction orders were numerically calculated using the given geometrical parameters of the fabricated nanostructures and the imaging system (angle of incidence, collection angle). In the case of the 1st diffraction order, *OF* was found to range from 0.45 at a wavelength of 400 nm to 0.045 at 800 nm. Overlap factors for higher diffraction orders were zero except for negligible values at the shortest wavelengths and, therefore, were not considered in further calculations. The reflection at the pillar-substrate interface was calculated by summing 100% of the reflectivity values obtained for the 0th diffraction order and the values of the 1st diffraction order multiplied by $4OF_\lambda$, where 4 is the number of contributing 1st order diffraction cones, and OF_λ the overlap for the respective wavelength. The reflection at the backside of the substrate was calculated in the same way but using transmissive diffraction efficiencies. For each grating, both contributions were added to a final reflection spectrum.

4. Results and Discussion

4.1. Nanopillar Sizes

Sizes of the nanopillars fabricated with laser powers varying from 10 to 23 mW were obtained from SEM-micrographs (Figure 1a). The pillar diameters ranged from 120 nm to 430 nm (measured at the bottom of the pillars). The heights of the pillars extended from 330 to 1315 nm. The structures exhibited an almost constant aspect ratio (height over diameter) of about 3 (Figure 1b). Evaluation of the defect rate revealed high quality of the obtained optical gratings with 100% upright standing pillars for all gratings with pillar diameters down to 184 nm (Figure 1c). Gratings fabricated with the lowest laser power of 10 mW and thus, closest to the polymerization threshold, exhibited 30% freestanding pillars and 70% collapsed structures. With a diameter of 120 nm, these pillars were the smallest high aspect ratio pillars fabricated with TPL to our best knowledge so far. The collapse of the pillars is most likely induced by capillary forces during drying upon development and post-curing. These collapses occur when the capillary forces exceed the elastic restoring forces of the pillars [36]. As the latter decrease with pillar diameter, smaller structures tend to collapse more easily. For the sake of completeness and to demonstrate the importance of using a UV-post-curing during development, similar pillar structures were fabricated without the additional UV-exposure, exhibiting much more defects even for larger structures (see Figure S2 in the Supplementary Materials).

Theoretical voxel sizes obtained from numerical simulations ranged from 144 to 503 nm in diameter and from 531 to 1873 nm in height with a mean aspect ratio of 3.6 ± 0.1. As a result, the theoretical height of the nanopillars (i.e., four vertically stacked voxels in accordance to Equation (7)) ranges from 965 to 1637 nm as shown in Figure 1b. All values are summarized in Table S1 in the Supplementary Materials. For the diameter, the numerical simulations overestimate the experimental data by 18 ± 3% for all applied laser powers. For the heights, the numerical values overestimate the experimental data by 65% for 10 mW, but only 20% for 23 mW. These discrepancies are most likely related to the shrinkage of the nanopillars, their adhesion to the substrate, and surface tension [36,41]. The degree of shrinkage is mostly related to the removal of unreacted molecules during development. With the reduction of the laser power, the amount of unreacted and not covalently bound oligomers and fragments increased, which in turn did not contribute to the formation of the nanostructures. Furthermore, the distance of the stacked voxels was kept constant for all laser powers, so that voxels overlapped less at lower laser powers, which amplified the effect of an incomplete cross-linking reaction. These unreacted and non-crosslinked fragments were removed during development, which explains higher shrinkage with lower laser power. Upon development, the nanostructures remained in isopropanol and were exposed to UV again. This post-curing led to an additional crosslinking reaction providing enhanced mechanical stability and higher resistance against a capillary forces during drying as is evident from the comparison with structures developed without the additional cross-linking (Figure S2 in the Supplementary Materials). Furthermore, the shrinkage close to the substrate is limited by the adhesion of the nanopillars to the rigid substrate, which led to low shrinkage of the nanopillar diameters. The shrinkage of the pillars heights, in contrast, is not constrained and thus much stronger. This anisotropic shrinkage as well as surface tension effects are assumed to lead to the conical shape of the nanopillars.

4.2. Optical Properties

Figure 2 shows optical micrographs of the nanopillar gratings. The divergence angle of the illuminating light as well as the collection angle of the microscope were both 27°. The optical micrographs revealed a colored reflection of the gratings, which is based mainly on diffraction and interference effects. Absorption can be neglected due to the low extinction coefficients of all materials involved (Figure S1 in the Supplementary Materials). The color of the gratings changed with the laser power from slight brownish (11 mW) to blue-green (23 mW). For 10 mW, the surface appeared colorless (Figure 2a). To quantify the optical appearance of the gratings, the optical micrographs were compared with numerical simulations that provide expected optical spectra for the experimental set-up used (Figure 3). The simulations were performed for six gratings corresponding to 10 mW (colorless), 11 mW (slightly yellow), 14 mw (brown), 17 mW (blue-brown), 20 mW (blue), and 23 mW (blue-green).

The obtained spectra did not fully agree with the colors recorded by optical microscopy. Minor deviations can be expected from using diffraction efficiencies for normal incidence to approximate the whole cone of incident light. Moreover, considerable differences could be attributed to variations of the pillar shape. In the fabricated gratings, the shape of the pillars varied between cylinders and cones (compare Figure 1a), whereas cylindrical pillars with elliptical tips were assumed in the simulations. The shape, though, is important for the choice of an appropriate pillar diameter for simulations. This argument was confirmed by simulations with 20% smaller pillar diameters, which led to a significant blue-shift of the spectra (dashed lines in Figure 3) and a better correspondence with the colors observed. From this, it can be concluded that even small variations in the shape of the pillars dramatically affect the optical appearance of the gratings, and that it is therefore not sufficient to evaluate diameters at the pillars' bases only. This outcome supports our arguments that good understanding of occurring effects, such a shrinkage, surface tension, and related mechanisms, as well as their influence on the feature shape and sizes are important for a precise prediction of the optical properties of pillar gratings fabricated via TPL.

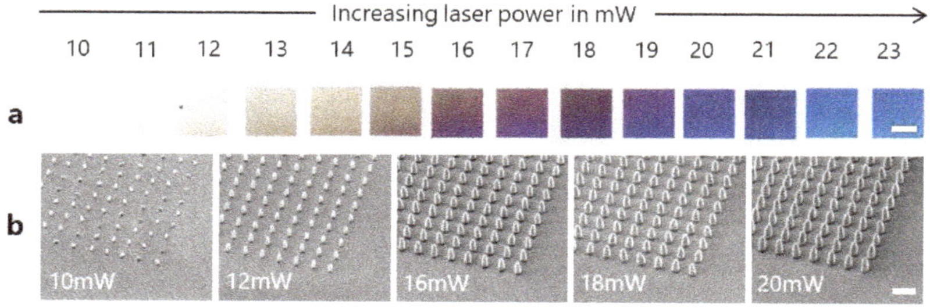

Figure 2. Optical appearance of the nanopillar gratings in dependence on the laser power. (**a**) Optical micrographs of 50 × 50 µm nanopillar gratings on a fused silica substrate. Scale bar is 25 µm. (**b**) Scanning electron micrographs showing the corresponding nanopillars. Scale bar is 1 µm.

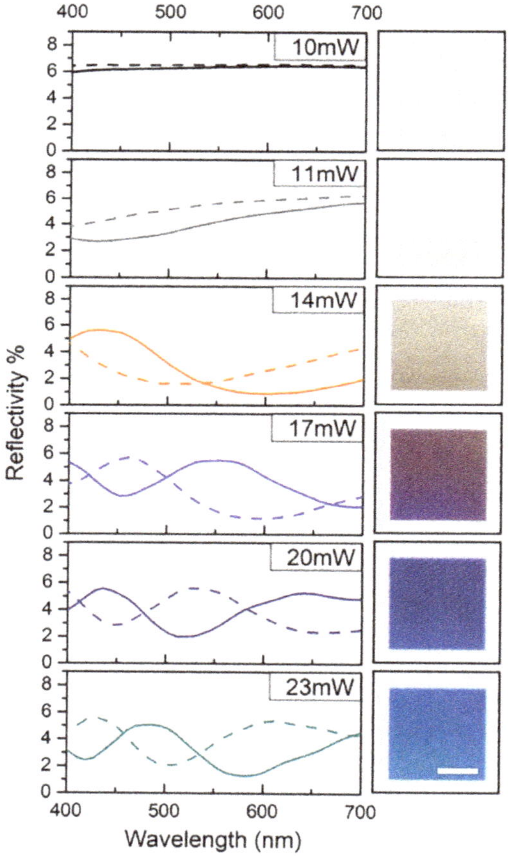

Figure 3. Optical properties of the nanopillar gratings. Reflectivity in dependence on the wavelength obtained from numerical simulations for structure sizes as measured by SEM (solid line) and pillars assuming 20% smaller diameters (dashed line) (left) compared to optical micrographs of nanopillar gratings fabricated with different laser powers (right). The scale bar is 25 µm.

The total reflectivities shown in Figure 3 comprise the reflected diffraction at the pillar-substrate interface as well as the transmissive diffraction of light reflected at the backside of the substrate. As these two are expected to differ strongly in their minima due to variation in the optical path, minima as low as 1% predicted by simulation were surprising. Nevertheless, their presence is confirmed by the intense colors of the gratings. We assume that this could be caused by Mie resonances [42]. This assumption is supported by the strong dependence of the spectra on the pillar diameter. Small pillars obtained at low laser powers interact predominantly with ultraviolet to blue light, leading to a brownish hue. As the pillar size increases with increasing laser power, the resonance shifts towards red wavelengths, resulting in a blueish hue. This variability in color due to the size and shape of nanopillars gratings allows for efficient diffractive color filters.

5. Conclusions

Optical pillar gratings were successfully fabricated via TPL at different laser powers close to the polymerization threshold of the photoresist and investigated with respect to their sizes and optical properties using imaging techniques and numerical simulations. The following conclusions can be drawn:

(1) Defect-free nanopillar gratings were fabricated down to pillar diameters of 184 nm and aspect ratios about 3. The smallest pillar diameters achieved were 120 nm, but on imperfect arrays, and therefore would require further optimization in fabrication.

(2) Simulations of the voxel sizes overestimated experimental pillar sizes by 20% in lateral and up to 65% in the vertical direction. This effect can be rationalized by shrinkage that differs due to varying amounts of unreacted oligomers, different overlaps between adjacent voxels, substrate adhesion, and probably surface tension effects.

(3) The nanopillar gratings interfered with visible wavelengths and varied in their optical properties depending on the pillar sizes tuned by TPL. The simulation of the optical spectra confirmed that the coloration originates indeed from the diffraction of reflected light, but also prompts the notion that the size and shape of the nanopillars strongly influence the optical appearance.

In summary, optical gratings based on different pillar sizes can be manufactured by varying the laser power in TPL in a single process step on one surface. However, the fabrication of precise optical gratings close to the polymerization threshold of the photoresist requires a deep understanding of the involved processes.

Supplementary Materials: The following are available online at http://www.mdpi.com/2079-4991/9/10/1495/s1, Table S1: Summary of pillar dimensions as a function of the laser power. Figure S1: Optical properties of the cured photoresist and the fused silica substrate obtained by ellipsometry. Figure S2: Optical pillar gratings when fabricated without UV-post-curing.

Author Contributions: Conceptualization, J.P. and R.H.; methodology, J.P. and P.R.; validation, A.V., P.R., R.H.; formal analysis, J.P., P.R., A.V., V.E.J.; investigation, J.P.; resources, R.H.; data curation, J.P. and P.R.; writing, J.P., P.R., A.V., and R.H.; visualization, J.P. and R.H.; project administration, J.P. and R.H.; funding acquisition, V.E.J. and R.H.

Funding: This research was funded by the European Commission (Marie Curie Fellowship Looking Through Disorder (LODIS), 701455) as well as the European Research Council under the European Union's Seventh Framework Program (FP/2007-2013)/ERC Advanced Grant no. 340929.

Acknowledgments: The authors thank Reza Hosseinabadi for his help with the TPL-fabrication and the optical imaging as well as Bruno Schäfer for his support during the ellipsometry measurements.

Conflicts of Interest: The authors declare no conflict of interest.

References

1. Catchpole, K.R.; Green, M.A. A conceptual model of light coupling by pillar diffraction gratings. *J. Appl. Phys.* **2007**, *101*, 063105. [CrossRef]
2. Chong, T.K.; Wilson, J.; Mokkapati, S.; Catchpole, K.R. Optimal wavelength scale diffraction gratings for light trapping in solar cells. *J. Opt.* **2012**, *14*, 024012. [CrossRef]

3. Yan, H.; Huang, L.; Xu, X.; Chakravarty, S.; Tang, N.; Tian, H.; Chen, R.T. Unique surface sensing property and enhanced sensitivity in microring resonator biosensors based on subwavelength grating waveguides. *Opt. Express* **2016**, *24*, 29724–29733. [CrossRef] [PubMed]
4. Mao, X.; Zeng, L. Design and fabrication of crossed gratings with multiple zero-reference marks for planar encoders. *Meas. Sci. Technol.* **2018**, *29*, 025204. [CrossRef]
5. Kim, D.Y.; Tripathy, S.K.; Li, L.; Kumar, J. Laser-induced holographic surface relief gratings on nonlinear optical polymer films. *Appl. Phys. Lett.* **1995**, *66*, 1166–1168. [CrossRef]
6. Gale, M.T.; Knop, K.; Morf, R.H. *Zero-Order Diffractive Microstructures for Security Applications*; SPIE: Bellingham, WA, USA, 1990; Volume 1210.
7. Wang, Y.; Lu, N.; Xu, H.; Shi, G.; Xu, M.; Lin, X.; Li, H.; Wang, W.; Qi, D.; Lu, Y.; et al. Biomimetic corrugated silicon nanocone arrays for self-cleaning antireflection coatings. *Nano Res.* **2010**, *3*, 520–527. [CrossRef]
8. Dewan, R.; Fischer, S.; Meyer-Rochow, V.B.; Özdemir, Y.; Hamraz, S.; Knipp, D. Studying nanostructured nipple arrays of moth eye facets helps to design better thin film solar cells. *Bioinspiration Biomim.* **2011**, *7*, 016003. [CrossRef]
9. Ji, S.; Park, J.; Lim, H. Improved antireflection properties of moth eye mimicking nanopillars on transparent glass: Flat antireflection and color tuning. *Nanoscale* **2012**, *4*, 4603–4610. [CrossRef]
10. Lora Gonzalez, F.; Chan, L.; Berry, A.; Morse, D.E.; Gordon, M.J. Simple colloidal lithography method to fabricate large-area moth-eye antireflective structures on Si, Ge, and GaAs for IR applications. *J. Vac. Sci. Technol. B* **2014**, *32*, 051213. [CrossRef]
11. Chigrin, D.N.; Lavrinenko, A.V. *Nanopillar Coupled Periodic Waveguides: From Basic Properties to Applications*; SPIE: Bellingham, WA, USA, 2006; Volume 6393.
12. Farsari, M.; Chichkov, B.N. Materials processing: Two-photon fabrication. *Nat. Photonics* **2009**, *3*, 450–452. [CrossRef]
13. Gissibl, T.; Thiele, S.; Herkommer, A.; Giessen, H. Two-photon direct laser writing of ultracompact multi-lens objectives. *Nat. Photonics* **2016**, *10*, 554–560. [CrossRef]
14. Hohmann, J.K.; Renner, M.; Waller, E.H.; von Freymann, G. Three-Dimensional μ-Printing: An Enabling Technology. *Adv. Opt. Mater.* **2015**, *3*, 1488–1507. [CrossRef]
15. Kawata, S.; Sun, H.B.; Tanaka, T.; Takada, K. Finer features for functional microdevices. *Nature* **2001**, *412*, 697. [CrossRef] [PubMed]
16. Waheed, S.; Cabot, J.M.; Macdonald, N.P.; Lewis, T.; Guijt, R.M.; Paull, B.; Breadmore, M.C. 3D printed microfluidic devices: Enablers and barriers. *Lab Chip* **2016**, *16*, 1993–2013. [CrossRef] [PubMed]
17. Fischer, S.C.L.; Groß, K.; Torrents Abad, O.; Becker, M.M.; Park, E.; Hensel, R.; Arzt, E. Funnel-Shaped Microstructures for Strong Reversible Adhesion. *Adv. Mater. Interfaces* **2017**, *4*, 1700292. [CrossRef]
18. Hensel, R.; Moh, K.; Arzt, E. Engineering Micropatterned Dry Adhesives: From Contact Theory to Handling Applications. *Adv. Funct. Mater.* **2018**, *28*, 1800865. [CrossRef]
19. Marino, A.; Filippeschi, C.; Mattoli, V.; Mazzolai, B.; Ciofani, G. Biomimicry at the nanoscale: Current research and perspectives of two-photon polymerization. *Nanoscale* **2015**, *7*, 2841–2850. [CrossRef]
20. Wolfenson, H.; Meacci, G.; Liu, S.; Stachowiak, M.R.; Iskratsch, T.; Ghassemi, S.; Roca-Cusachs, P.; O'Shaughnessy, B.; Hone, J.; Sheetz, M.P. Tropomyosin controls sarcomere-like contractions for rigidity sensing and suppressing growth on soft matrices. *Nat. Cell Biol.* **2015**, *18*, 33. [CrossRef]
21. Marino, A.; Ciofani, G.; Filippeschi, C.; Pellegrino, M.; Pellegrini, M.; Orsini, P.; Pasqualetti, M.; Mattoli, V.; Mazzolai, B. Two-Photon Polymerization of Sub-micrometric Patterned Surfaces: Investigation of Cell-Substrate Interactions and Improved Differentiation of Neuron-like Cells. *ACS Appl. Mater. Interfaces* **2013**, *5*, 13012–13021. [CrossRef]
22. Klein, F.; Striebel, T.; Fischer, J.; Jiang, Z.; Franz, C.M.; von Freymann, G.; Wegener, M.; Bastmeyer, M. Elastic Fully Three-dimensional Microstructure Scaffolds for Cell Force Measurements. *Adv. Mater.* **2010**, *22*, 868–871. [CrossRef]
23. Thiele, S.; Arzenbacher, K.; Gissibl, T.; Giessen, H.; Herkommer, A.M. 3D-printed eagle eye: Compound microlens system for foveated imaging. *Sci. Adv.* **2017**, *3*. [CrossRef] [PubMed]
24. von Freymann, G.; Ledermann, A.; Thiel, M.; Staude, I.; Essig, S.; Busch, K.; Wegener, M. Three-Dimensional Nanostructures for Photonics. *Adv. Funct. Mater.* **2010**, *20*, 1038–1052. [CrossRef]
25. Nawrot, M.; Zinkiewicz, Ł.; Włodarczyk, B.; Wasylczyk, P. Transmission phase gratings fabricated with direct laser writing as color filters in the visible. *Opt. Express* **2013**, *21*, 31919–31924. [CrossRef] [PubMed]

26. Mueller, J.B.; Fischer, J.; Mayer, F.; Kadic, M.; Wegener, M. Polymerization Kinetics in Three-Dimensional Direct Laser Writing. *Adv. Mater.* **2014**, *26*, 6566–6571. [CrossRef] [PubMed]
27. Sun, H.B.; Maeda, M.; Takada, K.; Chon, J.W.M.; Gu, M.; Kawata, S. Experimental investigation of single voxels for laser nanofabrication via two-photon photopolymerization. *Appl. Phys. Lett.* **2003**, *83*, 819–821. [CrossRef]
28. Fischer, J.; Wegener, M. Three-dimensional optical laser lithography beyond the diffraction limit. *Laser Photonics Rev.* **2013**, *7*, 22–44. [CrossRef]
29. Xing, J.F.; Dong, X.Z.; Chen, W.Q.; Duan, X.M.; Takeyasu, N.; Tanaka, T.; Kawata, S. Improving spatial resolution of two-photon microfabrication by using photoinitiator with high initiating efficiency. *Appl. Phys. Lett.* **2007**, *90*, 131106. [CrossRef]
30. Tanaka, T.S.; Sun, H.B.; Kawata, S. Rapid sub-diffraction-limit laser micro/nanoprocessing in a threshold material system. *Appl. Phys. Lett.* **2002**, *80*, 312–314. [CrossRef]
31. Fischer, J.; Mueller, J.B.; Quick, A.S.; Kaschke, J.; Barner-Kowollik, C.; Wegener, M. Exploring the Mechanisms in STED-Enhanced Direct Laser Writing. *Adv. Opt. Mater.* **2015**, *3*, 221–232. [CrossRef]
32. Sun, H.B.; Takada, K.; Kim, M.S.; Lee, K.S.; Kawata, S. Scaling Laws of Voxels in Two-Photon Photopolymerization Nanofabrication. *Appl. Phys. Lett.* **2003**, *83*, 1104–1106. [CrossRef]
33. Sun, H.B.; Suwa, T.; Takada, K.; Zaccaria, R.P.; Kim, M.S.; Lee, K.S.; Kawata, S. Shape precompensation in two-photon laser nanowriting of photonic lattices. *Appl. Phys. Lett.* **2004**, *85*, 3708–3710. [CrossRef]
34. LaFratta, C.; Baldacchini, T. Two-Photon Polymerization Metrology: Characterization Methods of Mechanisms and Microstructures. *Micromachines* **2017**, *8*, 101. [CrossRef]
35. Hu, Y.; Lao, Z.; Cumming, B.P.; Wu, D.; Li, J.; Liang, H.; Chu, J.; Huang, W.; Gu, M. Laser printing hierarchical structures with the aid of controlled capillary-driven self-assembly. *Proc. Natl. Acad. Sci. USA* **2015**, *112*, 6876–6881. [CrossRef] [PubMed]
36. Purtov, J.; Verch, A.; Rogin, P.; Hensel, R. Improved development procedure to enhance the stability of microstructures created by two-photon polymerization. *Microelectron. Eng.* **2018**, *194*, 45–50. [CrossRef]
37. Roca-Cusachs, P.; Rico, F.; Martínez, E.; Toset, J.; Farré, R.; Navajas, D. Stability of Microfabricated High Aspect Ratio Structures in Poly(dimethylsiloxane). *Langmuir* **2005**, *21*, 5542–5548. [CrossRef] [PubMed]
38. Chandra, D.; Yang, S. Capillary-Force-Induced Clustering of Micropillar Arrays: Is It Caused by Isolated Capillary Bridges or by the Lateral Capillary Meniscus Interaction Force? *Langmuir* **2009**, *25*, 10430–10434. [CrossRef]
39. Liu, V.; Fan, S. S4: A free electromagnetic solver for layered periodic structures. *Comput. Phys. Commun.* **2012**, *183*, 2233–2244. [CrossRef]
40. Available online: https://web.stanford.edu/group/fan/S4/ (accessed on 24 Nov 2018).
41. Lim, T.W.; Son, Y.; Yang, D.Y.; Pham, T.A.; Kim, D.P.; Yang, B.I.; Lee, K.S.; Park, S.H. Net Shape Manufacturing of Three-Dimensional SiCN Ceramic Microstructures Using an Isotropic Shrinkage Method by Introducing Shrinkage Guiders. *Int. J. Appl. Ceram. Technol.* **2008**, *5*, 258–264. [CrossRef]
42. Kuznetsov, A.I.; Miroshnichenko, A.E.; Brongersma, M.L.; Kivshar, Y.S.; Luk'yanchuk, B. Optically resonant dielectric nanostructures. *Science* **2016**, *354*, aag2472. [CrossRef]

nanomaterials

Article

Picosecond Laser-Induced Hierarchical Periodic Near- and Deep-Subwavelength Ripples on Stainless-Steel Surfaces

Shijie Ding [1,†], Dehua Zhu [2,†], Wei Xue [1], Wenwen Liu [1] and Yu Cao [2,*]

1 College of Mechanical & Electrical Engineering, Wenzhou University, Wenzhou 325035, China; dingshijie1201@163.com (S.D.); xw@wzu.edu.cn (W.X.); sophialww@163.com (W.L.)

2 Institute of Laser and Optoelectronic Intelligent Manufacturing, Wenzhou University, Wenzhou 325035, China; zhu_556@163.com

* Correspondence: yucao@wzu.edu.cn

† These authors contributed equally to this work.

Received: 5 December 2019; Accepted: 23 December 2019; Published: 26 December 2019

Abstract: Ultrafast laser-induced periodic surface subwavelength ripples, categorized based on the ripple period into near-subwavelength ripples (NSRs) and deep-subwavelength ripples (DSRs), are increasingly found in the variety of materials such as metals, semiconductors and dielectrics. The fabrication of hierarchical periodic NSRs and DSRs on the same laser-irradiated area is still a challenge since the connection between the two remains a puzzle. Here we present an experimental study of linearly polarized picosecond laser-induced hierarchical periodic NSRs and DSRs on stainless-steel surfaces. While experiencing peak power density higher than a threshold value of 91.9 GW/cm^2, in the laser-scanned area appear the hierarchical periodic NSRs and DSRs (in particular, the DSRs are vertically located in the valley of parallel NSRs). A large area of the uniformly hierarchical periodic NSRs and DSRs, with the spatial periods 356 ± 17 nm and 58 ± 15 nm, respectively, is fabricated by a set of optimized laser-scanning parameters. A qualitative explanation based on the surface plasmon polariton (SPP) modulated periodic coulomb explosion is proposed for unified interpretation of the formation mechanism of hierarchical periodic NSRs and DSRs, which includes lattice orientation of grains as a factor at low peak power density, so that the initial DSRs formed have a clear conformance with the metallic grains.

Keywords: laser induced subwavelength ripples; surface plasmon polaritons; coulomb explosion; stainless steel

1. Introduction

The laser-induced periodic surface structure (LIPSS) [1,2], a by-product effect of laser material processing applications is at its inception, but in the past few years has been a research hotspot that involves multiple branches of physics such as wave optics, non-linear optics, fluid dynamics and thermodynamics. With the rapid development in particular of ultrafast laser-matter interaction in both science and engineering, the ultrafast laser-induced periodic surface subwavelength ripples [3–5] which can be categorized based on the ripple period (Λ) into near-subwavelength ripples (NSRs, $0.4 < \Lambda/\lambda < 1$) and deep-subwavelength ripples (DSRs, $\Lambda/\lambda < 0.4$) [3], have been reported increasingly on a variety of materials such as metals [5,6], semiconductors [7,8] and dielectrics [3,9].

However, the existing formation mechanisms for construction of NSRs and DSRs are considered completely different, and their theoretical models have been in a state of debate and development [3,10,11]. The NSRs (sometimes called classical ripples) are found to be formed by a wide range of pulse durations from a few hundred nanosecond (ns) to femtosecond (fs) laser irradiation on the solid material surface, which makes it seem like a universal pulse laser-induced material response. The orientation of NSRs, which were found always parallel or perpendicular to the incident laser polarization, is usually ascribed to the periodic electric field modulation by the incident laser beam interference with surface scattering laser wave due to the sample surface roughness [12,13]. The excitation of surface plasmon polaritons (SPPs) by incident laser pulses is also widely used for the explanation of NSR formation [3]. Huang et al. [3] compared experiments with metal, semiconductor and dielectric and assumed the grating-assisted surface plasmon (SP)-laser coupling should be responsible for the origination of NSRs. Reif et al. [14,15] considered the ion sputtering and thin liquid film, and proposed the self-organized effect of the non-stable material to explain the formation of subwavelength ripples. Instead of easy production and observation of the NSRs, DSRs are only rarely found in ultra-short laser pulse (duration less than 100 picoseconds) irradiation of some selected materials, including fused silica [16], silicon [8,10,17], graphite [11] and some other materials [10,18,19]. Diverse mechanisms are proposed for explanation of the origins of DSRs, such as second and higher harmonic generation [10,20], SPPs [11] and self-organized effect [14,21].

The relationship between the formation mechanisms of NSRs and DSRs still remains a puzzle. The interference model of incident and surface scattered laser beam can effectively explain the correlation between the spatial period of NSRs and the incident angle, wavelength of laser irradiation, but not fit the period of the DSRs. The SPPs model excited from the rough surface by scattering with local defects is doubted as its intensity distribution is not at regularity but intricacy [22]. Rudenko A. gave a reasonable opinion that the surface scatter electromagnetic waves have been demonstrated to consist of the SPP, an evanescent cylindrical wave, and a Norton wave [23]. The DSRs models like second and higher harmonic generation [10] and grating-assisted SP-laser coupling [3,10] are not compatible with the spatial period of NSRs. The self-organized theory based on the ion sputtering and instability of thin liquid films is theoretically compatible with various ripple periods for any NSRs and DSRs, which attribute the directional asymmetry of laser-induced ripples to a spatial asymmetry of the excited-electron kinetic energies [21]. But it can only meet with the subwavelength periodic ripples that are perpendicular to the incident laser electric polarization. This does not agree with the scenario of ripples' orientation parallel with laser polarization that is observed in dielectric materials [24], metals and semiconductor materials [10,25,26].

Another interesting subject of debate is whether the hierarchical periodic NSRs and DSRs can be constructed simultaneously on the same laser-irradiated area, which should be helpful to reveal the connection between the two, and to discover a unified theoretical model. Rare reports took notice of the mechanism and regularity of hierarchical periodic NSRs and DSRs formation. Yao et al. [27] used femtosecond laser scanning to fabricate large area subwavelength ripples on stainless-steel surfaces for a reinforced antireflection property. The induced ripples included both NSRs and DSRs, but the DSRs were not noted by the authors since the DSRs were not clear. Ji et al. [28] reported cross-periodic structures (NSRs and DSRs) on the silicon surface that the DSRs induced by a low laser pulse repetition rate of 1 kHz were not distinct and the formation mechanism remains in doubt. In contrast with the fabrication of NSRs, the DSRs usually need a higher repetition rate of the ultrashort laser pulse (above 76 MHz) [10] or an underwater environment [29,30]. Ahsan et al. [31] and Romano et al. [32] reported that the laser irradiation dose and the transient process had a significant influence on the ripple formation, suitable adjustment of the ultrafast laser-induced surface subwavelength ripples could offer a variety of colors. Gurevich et al. [33] proposed that the ripples formation cannot be explained in the frames of pure plasmonic theory and hydrodynamic instabilities are an important influence in the ripple formation. Liao et al. [34] reported that excitation of standing plasma waves at the interfaces between areas modified and unmodified by the femtosecond laser irradiation plays a crucial role in promoting the growth of periodic nanoripples. In fact, there are many diverse mechanisms proposed for the origins of laser-induced DSRs and NSRs, such as the second and higher harmonic generation [10,20], SPPs [11] and self-organized effect [14,21], so far no model can satisfy all the phenomena observed. Hence, the fabrication of in situ hierarchical periodic NSRs and DSRs remains a challenge when exploring their physical mechanisms and potential applications.

In this work, we present a picosecond (ps) laser induced hierarchical periodic NSRs and DSRs on stainless-steel surfaces. The evolution of DSRs to NSRs with laser peak power density deposition was observed, and the lattice orientation as the intrinsic characteristic of the stainless-steel material shows a significant role in the evolutionary process of DSRs to NSRs. A qualitative explanation of the ps laser-induced hierarchical periodic DSR and NSR formation mechanism is discussed and compared with a self-organized effect and the grating-assisted SP-laser coupling theory.

2. Materials and Methods

Stainless steel is one of the most extensively studied metals because of its versatility in industry and daily life. It has been reported that periodic NSRs are induced by the ultrashort pulse laser irradiation on stainless steel surface [35], but so far no DSRs. AISI 316L stainless steel sheets with dimensions of 50 mm × 50 mm and 1 mm of thickness were used in the experiment. Samples were mechanically polished and divided into two groups A and B. The group A samples were directly subject to the laser irradiation and the group B samples were firstly corroded by the surface erosion agent (HF 20 %vol. + HNO_3 10 %vol. + H_2O 70 %vol.) before the subsequent laser irradiation. After the corrosion, the group B samples revealed the metallographic lattice structure on the surface. Before the laser irradiation, all the samples were cleaned by ultrasonic ethanol bath for 10 min. The same laser processing parameters were used for the fabrication of the hierarchical periodic NSRs and DSRs on the non-corroded (group A) and corroded (group B) sample surfaces.

The schematic setup of the experimental ps laser-scanning system is shown in Figure 1. The as-prepared samples were irradiated with a linearly polarized ps laser (Trumicro5050, TRUMPF Company, Ditzingen, Germany) with the repetition rate of 400 kHz and the pulse duration of 10 ps. The BBO (Beta barium borate, β-BaB2O4) crystal was utilized for frequency doubling of the incident laser beam from the wavelength of 1030 nm to 515 nm. Also, we set up the beam expander in order to achieve smaller focal spot size and higher laser fluence. A 2-D galvanometric scanner (SCANLab Company, Puchheim, Germany) with F-theta objective lens was used for the focused laser beam scanning in x-y directions. A z-direction stage was used to control the focal length of the laser beam spot since the laser beam was set perpendicular to the surface of samples. The focal spot diameter, defined as the gaussian-profile laser beam at $1/e^2$ of its maximum intensity, was approximately 16 µm. Different scanning speeds and power density were selected for fabrication of the hierarchical periodic dual subwavelength ripples. Large-area hierarchical periodic NSRs and DSRs were prepared by laser parallel line scanning on the sample surfaces.

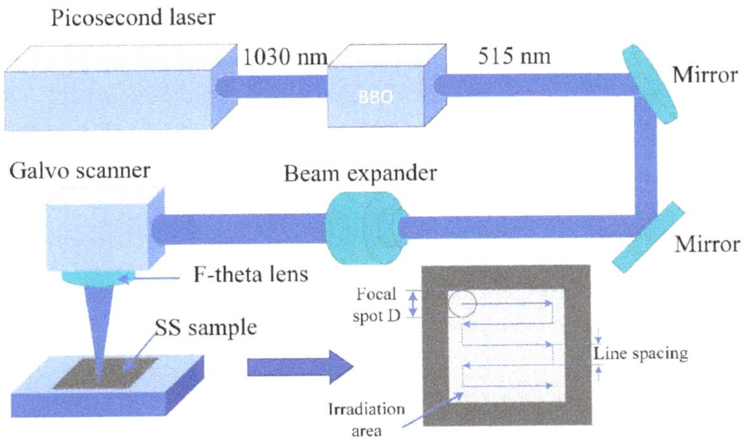

Figure 1. Schematic setup of the picosecond (ps) laser scanning system for fabrication of hierarchical periodic near-subwavelength ripples (NSRs) and deep-subwavelength ripples (DSRs).

The micro-morphologies of the prepared stainless-steel sample surfaces were investigated by laser-scanning confocal microscopy (OSL4100, OLYMPUS, Tokyo, Japan) and a scanning electron microscope (SEM, QUANTA 200F, FEI, Hillsboro, OR, USA), and the spatial period of the NSRs and DSRs were estimated from the high-magnification morphology images. Since there are reports that the single ripple line seems not always to be straight and perpendicular or parallel to laser polariton [36], which means the laser induced subwavelength ripples have bending and bifurcation, the orientation variation of ripple formation were also investigated.

3. Results and Discussion

3.1. Preparation of Hierarchical Periodic Near-Subwavelength Ripples (NSRs) and Deep-Subwavelength Ripples (DSRs)

During the interaction between pulsed laser and solid material, the cumulative history of receiving laser irradiation by a solid surface is essentially affected by the transient laser power density of single laser pulse (space factor) and the superposition rate of multiple laser pulses (time factor). Therefore, our experimental exploration was started by counted ps laser pulses stacking hitting a fixed point on the stainless-steel sample surface for investigation of the influences of space-time characteristics on laser-induced NSRs/DSRs, then expanded to line scanning for fabrication of large area hierarchical periodic NSRs and DSRs.

Figure 2 shows the SEM images of the ps laser induced surface subwavelength ripples at fixed points, each irradiated by 4 laser pulses with the single pulse energy of 1.9 μJ. The average laser power adopted in the experiment was 0.76 W. The schematic power density correlation of two adjacent ablation spots is illustrated in Figure 2h, where L is the center distance between two adjacent pulse points which equals 20 μm, 10 μm and 5 μm in Figure 2a–c, respectively. The laser power density *I(r)* at a certain point in the focal spot associating the material thermo-physical properties can be defined as the following equation:

$$\mathrm{P} = \int_{0}^{2\pi}\int_{0}^{+\infty} I_0 e^{-2r^2/w_0^2} r dr d\theta = \pi w_0^2 I_0/2 \tag{1}$$

$$I(r) = I_0 \times e^{-2r^2/w_0^2} \tag{2}$$

where $p = E/t_p$ is the peak laser power, E is the single pulse energy, t_p is the pulse duration, I_0 is the maximum power density in the center of the focal spot and w_0 is the waist radius of the laser beam.

The results clearly suggest the peak power density plays an important role in the formation of subwavelength ripples. Figure 2a shows the surface morphologies of two separate ablation craters with hatch distance L = 20 μm. The ablated crater presents a central, circular region with a radius of 8 μm can be characterized by two concentric rings, namely the dual DSRs + NSRs inner zone I and the single DSRs outer ring II, as shown in Figure 2d. In the inner zone I, the NSRs preferentially aligned perpendicular with the incident laser polarization (marked with the arrow *E*), which has a spatial period of 381 ± 23 nm as measured in the magnified Figure 2d. The DSRs situate in the valley of NSRs with orientation parallel to the incident laser polarization has a spatial period of 381 ± 23 nm. Considering the gaussian profile of the focal spot energy distribution, the laser peak power density threshold value of the inner zone I was calculated to be 91.9 GW/cm^2. The ridges of the NSRs were steep and there were nano-particles on both the valleys and ridges. The outer ring II presents the characteristic texture of DSRs with a thickness of about 2 μm aligned along the laser polarization as shown in Figure 2g. The laser peak power density threshold value of the outer ring II with DSRs was calculated as 55 GW/cm^2. As shown in Figure 2b, the hatch distance L is set to 10 μm and the overlapping rate is 33% (only the outer ring II of two ablated craters overlapped). In Figure 2e, it is clearly shown that more nanoparticles are dispersed on the DSRs and more DSRs formed in the valley of NSRs close to the overlapping area. With high laser peak power density in the center of the focal spot, the NSRs can be fabricated without pulse overlapping. Due to the low power density in the periphery of the focal spot, the DSRs formed without pulse overlapping while the NSRs formed with pulse overlapping. These phenomena are in accord with the reported theory that the threshold of multiple laser pulse material ablation is less than of single laser pulse ablation [37]. As shown in Figure 2c, the hatch distance L is set to 5 μm and the overlapping rate is 66%. The overlapping area consisted of the outer ring II and the inner zone I of the two ablated craters. A similar morphology can be seen in the non-overlapped area in Figure 2f, but the overlapping area of two adjacent craters were observed with the higher laser peak power density irradiation and the appearance of the hierarchical periodic NSRs and DSRs (in particular, the DSRs were vertically located in the valley of parallel NSRs). The ridge profile became coarse and the groove became narrow accompanied with the aggregation of plenty of nanoparticles on the ridge, and the coarsened ridge period changed from 201 nm to 249 nm at different locations. For applications, the evolutionary morphology from the periphery to the center of the crater due to the gaussian beam energy distribution suggests a precision control of the laser peak power density is necessary for fabrication of distinct hierarchical periodic NSRs and DSRs.

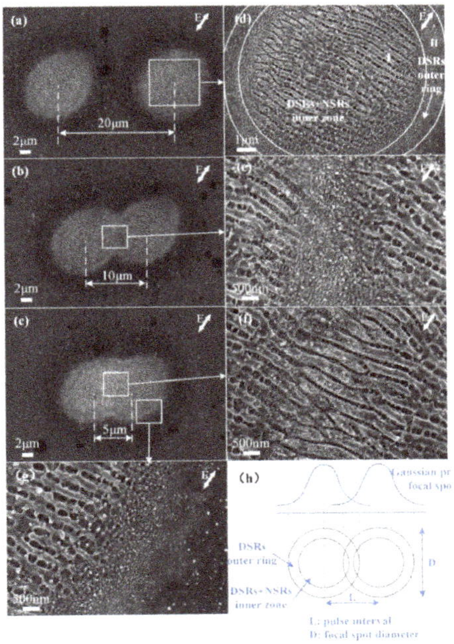

Figure 2. Scanning electron microscope (SEM) images of ps laser induced surface subwavelength ripples. Each of spot was carried out by 4 pulses with single pulse energy of 1.9 μJ and the hatch distance of the two spots was set as: (**a**) 20 μm, (**b**) 10 μm and (**c**) 5 μm. (**d–g**) The corresponding images of magnified areas separately. (**h**) The schematic of peak power density profile correlation with laser spots.

For the preparation of large-area hierarchical periodic NSRs and DSRs, it is worth exploring the relationship between the scanning speed and resultant morphology. The preparation of large-area hierarchical periodic NSRs and DSRs was implemented by a line-by-line laser scanning process. The spacing between the filled parallel lines was set to 10 μm. A series of laser scanning speeds from 50 mm/s to 1000 mm/s under the average laser power of 0.76 W were chosen to investigate the effect of space-time characteristic on large area laser-induced hierarchical periodic NSRs and DSRs. The numbers of overlapping laser pulses at any place of the laser scanning line path can be calculated by $N = D * f_p/v$, where D = 16 μm is the diameter of focused Gaussian spot, f_p = 400 kHz is the laser pulse repetition rate and v is the laser scanning speed as 50, 100, 300, 500, 800, 1000, 1300, and 1500 mm/s. Thus, the calculated overlapping pulse numbers (N) were 128, 64, 21, 13, 8, 6, 5 and 4, respectively.

The SEM images of the prepared hierarchical periodic NSRs and DSRs with increasing laser scanning speed are shown in Figure 3. As shown in Figure 3a, there were multiple nanoparticles redeposited on the ridge of the NSRs and, and the NSRs nearly disappeared for the high power density pulse repeated irradiation under a relatively low scanning speed of 50 mm/s. The DSRs in the valley of the NSRs cannot be observed or have vanished due to the absorption of exceeded pulse energy irradiation. When the laser scanning speed was increased to 100 mm/s, the uniformity NSRs and DSRs could be captured easily from the zoomed view of Figure 3b. With the scanning speed increased up to 500 mm/s, the smooth and clean NSRs and rarely nanoparticles redeposition fabricated accompany with more regular DSRs formed in the valley of the NSRs. When the scanning speed was higher than 500 mm/s and lower than 1000 mm/s, the continuity of the DSRs and NSRs became worse and the bending of the DSRs was pronounced. When the scanning speed increased to 1300 mm/s and 1500 mm/s, the nanoripples formed were more blurred. It can be seen from Figure 3g,h that although the periodic ripples formed were not strictly along one direction, but the disorderly DSRs cross with NSRs could still be observed.

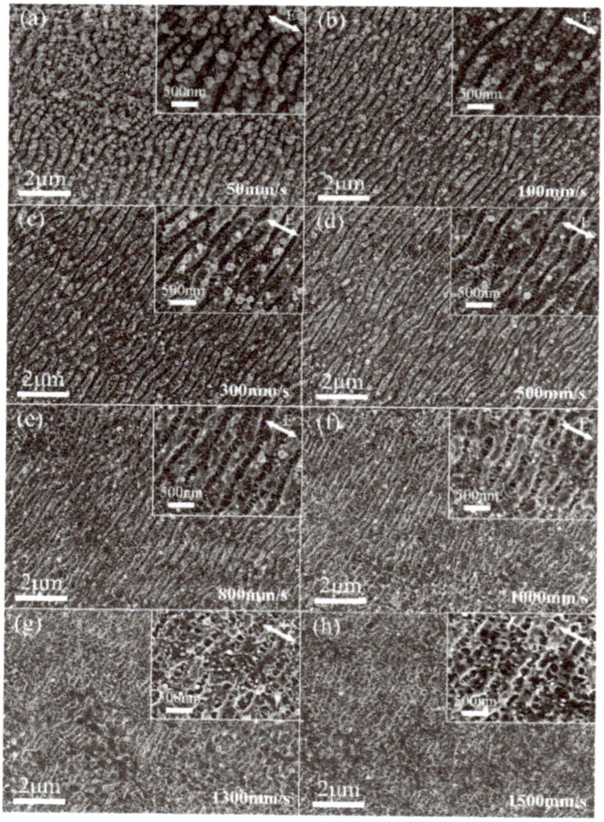

Figure 3. SEM images of subwavelength ripples prepared by different laser scanning speed and the fixed line spacing of 10 μm. The laser scanning speed of (**a**–**h**) is set to 50, 100, 300, 500, 800, 1000, 1300, and 1500 mm/s, respectively. The insets are zoomed views of the selected areas. The polarization of incident laser is indicated by the white double-headed arrow.

As shown in Figure 4, the spatial period of the NSRs and the DSRs which irradiated at average power of 0.76 W and various scanning speeds from 50 mm/s to 1000 mm/s were measured. The influence of pulse energy overlapping in the charge of the scanning speed was investigated by means of the line and symbol chart. The spatial period of the NSRs shows a small increment from 356.8 ± 17 nm to 411.4 ± 19 nm as the scanning speed increased. This scenario is similar to the previous findings that the spatial period of the NSRs decreases with increasing pulse energy [8,35]. An efficient amount of high power density and its distribution which deposited on the pre-existing ripple could decrease the spatial period and the ridge of the ripple could become coarser for its inevitable pulse overlapping offset. When the scanning speed was higher than 500 mm/s, the spatial period reached a higher limit value of 411.4 ± 19 nm. We assume that the few pulses overlapping that of the low power density in the periphery of the follow-up pulse, interferes with the NSRs formation under that of high power density. Moreover, the DSRs in the valley formed this superimposed effect in chaos, too. The spatial period of the DSRs shows no obvious trend and fluctuates within a range of 44 nm to 74.6 nm.

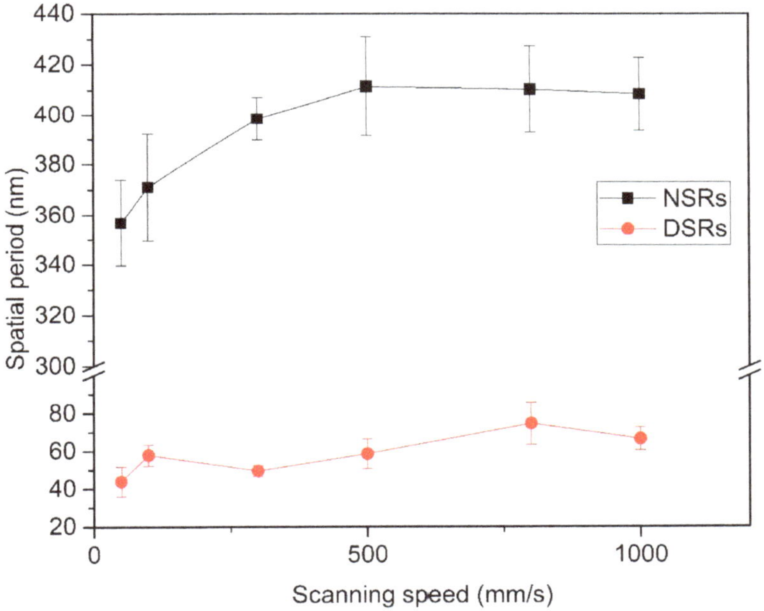

Figure 4. The spatial period of NSRs and DSRs irradiated at an average power of 0.76 W and various scanning speeds from 50 mm/s to 1000 mm/s.

Figure 5 shows the surface morphology evolution of ps laser-induced subwavelength ripples by increasing the average laser power from 0.49 W to 4.55 W at a fixed scanning speed of 500 mm/s, namely the peak power density increases from 60.9 GW/cm^2 to 565.7 GW/cm^2. As shown in Figure 5a, the surface is covered with uniform DSRs with orientation parallel to the incident laser polarization, since the peak power density is as low as 60.9 GW/cm^2 (average laser power of 0.49 W). It is interesting that the DSRs induced at low peak power density are not straight lines in succession and are similar to dendrite or well-known hydrodynamic instabilities of thin liquid films [21]. By increasing the laser peak power density to 94.5 GW/cm^2 (average laser power of 0.76 W), both the NSRs and DSRs were fabricated as illustrated in Figure 5b. More uniformity and regularity including continuous lines were observed on the samples when the laser peak power density was higher than 94.5 GW/cm^2. In agreement with the previous reports, a correlation between the electric field polarization direction and the NSRs orientation was observed in all cases [3,10]. Due to the ultrashort pulse laser irradiation, the time of the pulse duration is not enough for the excited electron transfer energy to lattice and thermalization. All the nano-ripples were fabricated by the "cold" process with few heating effects, which is why so much high power density without molten pool but only more and more nano-particle aggregation.

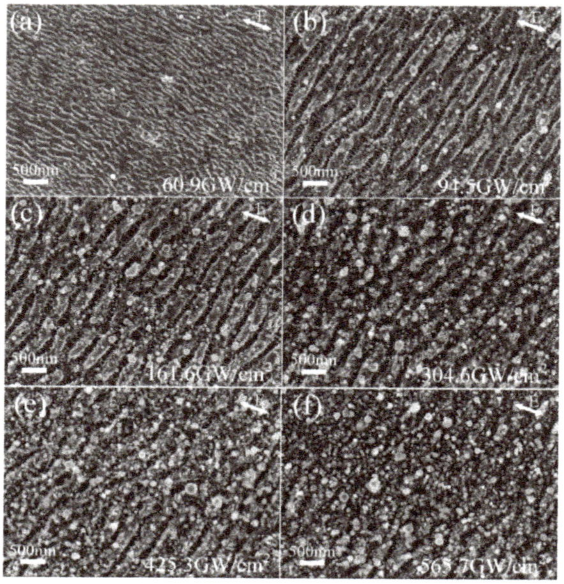

Figure 5. SEM image of the subwavelength ripples induced by different laser peak power density of (**a**) 60.9 GW/cm^2, (**b**) 94.5 GW/cm^2, (**c**) 161.6 GW/cm^2, (**d**) 304.6 GW/cm^2, (**e**) 425.3 GW/cm^2 and (**f**) 565.7 GW/cm^2, while the scanning speed is set to 500 mm/s and the line spacing is set to 10 μm.

As shown in Figure 5, the variation of laser peak power density has a great influence on the eventual morphology of the hierarchical periodic NSRs and DSRs. The spatial period is one of the most important topology parameters of the hierarchical periodic NSRs and DSRs, which may play a crucial role in its potential application such as superhydrophobicity, friction control and structural color.

The spatial period tendencies of NSRs and DSRs with incident laser peak power density and scanning speed are shown in Figure 6. The spatial period of the NSRs decreases as the incident laser peak power density increases. Under the scanning speed of 50 mm/s, when the laser peak power density increases from 60.9 GW/cm^2 to 161.6 GW/cm^2, the period of the NSRs decreases from 389 ± 11 nm to 352 ± 12 nm. When the laser peak power density is greater than 161.6 GW/cm^2, since the scanning speed is relatively low, the much greater pulse overlap results in the preformed periodic ripples' disappearance. A similar scenario can be observed when the scanning speed is 100 mm/s and the laser peak power density greater than 675.2 GW/cm^2. Moreover, the spatial period of the NSRs increases as the scanning speed increases, which agrees with the previous section. When the scanning speed is between 300 mm/s and 1000 mm/s, the NSRs were fabricated on the surface of the sample, and the period of the NSRs decreases from 430 ± 10 nm to 382 ± 14 nm with the laser peak power density increase.

On the other hand, the period of DSRs fluctuates at 65 ± 21 nm, and there is no obvious trend with the incident laser peak power density and scanning speed. The formation of the DSRs precedes the NSRs when the incident laser power density is slightly larger than the ablation threshold of the material. The periodic fluctuation is related to the local transient free electron density fluctuation of the material. The spatial period of the subwavelength ripples fabricated at different laser peak power density share a similar tendency with the scanning speed, which means the power density of incident laser has a decisive role in the light–matter interaction.

Finally, a large area of the hierarchical periodic NSRs and DSRs was fabricated on the stainless-steel surfaces by a set of optimized laser-scanning parameters. The scanning speed of 500 mm/s and the laser average power of 0.76 W were chosen. Figure 7 shows the SEM micrographs of the prepared large area hierarchical periodic NSRs and DSRs. The spatial period of obtained NSRs and DSRs were 356 ± 17 nm and 58 ± 15 nm, respectively.

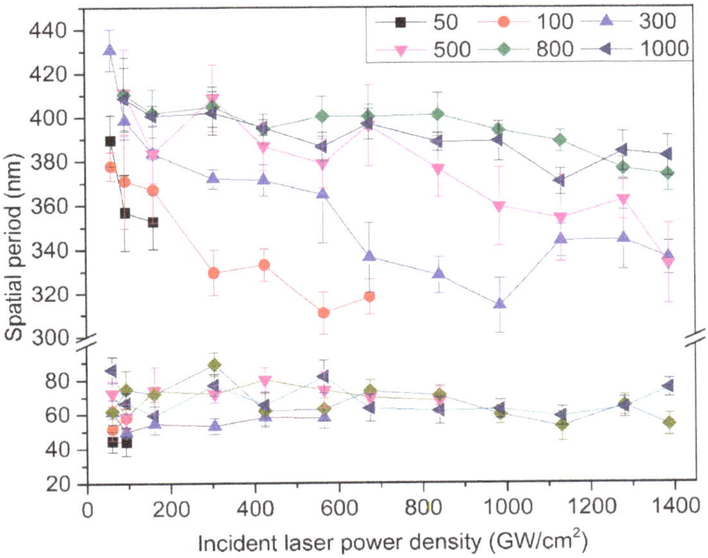

Figure 6. Comparison of the spatial period tendency of NSRs and DSRs with incident laser peak power density and scanning speed.

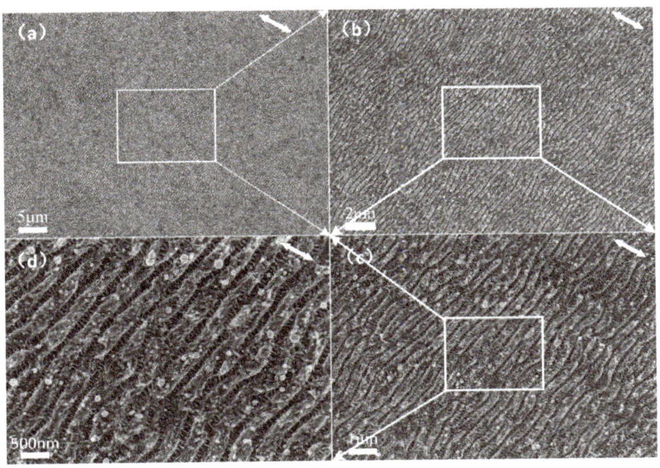

Figure 7. SEM micrographs of the prepared large-area hierarchical periodic NSRs and DSRs on the stainless-steel sample surface. (**a**–**d**) is the magnified picture of the selected area which indicated by a rectangular of white line.

3.2. Lattice Orientation of Grains as a Factor in the ps Laser-Induced Hierarchical Periodic NSRs and DSRs

An interesting phenomenon has been found in that the lattice orientation of grains is a factor in the ps laser-induced hierarchical periodic NSRs and DSRs on the stainless-steel surfaces. The evolutionary differences of the subwavelength periodic ripples with decreasing laser scanning speed under low peak power density (60.9 GW/cm^2, average laser power of 0.49 W) are revealed in Figure 8. By comparing Figure 8a,b we see that with the decrease of laser scanning speed (an increase of the overlapping pulse number N and laser energy absorption), the DSRs can be observed on stainless steel sample surface irradiated with all the different laser-scanning speeds (300~1000 mm/s), but the NSRs that are perpendicularly oriented only appear at low scanning speed (<300 mm/s, Figure 8d). With the high laser scanning speed of 1000 mm/s, the obscured DSRs were generated on the pristine surface and the spatial period of the nano-ripples was 58 nm. The ridges and valleys of the ripples are very smooth and these blurred DSRs show a regular arrangement that seems to be influenced by the lattice orientation of the material characteristics, as shown in Figure 8a,b. The initial DSRs formed at low peak power density have a conformance with the metallic grains, which has never been reported before. As shown in Figure 8a,b, the formation of DSRs has preferentially occurred in some "blocks", which is clearly related to the lattice orientation of the material itself. When the laser-scanning speed is decreased to 500 mm/s, the DSRs became distinct and the area induced by the lattice orientation became unclear because the power density from the laser pulses is high enough for the valence electrons excitation upon reaching the material ablation threshold [37]. When the laser-scanning speed is lower than 300 mm/s, the hierarchical periodic NSRs and DSRs are uniformly generated, as shown in Figure 8d. But at the scanning line overlapping area of Figure 8d, which is similar to the Figure 2b scenario as the two adjacent DSRs outer rings overlapped, only DSRs are fabricated because the peak power density is lower than the threshold value of 91.9 GW/cm^2.

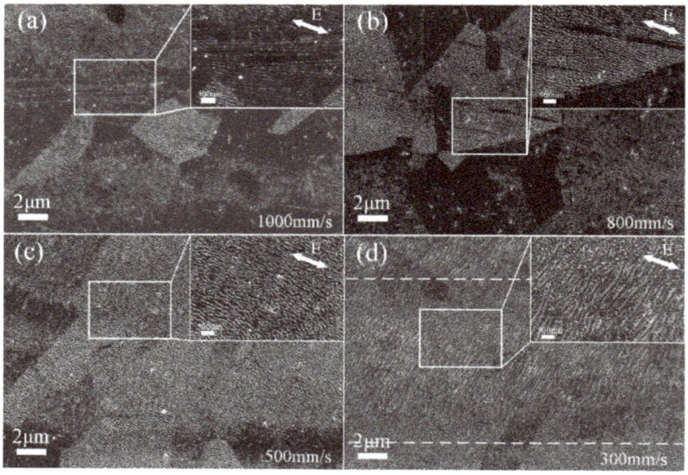

Figure 8. SEM images of subwavelength ripples prepared by different laser scanning speed and the fixed line spacing of 10 μm. The laser scanning speed is set to (**a**) 1000 mm/s, (**b**) 800 mm/s, (**c**) 500 mm/s, and (**d**) 300 mm/s, respectively. The insets are zoomed views of the selected areas. The polarization of the incident laser is indicated by the white double-headed arrow.

To further verify this hypothesis, we performed a pre-corrosion treatment on the stainless-steel samples in order to reveal the metallographic lattice structure on the surface. Hydrofluoric acid–nitric acid aqueous solution was selected on account of the grain of 304 stainless steel which is austenite grain. The metallographic lattice structure before and after the laser scanning treatment are shown in the confocal laser scanning microscope images of Figure 9a,b, respectively. Some grains turned a dark grey color after the ps laser irradiation, since the laser induced nano-scale ripples always present dark color in optical microscopes.

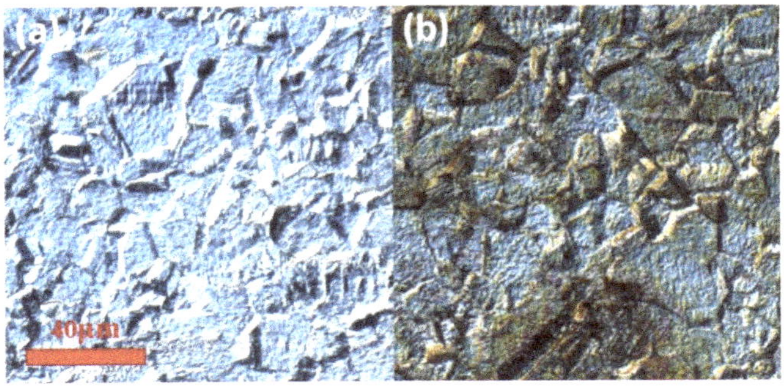

Figure 9. The metallographic lattice structure before and after the laser scanning treatment of 0.49 W and 1000 mm/s are shown in the confocal laser-scanning microscope images, (**a**): before, (**b**): after.

Figure 10 shows the SEM micrograph of the same sample of Figure 9b, where the DSRs similar to those in Figure 8a–c can be observed, and the formation of DSRs is affected by the lattice orientation of stainless steel. On the different grains of the stainless-steel sample, the DSRs are selectively generated. The DSRs are found in adjacent grains with little differences in orientation, as shown in Figure 8a,b. The lattice orientation variations, which are the manifestation of grain atomic bulk density differences, are believed to have an intrinsic fluence on the initial formation of DSRs. Moreover, the DSRs formation conformance with the grain structures is sensitive to the peak power density irradiation of the material. In this study, the energy threshold of the lattice orientation and the atomic bulk density affects the DSRs' formation. When the power density goes beyond the lattice binding energy of the stainless steel, the effect of the lattice on the initial formation behavior of DSRs disappears.

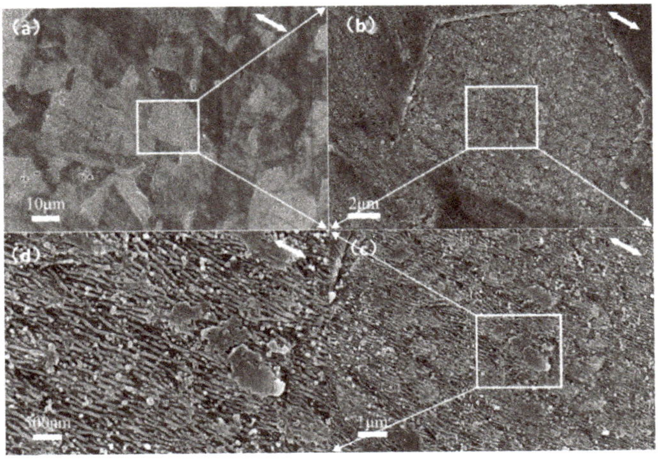

Figure 10. SEM images of the same sample of Figure 9b. (**a**–**d**) is the magnified picture of the selected area which indicated by a rectangular of white line.

3.3. Mechanism for the Picosecond (ps) Laser-Induced Hierarchical Periodic NSRs and DSRs

According to the above discussed facts, a feasible mechanism model should be compatible with the three clues that are discussed above: (a) DSRs can be independently generated at a relatively low peak power density, but the generation of NSRs need much higher peak power density and accompanied by DSRs in the valleys of the NSRs. (b) The orientation of the periodic DSRs is parallel to the polarization of incident laser and perpendicular to the orientation of the periodic NSRs which coincides with the laser-induced SPPs. (c) The formation of the DSRs at low peak power density has a conformance with the metallic grain structures, and preferentially occurs in the interior of the grains that have low surface atomic planar densities. Neither the classical scattering wave theory nor the SPPs excitation theory can explain the formation of DSRs in which spatial period much smaller than the incident laser wavelength [11]. The shortcoming of the self-organization theory is the influence of asymmetric ionization kinetic energy affected by the polarization of the laser electric field, which cannot explain the hierarchy and polarization dependence of different ripples in this experiment [21]. Second-harmonic generation (SHG) is only found during the irradiation of certain materials, the present experiments cannot clearly show that SHG is involved in the formation of DSRs on any surface [38].

A qualitative explanation based on SPP modulated periodic coulomb explosion is proposed for the formation mechanism of hierarchical periodic NSRs and DSRs. As shown in Figure 11, the atomic arrangement of the grains of 304 austenitic stainless steel is indicated in the dotted frame A. The white double-headed arrow shows the polarization of the incident laser. During a picosecond laser pulse duration, electron-absorbing photon ionization is always present, and the free electrons after ionization move in the direction of the electric field under the action of the laser electric field, forming a free electron gas limited to the surface movement. However, the concentration of free electrons on the surface of the sample is not uniformly distributed due to the surface roughness and defects, which lead to the initiation of the local non-thermal phase change coulomb explosion [39] at the place where local electron ionization intensity is large enough. As shown in Figure 11b, the material removal of coulomb explosion then makes the free electron gas movement blocked and concentrated at the adjacent place along the laser electric field, which results in the subsequent coulomb explosion chain that forms the DSRs. When the incident laser power density increases to a high threshold value, the laser electric field (Transverse Magnetic wave, TM wave) induced constraint SPPs will propagate along the metallic surface but perpendicular to the laser electric field orientation, and attenuate along the vertical direction of the metallic surface [40]. The dotted line frame B in Figure 11a shows the surface plasmon standing wave field induced at high energy density during laser irradiation. When the incident light is irradiated to a defect on the surface of the stainless steel, the SPPs are excited by the coupling of the surface electron plasma oscillation, and the surface plasmon standing wave is induced when the adjacent surface defects are separated by a certain distance.

Figure 11c shows that the surface energy field of the stainless-steel sample was being periodically modulated by the surface plasmon standing wave action, which made the free electron concentration increase and a strong coulomb explosion occurred. The subsequent free electrons under the SPP-modulated electric field are blocked and concentrated at the adjacent place along the SPPs' propagation direction, which results in the subsequent coulomb explosion chain that forms the NSRs with orientation vertical to the laser polarization directions. The weak electron concentration in the valley after the strong coulomb explosion is likely to form a micro coulomb explosion, thereby forming the DSRs in the valley. For the inevitable pulse overlapping offset, when the peak power density is too high, the hierarchical periodic NSRs and DSRs structures will be partly destroyed and formed in chaos.

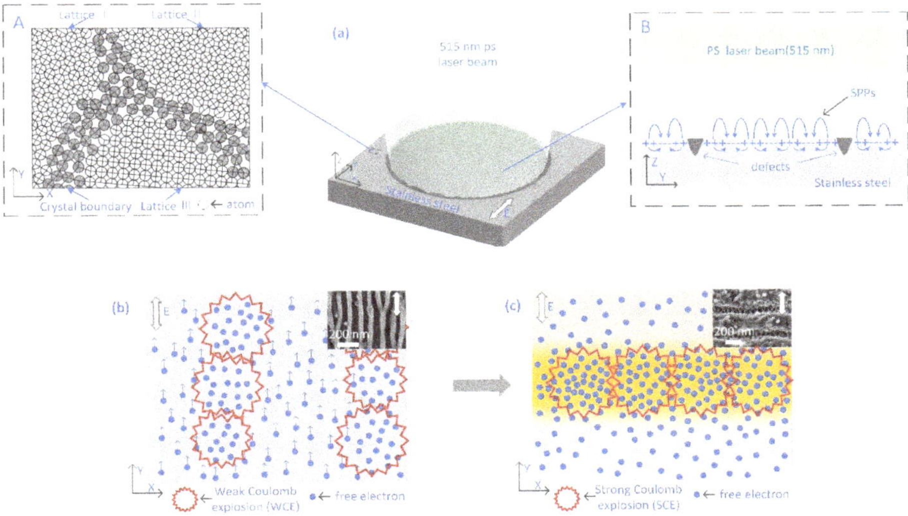

Figure 11. Schematic of the surface plasmon polariton (SPP)-modulated periodic coulomb explosion for the formation mechanism of hierarchical periodic NSRs and DSRs. (**a**) The picosecond laser irradiated on the stainless-steel surface. The dotted frame (**A**) indicated the atomic arrangement of the grains of 304 austenitic stainless steel. The dotted frame (**B**) shows the surface plasmon standing wave field induced at high energy density during laser irradiation. (**b**) The formation of DSRs under the material removal of coulomb explosion. (**c**) The NSRs induced by the periodically modulated by the surface plasmon standing wave.

4. Conclusions

Although many notable works have been undertaken over the years to investigate ultrafast laser-induced periodic surface subwavelength ripples (NSRs and DSRs), how to simultaneously construct large-scale hierarchical periodic NSRs and DSRs still remains a challenge. The preparation regulation of the picosecond laser induced hierarchical periodic NSRs and DSRs on stainless-steel surfaces is investigated, and a unified qualitative explanation based on a SPP-modulated periodic coulomb explosion is proposed in this work. The main conclusions obtained are as follows:

(1) The peak power density of incident laser makes a decisive role in the light–matter interaction which leads to the generation of hierarchical periodic NSRs and DSRs. The DSRs can be independently generated at a relatively low peak power density, whereas the generation of NSRs needs much higher peak power density and are accompanied by DSRs in the valleys of the NSRs. The orientation of the periodic DSRs is perpendicular to that of the periodic NSRs, which was always found to be perpendicular to the polarization of the incident laser and coincides with the laser-induced SPPs.

(2) The formation of the DSRs at low peak power density has a conformance with the metallic grain structures, and preferentially occurs in the interior of the grains that have low surface atomic planar densities. Moreover, the spatial period of the NSRs is determined by the peak power density absorption and the material intrinsic thermo-physical properties.

(3) A qualitative explanation based on SPP-modulated periodic coulomb explosion is proposed for the formation mechanism of hierarchical periodic NSRs and DSRs. During a picosecond laser pulse duration, the photon-absorbed free electrons' motion initiates the locally non-thermal phase change coulomb explosion and results in the subsequent coulomb explosion chain that forms the DSRs. The laser electric field (TM wave)-induced constraint on SPPs' propagation along the

metallic surface makes the free electron concentration increase and a strong coulomb explosion chain occurs that forms the NSRs with orientation vertical to the laser polarization direction.

(4) The preparation of large-area hierarchical periodic NSRs and DSRs was implemented by a line-by-line laser scanning process within which either the incident laser power or scanning speed can be used as the control variable. The spatial periods of the obtained NSRs and DSRs were 356 ± 17 nm and 58 ± 15 nm, respectively.

(5) Further theoretical calculation and simulation are needed to verify the qualitative explanation and find routes to improve continuity and strict directionality of the hierarchical periodic NSRs and DSRs. The potential applications such as wettability, tribology, structural color are promising since the laser-scanning preparation method is simple and precisely controllable.

Author Contributions: All authors have read and agreed to the published version of the manuscript. Formal analysis, S.D., D.Z. and W.L.; Funding acquisition, W.X. and Y.C.; Investigation, S.D.; Methodology, D.Z., W.L. and Y.C.; Project administration, W.X. and Y.C.; Writing—original draft, S.D. and D.Z.; Writing—review & editing, Y.C.

Funding: This research was funded by National Natural Science Foundation of China (Grant numbers U1609209,11704285) and the Zhejiang Provincial Natural Science Foundation (Grant numbers LZ20E050003, LQ17F050001).

Conflicts of Interest: The authors declare no conflict of interest.

References

1. Bonse, J.; Hohm, S.; Kirner, S.V.; Rosenfeld, A.; Kruger, J. Laser-induced periodic surface structures—A scientific evergreen. *IEEE J. Sel. Top. Quantum Electron.* **2016**, *23*, 1. [CrossRef]
2. Muller, F.A.; Kunz, C.; Graf, S. Bio-Inspired Functional Surfaces Based on Laser-Induced Periodic Surface Structures. *Materials* **2016**, *9*, 476. [CrossRef] [PubMed]
3. Huang, M.; Zhao, F.L.; Cheng, Y.; Xu, N.S.; Xu, Z.Z. Origin of Laser-Induced Near-Subwavelength Ripples: Interference between Surface Plasmons and Incident Laser. *ACS Nano* **2009**, *3*, 4062–4070. [CrossRef] [PubMed]
4. Shugaev, M.V.; Gnilitskyi, I.; Bulgakova, N.M.; Zhigilei, L.V. Mechanism of single-pulse ablative generation of laser-induced periodic surface structures. *Phys. Rev. B* **2017**, *96*, 205429. [CrossRef]
5. Gurevich, E.L. Mechanisms of femtosecond LIPSS formation induced by periodic surface temperature modulation. *Appl. Surf. Sci.* **2016**, *374*, 56–60. [CrossRef]
6. Gnilitskyi, I.; Derrien, T.J.; Levy, Y.; Bulgakova, N.M.; Mocek, T.; Orazi, L. High-speed manufacturing of highly regular femtosecond laser-induced periodic surface structures: Physical origin of regularity. *Sci. Rep.* **2017**, *7*, 8485. [CrossRef]
7. Fuentes-Edfuf, Y.; Garcia-Lechuga, M.; Puerto, D.; Florian, C.; Garcia-Leis, A.; Sanchez-Cortes, S.; Solis, J.; Siegel, J. Coherent scatter-controlled phase-change grating structures in silicon using femtosecond laser pulses. *Sci. Rep.* **2017**, *7*, 4594. [CrossRef]
8. Nivas, J.J.J.; He, S.; Song, Z.; Rubano, A.; Vecchione, A.; Paparo, D.; Marrucci, L.; Bruzzese, R.; Amoruso, S. Femtosecond laser surface structuring of silicon with Gaussian and optical vortex beams. *Appl. Surf. Sci.* **2017**, *418*, 565–571. [CrossRef]
9. Huang, M.; Zhao, F.L.; Jia, T.Q.; Cheng, Y.; Xu, N.S.; Xu, Z.Z. A uniform 290 nm periodic square structure on ZnO fabricated by two-beam femtosecond laser ablation. *Nanotechnology* **2007**, *18*, 505301. [CrossRef]
10. Lin, X.; Li, X.; Zhang, Y.; Xie, C.; Liu, K.; Zhou, Q. Periodic structures on germanium induced by high repetition rate femtosecond laser. *Opt. Laser Technol.* **2018**, *101*, 291–297. [CrossRef]
11. Huang, M.; Zhao, F.; Cheng, Y.; Xu, N.; Xu, Z. Mechanisms of ultrafast laser-induced deep-subwavelength gratings on graphite and diamond. *Phys. Rev. B* **2009**, *79*, 125436. [CrossRef]
12. Sipe, J.E.; Young, J.F.; Preston, J.S.; van Driel, H.M. Laser-induced periodic surface structure. I. Theory. *Phys. Rev. B* **1983**, *27*, 1141–1154. [CrossRef]
13. Young, J.F.; Preston, J.S.; van Driel, H.M.; Sipe, J.E. Laser-induced periodic surface structure. II. Experiments on Ge, Si, Al, and brass. *Phys. Rev. B* **1983**, *27*, 1155–1172. [CrossRef]
14. Reif, J.; Varlamova, O.; Costache, F. Femtosecond laser induced nanostructure formation: Self-organization control parameters. *Appl. Phys. A* **2008**, *92*, 1019–1024. [CrossRef]

15. Reif, J.; Varlamova, O.; Uhlig, S.; Varlamov, S.; Bestehorn, M. On the physics of self-organized nanostructure formation upon femtosecond laser ablation. *Appl. Phys. A* **2014**, *117*, 179–184. [CrossRef]
16. Shi, X.; Jiang, L.; Li, X.; Wang, S.; Yuan, Y.; Lu, Y. Femtosecond laser-induced periodic structure adjustments based on electron dynamics control: From subwavelength ripples to double-grating structures. *Opt. Lett.* **2013**, *38*, 3743–3746. [CrossRef]
17. Feng, L.; Réal, V. Femtosecond laser-induced ultra-fine nanostructures on silicon surface. *Opt. Mater. Express* **2016**, *6*, 3330–3338.
18. Gemini, L.; Hashida, M.; Miyasaka, Y.; Inoue, S.; Limpouch, J.; Mocek, T.; Sakabe, S. Periodic surface structures on titanium self-organized upon double femtosecond pulse exposures. *Appl. Surf. Sci.* **2015**, *336*, 349–353. [CrossRef]
19. Jia, T.; Baba, M.; Suzuki, M.; Ganeev, R.A.; Kuroda, H.; Qiu, J.; Wang, X.; Li, R.; Xu, Z. Fabrication of two-dimensional periodic nanostructures by two-beam interference of femtosecond pulses. *Opt. Express* **2008**, *16*, 1874–1878.
20. Jia, T.Q.; Chen, H.X.; Huang, M.; Zhao, F.L.; Qiu, J.R.; Li, R.X.; Xu, Z.Z.; He, X.K.; Zhang, J.; Kuroda, H. Formation of nanogratings on the surface of a ZnSe crystal irradiated by femtosecond laser pulses. *Phys. Rev. B* **2005**, *72*, 125429. [CrossRef]
21. Reif, J.; Varlamova, O.; Varlamov, S.; Bestehorn, M. The role of asymmetric excitation in self-organized nanostructure formation upon femtosecond laser ablation. *Appl. Phys. A* **2011**, *104*, 969–973. [CrossRef]
22. Zayats, A.V.; Smolyaninov, I.I.; Maradudin, A.A. Nano-optics of surface plasmon polaritons. *Phys. Rep.* **2005**, *408*, 131–314. [CrossRef]
23. Rudenko, A.; Mauclair, C.; Garrelie, F.; Stoian, R.; Colombier, J.P. Self-organization of surfaces on the nanoscale by topography-mediated selection of quasi-cylindrical and plasmonic waves. *Nanophotonics* **2019**, *8*, 459–465. [CrossRef]
24. Bonse, J.; Krüger, J.; Höhm, S.; Rosenfeld, A. Femtosecond laser-induced periodic surface structures. *J. Laser Appl.* **2012**, *24*, 042006. [CrossRef]
25. Wang, L.; Chen, Q.D.; Cao, X.W.; Buividas, R.; Wang, X.; Juodkazis, S.; Sun, H.B. Plasmonic nano-printing: Large-area nanoscale energy deposition for efficient surface texturing. *Light Sci. Appl.* **2017**, *6*, e17112. [CrossRef]
26. Bonse, J.; Rosenfeld, A.; Krüger, J. On the role of surface plasmon polaritons in the formation of laser-induced periodic surface structures upon irradiation of silicon by femtosecond-laser pulses. *J. Appl. Phys.* **2009**, *106*, 104910. [CrossRef]
27. Yao, C.; Ye, Y.; Jia, B.; Li, Y.; Ding, R.; Jiang, Y.; Wang, Y.; Yuan, X. Polarization and fluence effects in femtosecond laser induced micro/nano structures on stainless steel with antireflection property. *Appl. Surf. Sci.* **2017**, *425*, 1118–1124. [CrossRef]
28. Ji, X.; Jiang, L.; Li, X.; Han, W.; Liu, Y.; Wang, A.; Lu, Y. Femtosecond laser-induced cross-periodic structures on a crystalline silicon surface under low pulse number irradiation. *Appl. Surf. Sci.* **2015**, *326*, 216–221. [CrossRef]
29. Hamad, S.; Podagatlapalli, G.K.; Vendamani, V.S.; Nageswara Rao, S.V.S.; Pathak, A.P.; Tewari, S.P.; Venugopal Rao, S. Femtosecond Ablation of Silicon in Acetone: Tunable Photoluminescence from Generated Nanoparticles and Fabrication of Surface Nanostructures. *J. Phys. Chem. C* **2014**, *118*, 7139–7151. [CrossRef]
30. Derrien, T.J.Y.; Koter, R.; Krüger, J.; Höhm, S.; Rosenfeld, A.; Bonse, J. Plasmonic formation mechanism of periodic 100-nm-structures upon femtosecond laser irradiation of silicon in water. *J. Appl. Phys.* **2014**, *116*, 074902. [CrossRef]
31. Ahsan, M.S.; Ahmed, F.; Kim, Y.G.; Lee, M.S.; Jun, M.B.G. Colorizing stainless steel surface by femtosecond laser induced micro/nano-structures. *Appl. Surf. Sci.* **2011**, *257*, 7771–7777. [CrossRef]
32. Romano, J.M.; Garcia-Giron, A.; Penchev, P.; Dimov, S. Triangular Laser-Induced submicron textures for functionalising stainless steel surfaces. *Appl. Surf. Sci.* **2018**, *440*, 162–169. [CrossRef]
33. Gurevich, E.L.; Gurevich, S.V. Laser Induced Periodic Surface Structures induced by surface plasmons coupled via roughness. *Appl. Surf. Sci.* **2014**, *302*, 118–123. [CrossRef]
34. Liao, Y.; Ni, J.; Qiao, L.; Huang, M.; Bellouard, Y.; Sugioka, K.; Cheng, Y. High-fidelity visualization of formation of volume nanogratings in porous glass by femtosecond laser irradiation. *Optica* **2015**, *2*, 329. [CrossRef]

35. Razi, S.; Varlamova, O.; Reif, J.; Bestehorn, M.; Varlamov, S.; Mollabashi, M.; Madanipour, K.; Ratzke, M. Birth of periodic Micro/Nano structures on 316L stainless steel surface following femtosecond laser irradiation; single and multi scanning study. *Opt. Laser Technol.* **2018**, *104*, 8–16. [CrossRef]
36. He, S.; Nivas, J.J.J.; Anoop, K.K.; Vecchione, A.; Hu, M.; Bruzzese, R.; Amoruso, S. Surface structures induced by ultrashort laser pulses: Formation mechanisms of ripples and grooves. *Appl. Surf. Sci.* **2015**, *353*, 1214–1222. [CrossRef]
37. Eichstädt, J.; Römer, G.R.B.E.; Huis in 't Veld, A.J. Determination of irradiation parameters for laser-induced periodic surface structures. *Appl. Surf. Sci.* **2013**, *264*, 79–87. [CrossRef]
38. Buividas, R.; Rosa, L.; Šliupas, R.; Kudrius, T.; Šlekys, G.; Datsyuk, V.; Juodkazis, S. Mechanism of fine ripple formation on surfaces of (semi)transparent materials via a half-wavelength cavity feedback. *Nanotechnology* **2010**, *22*, 055304. [CrossRef]
39. Li, S.; Li, S.; Zhang, F.; Tian, D.; Li, H.; Liu, D.; Jiang, Y.; Chen, A.; Jin, M. Possible evidence of Coulomb explosion in the femtosecond laser ablation of metal at low laser fluence. *Appl. Surf. Sci.* **2015**, *355*, 681–685. [CrossRef]
40. Maier, S.A. *Plasmonics: Fundamentals and Applications*; Springer: Berlin, Germany, 2007; pp. 49–74.

nanomaterials

MDPI

Article

Miniaturized GaAs Nanowire Laser with a Metal Grating Reflector

Wei Wei [1,2,*], Xin Yan [3] and Xia Zhang [3]

[1] School of Mechanical and Electric Engineering, Guangzhou University, Guangzhou 510006, China
[2] Photonics Research Centre, Department of Electronic and Information Engineering, The Hong Kong Polytechnic University, Hung Hom, Kowloon, Hong Kong, China
[3] State Key Laboratory of Information Photonics and Optical Communications, Beijing University of Posts and Telecommunications, Beijing 100876, China; xyan@bupt.edu.cn (X.Y.); xzhang@bupt.edu.cn (X.Z.)
* Correspondence: wei.wei@outlook.ie

Received: 26 February 2020; Accepted: 30 March 2020; Published: 4 April 2020

Abstract: This work proposed a miniaturized nanowire laser with high end-facet reflection. The high end-facet reflection was realized by integrating an Ag grating between the nanowire and the substrate. Its propagation and reflection properties were calculated using the finite elements method. The simulation results show that the reflectivity can be as high as 77.6% for a nanowire diameter of 200 nm and a period of 20, which is nearly three times larger than that of the nanowire without a metal grating reflector. For an equal length of nanowire with/without the metal grating reflector, the corresponding threshold gain is approximately a quarter of that of the nanowire without the metal grating reflector. Owing to the high reflection, the length of the nanowire can be reduced to 0.9 μm for the period of 5, resulting in a genuine nanolaser, composed of nanowire, with three dimensions smaller than 1 μm (the diameter is 200 nm). The proposed nanowire laser with a lowered threshold and reduced dimensions would be of great significance in on-chip information systems and networks.

Keywords: nanowire; nanolaser; metal grating

---

1. Introduction

Nanolasers hold the key position in on-chip sensing, optical interconnection and computing systems [1–5]. With cylindrical geometry and strong two-dimensional confinement of electrons, holes and photons, the independent semiconductor nanowire is one of the ideal candidates for nanolasers [6–8]. To date, room-temperature lasing emission has been realized in ZnO, GaN, CdS and GaAs nanowires, covering the optical spectrum from ultra-violet to near-infrared [9–14]. Among them, GaAs nanowires, with a direct band gap and high electron mobility, are considered to be a prime candidate for advanced opto-electronic devices [15]. However, due to the nanowires' small size compared with their emission wavelength, ranging from 800 to 900 nm, the end facets of the nanowire display weak reflection at the near-infrared region, resulting in significant mirror losses [16]. Therefore, the light propagation in the nanowire has to be long enough to overcome a threshold loss induced by the weak facet reflectivities. The distinct index contrast reduces the diameter of the nanowires down to 200 nm, and even below the diffraction limit with the contribution of surface plasmons (collective electron oscillations) [17–19]. To keep the lasing emission from the nanowires at a relatively low threshold level, the nanowire's length becomes too long compared with its diameter, which may impede its potential applications in nanophotonics and high-density photonics integration.

In this paper, an Ag grating with a short period number is proposed to enhance the end-facet reflection and lower the mirror loss of GaAs nanowires. The grating is employed as a high-quality reflector and placed at only one end facet to shrink the length of the nanowires. The other end-facet reflection is provided by the Au cap. The Ag grating and Au cap compose the optical resonant cavity. In that cavity, two mirrors of Ag grating and Au cap face each other, forming an optical resonator in which a light wave can be resonantly enhanced. To reveal the mechanism of the enhanced reflection by the short-period Ag grating, the finite elements method (FEM) is applied to a numerical simulation of the reflection and cavity properties. With the optimized structural parameters of the grating, the reflected spectrum can be located within the gain spectrum of the GaAs nanowires to generate an increase in feedback. Consequently, the lasing threshold will be lower than that without active feedback, and the nanowire's length can be shortened while maintaining the laser's power.

2. Structure of GaAs Nanowire with Ag Grating

The schematic diagram of the GaAs nanowire with the Ag grating is shown in Figure 1. On the silica substrate there is an Ag grating, the permittivity of which is described by the Drude-Lorentz model

$$\varepsilon(\omega) = 1 + \sum_k \frac{\Delta\varepsilon_k}{-\omega^2 - a_k(i\omega) + b_k} \tag{1}$$

where $\Delta\varepsilon_k$, a_k and b_k are constants that provide the best fit for silver, when compared with the optical constant data of silver given by Palik et al. [20]. The terms a_k and b_k denote the damping frequency of the electron gas and the effective electron collision frequency, respectively. The Drude-Lorentz model is an extended Drude model, incorporating additional Lorentz terms [21–23]. Both of the models are classic models which describe the transport properties of electrons in metals. Figure 1 illustrates that Λ is the grating period; the width and height of the metal grating are denoted by W_g and H_g, respectively. The duty cycle of the grating was fixed at 50% during the simulation. The GaAs nanowire is placed on the top of the grating. The part close to the dielectric interface is on the grating, to make the light propagated inside the nanowire interact with the metal grating. Meanwhile, the other part close to the Au cap is left suspended in the air. The Au cap on one top end of the nanowire is the Au particle, which is used as a catalyst during the growth of GaAs nanowires [14,24,25]. Its diameter is approximately equal to that of the nanowire. Due to the high reflection of the metal/dielectric interface, the Au cap is employed here as a high-quality reflector. Its reflectivity can be higher than 70%, depending on the nanowire's diameter [13]. The Au cap naturally decreases the threshold and the nanowire's length. The other end of the nanowire is cleaved, and the mirror is formed by the GaAs/air interface. The reflectivity of the dielectric interface for the fundamental mode is lower than 50%, and decreases with the shrinking nanowire diameter. It becomes less than 30% when the diameter is 200 nm. Therefore, only one end of the nanowire is located on the grating and the other end is left. Owing to the high modal confinement provided by the unique 1D geometry, the nanowire is employed both as the waveguide and gain medium. The Au cap and metal grating act as reflectors, forming the optical resonant cavity.

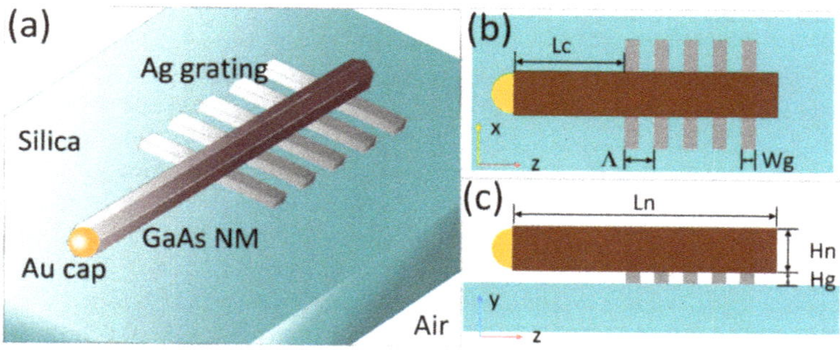

Figure 1. Schematic diagrams of the nanowire with the Ag grating reflector. 3D model (**a**), lateral view (**b**) and top view (**c**) of the proposed structure.

3. Results and Discussion

When the nanowire is placed on the Ag grating, the light wave will be reflected by the grating, forming the gain feedback, which is shown in Figure 2. The large index difference of the waveguide mode HE_{11a} for the nanowire in the air and that on the grating tooth enables the reflection of the light wave by the grating. The reflected spectrum of the grating is shown in Figure 2c, where the period and period number are 140 nm and 20, which makes the spectrum within the gain spectrum of the GaAs nanowire. The reflected spectrum has a wide bandwidth of 64 nm. Owing to the high index contrast, the grating, with a short period number of 20, can provide a maximum reflectivity of 74.5%. As shown in Figure 2d, the profile of H_z of the propagating mode is highly reflected by the Ag grating, and very little electromagnetic energy is transmitted through the nanowire.

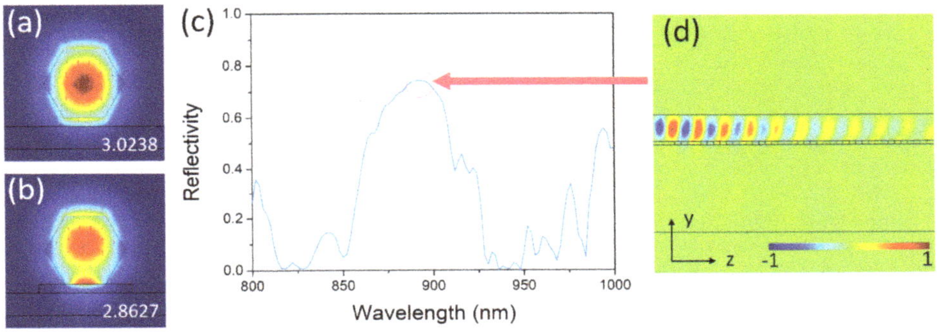

Figure 2. Modal profiles of HE_{11a} for the nanowire in the air (**a**) and on the grating tooth (**b**). (**c**) Reflected spectrum of the grating. (**d**) Profile of H_z of the propagating mode. Numbers in (**a**) and (**b**) are modal effective indices. The pseudo-colors indicate the intensity of the magnetic field of H_z.

The reflected spectrum, including bandwidth and peak, is dependent on the structural parameters of the grating. As depicted in Figure 3a, the grating height varies from 10 to 50 nm. The reflection is weakest for a grating height of 10 nm. It is significant to be note that due to the limitation of the classic Drude model and its extensions of the adopted Drude-Lorentz model at the order of ~10 nm, the simulation result values may not be very accurate [26]. The reflection curve of 10 nm here is only to demonstrate the varying trend as the grating height decreases. With the increasing height, the reflection becomes higher, and the bandwidth of the reflected spectrum becomes wider. However, reflection stays nearly similar when the height increases from 30 to 40 nm, and then becomes weaker at 50 nm. This can be attributed to the effective index contrast. The index contrast increases along with the height, resulting in higher reflection. However, when the grating height gets too large, the very high index contrast and the metal tooth are similar to a cliff or wall for the guided mode inside the nanowire, impeding the propagation and reflection of the mode. The period number is a direct factor in deciding the reflectivity of the grating. In Figure 3b, the reflection gets stronger with the increasing period number, from 10 to 30. When the period number increases from 25 to 30, the reflectivity increases by a very small amount. To be different from the dielectric grating with a small index contrast, the metal grating only needs a short period number to realize high reflection. The reflection capability of the grating gets saturated at a period number of 25. Furthermore, a long period number brings a long grating length, which also increases the nanowire's cavity length. This is contradictory to the purpose of the integration with the metal grating. Thus, the period number would be 20 or 25. In the following calculation, all the period numbers are assumed to be 20. The grating period Λ, or width, is crucial to the peak of the reflected spectrum. As depicted in Figure 3c, the reflection peak displays red-shift with the increasing Λ By optimizing the period Λ or width, the reflected spectrum of the grating can be adjusted within, or to cover, the gain spectrum of the gain medium. The wavelength of the lasing emission from a GaAs nanowire most probably ranges from 850 to 880 nm [14]. Therefore, all of the structural parameters were selected to cover or partly cover that spectral region.

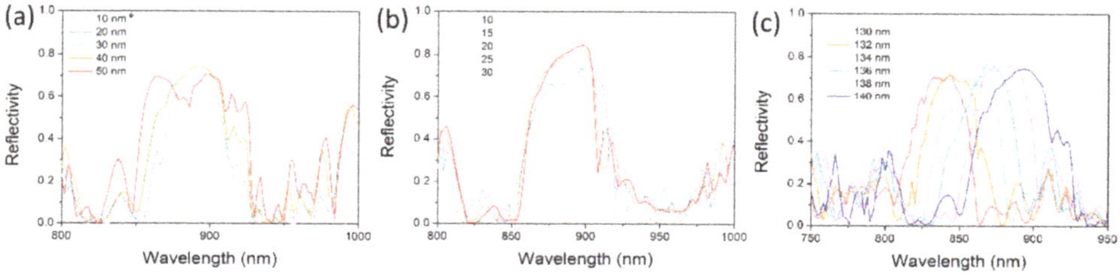

Figure 3. Reflected spectra of the grating for different grating heights (**a**), period numbers (**b**) and Λ (**c**). * Due to the limitation of classic model of Drude-Lorentz at the order of ~10 nm, the simulation result values may not be very accurate. The results given here are only to show the varying trend of reflection intensity as the grating height decreases.

The photonic integration is going towards high density, requiring reduced dimensions of photonic components. As the nanowire diameter decreases to the limit of fundamental modes, the reflectivity decreases dramatically, together with the threshold gain. The metal grating reflector is more meaningful for nanowires with smaller diameters. The reflected spectra for nanowire diameters of 300 nm, 250 nm, 230 nm, and 200 nm are shown in Figure 4, where the other structural parameters of the grating height, duty cycle and period number are the same. The grating period or width are slightly tuned, to keep the spectra within the gain spectrum of the GaAs nanowire. The bandwidth and intensity of the reflection spectrum have negative relationships with the diameter, due to the index contrast increasing with the decreasing diameter. As the diameter decreases, the electromagnetic wave guided inside the nanowire has a more intense interaction with the metal grating tooth, resulting in a larger modal effective index, together with index contrast.

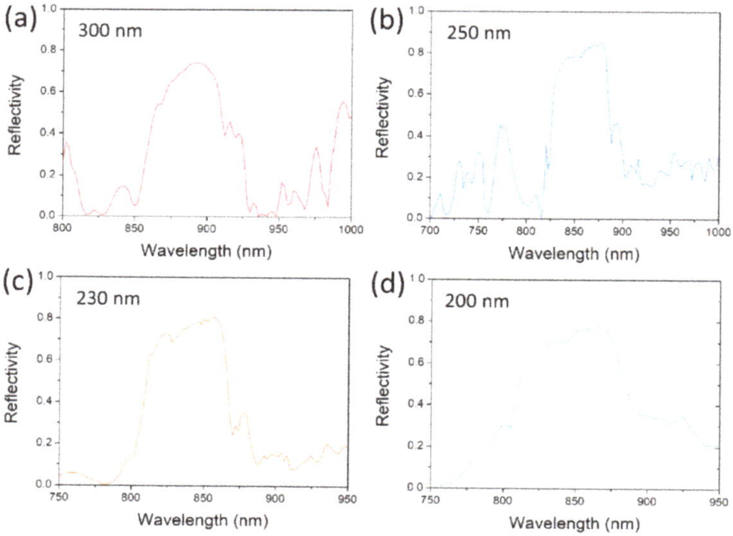

Figure 4. Reflected spectra for nanowire diameters of 300 nm (**a**), 250 nm (**b**), 230 nm (**c**) and 200 nm (**d**).

To compare the reflection between the Ag grating, end facet, and Au cap, functions of the Ag-grating reflectivity, end-facet reflectivity and Au-cap reflectivity on nanowire diameter are depicted in Figure 5a. The end-facet reflection is very weak and gets weaker for diameters of 200 nm and 220 nm. At a diameter of 200 nm, the end-facet reflectivity is below 30%. In contrast, for the Au-cap reflection, it is very strong even at a diameter of 200 nm. Its reflectivity can be 80% at larger diameters, and more than 65% at a diameter of 200 nm. Thus, the Au cap can provide excellent reflection for the propagated light wave with extremely small dimensions. This can be an additional advantage of the Au-catalyst growth method. For the reflection of the Ag grating, it is much stronger than the end-facet reflection. Its reflectivity is two times larger than its end-facet reflectivity. For small nanowire diameters below 260 nm, the reflectivity of Ag grating is even greater than Au-cap reflectivity. The introduction of the metal grating reflector is to compensate for the end-facet reflection, resulting in the reduced nanowire length or lowered threshold gain. The lasing threshold is the lowest excitation level, at which laser output is dominated by stimulated emission rather than spontaneous emission. The threshold gain g_{th}, which describes the required gain per unit length for lasing, is defined as [27]

$$g_{th} = \frac{1}{\Gamma_{wg}}\left[\alpha_i + \frac{1}{L}\ln\left(\frac{1}{R}\right)\right] \tag{2}$$

where R denotes the geometric mean of the reflectivity of the end facets of the nanowire, and L is the length of the nanowire's optical resonant cavity. Γ_{wg} is the modal confinement factor, which is an indicator of how well the mode overlaps with the gain medium, and is defined as the ratio between the modal gain and material gain in the active region [28–30]

$$\Gamma_{wg} = \frac{\frac{n_a}{2\eta_0}\int_{A_a} d\rho\left|E(\rho)\right|^2}{\int_{A_a} d\rho\frac{1}{2}Re[E(\rho) \times H^*(\rho)]\cdot\hat{z}} \tag{3}$$

where η_0 is the intrinsic impedance, n_a is the index of the active region, A_a is the cross section of the active region, A is the whole cross section ideally extending to infinity, and E and H are the complex electric and magnetic fields of the guided modes. The threshold gains of a nanowire laser with and without metal grating for diameters varying from 200 to 300 nm are shown in Figure 5b. Three lengths were chosen to demonstrate the threshold performance of the nanowire laser. As lasing emission is easily output from nanowires with lengths from 5–10 µm, lengths of 6 and 9 µm are typical parameters for a nanowire laser. A nanowire of 6 µm length without grating has a relatively high threshold gain, around 800 cm^{-1}, and increases to about 1000 cm^{-1} when the diameter decreases to 200 nm. This is not beneficial for lasing emission from nanowire and requires high-power pump. If the length is changed to 9 µm, the threshold gain becomes moderate, ranging from 600 to 800 cm^{-1}, which is helpful for lasing emission. However, for nanowire with metal grating, threshold gains for lengths of 6 and 9 µm are both low. The threshold gain of nanowire with a length of 6 µm ranges from 300 to 400 cm^{-1}. For nanowire with a length of 9 µm, threshold gain can go down to 200 to 250 cm^{-1}. At this threshold level, the laser can be easily lased without a high-power pump. To shrink the nanowire's length, we make the length 3 µm, at which lasing emission requires a high-power pump. Its threshold gain ranges from 1600 to 2000 cm^{-1}. With the metal grating, the threshold gain can go down to 600 to 800 cm^{-1}, decreasing threshold gain by more than 2 times. At this threshold level, lasing is not hard to realize. Thus, the length of the nanowire laser can be shortened to 3 µm under moderate pump power. In high-density photonic integration, a nanolaser with three dimensions at nanoscale is preferred. Therefore, we make the length of the nanowire laser 900 nm by shortening the period number of the metal grating to 5. As shown in Figure 5c, the bandwidth becomes wider and the reflection becomes weaker than nanowire of 3 µm length. The maximum reflectivity decreases from 77.6% to 69.2%, and the corresponding threshold gain increases from 756 to 2978 cm^{-1}. Lasing at a threshold gain of ~3000 cm^{-1} requires a high-power pump. Although the threshold gain is not low, a nanowire laser can be shortened to within 1 micron and is promising to lase under a high-power pump. In future research, if some additional techniques, like surface plasmons, were added onto the nanowire with metal grating, the confinement factor would be increased by more than 1.5 times, lowering the threshold gain further [31,32].

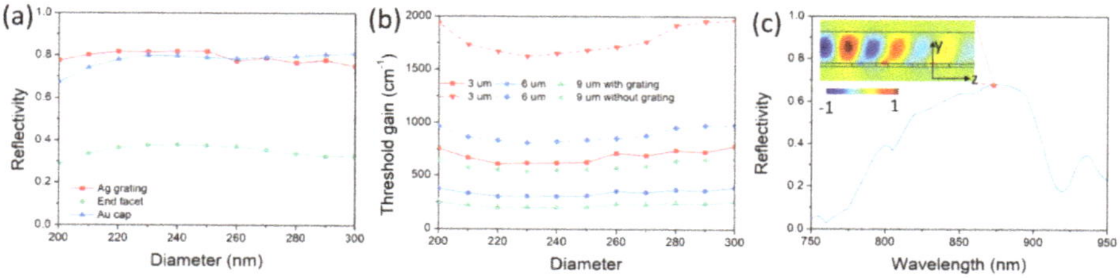

Figure 5. (**a**) Reflected spectra of the Ag grating, end facet and Au cap for nanowire diameter vary from 200 to 300 nm. (**b**) Threshold gain of nanowire with and without Ag grating for nanowire diameters varying from 200 to 300 nm. (**c**) Reflected spectrum for a period number of 5. Pseudo-colors indicate the intensity of the magnetic field of H_z.

4. Conclusions

In summary, we proposed a metal grating with a short period to provide high reflection for the light wave propagating inside the nanowire. The metal grating is placed between the silica substrate and the nanowire. To shorten the nanowire's length, the metal grating is only integrated with the end facet of the dielectric interface. The other end-facet reflection is provided by the Au cap, with a high reflectivity of around 70%. When the period number of the metal grating is set to 20, the grating can realize a maximum reflectivity of 81.8%. Moreover, a high reflectivity of 77.6% was realized for the nanowire diameter of 200 nm. The reflectivity is nearly three times larger than that without metal grating. Owing to the high reflection of the metal grating, the threshold gain of the nanowire laser can be decreased by more than 2.5 times. The length of the nanowire laser can be shortened to 3 μm under moderate pump power. With a high-power pump, the length of the nanowire laser could be shortened down to 900 nm, which is promising for the realization of lasing emission at a threshold gain of 2978 cm^{-1}. Consequently, all of the three dimensions of the nanowire laser, especially the length, could be reduced below 1 micron. The proposed miniaturized nanowire laser with metal grating would be promising for use in on-chip sensing, optical interconnection and computing systems.

Author Contributions: Conceptualization, W.W.; methodology, W.W. and X.Y.; software, W.W.; writing—original draft preparation, W.W. and X.Y.; writing—review and editing, W.W. and X.Z. All authors have read and agreed to the published version of the manuscript.

Funding: This work was funded by the National Natural Science Foundation of China (61905045, 61774021 and 61911530133), the Hong Kong Scholar Program (XJ2018052) and the Fund of State Key Laboratory of Information Photonics and Optical Communications (Beijing University of Posts and Telecommunications), China (IPOC2019ZT07).

Conflicts of Interest: The authors declare no conflict of interest.

References

1. Ning, C.Z. Semiconductor nanolasers. *Phys. Status Solidi B* **2010**, *247*, 774–788. [CrossRef]
2. Lu, Y.; Kim, J.; Chen, H.; Wu, C.; Dbidian, N.; Sander, C.; Wang, C.; Lu, M.; Li, B.; Qiu, X.; et al. Plasmonic nanolaser using epitaxially grown silver film. *Science* **2012**, *337*, 450–453. [CrossRef] [PubMed]
3. Wang, X.; Wang, Y.; Wang, S.; Li, B.; Zhang, X.; Dai, L.; Ma, R. Lasing enhanced surface plasmon resonance sensing. *Nanophotonics* **2016**, *5*, 52–58. [CrossRef]
4. Gwo, S.; Shih, C. Semiconductor plasmonic nanolasers: Current status and perspectives. *Rep. Prog. Phys.* **2016**, *79*, 086501. [CrossRef] [PubMed]
5. Ma, R.; Oulton, R.F. Applications of nanolasers. *Nat. Nanotech.* **2019**, *14*, 12–22. [CrossRef] [PubMed]
6. Panzauskie, P.J.; Yang, P. Nanowire photonics. *Phys. Mater. Today* **2006**, *9*, 36–45. [CrossRef]
7. Yan, R.; Gargass, D.; Yang, P. Nanowire photonics. *Nat. Photon.* **2009**, *3*, 569–576. [CrossRef]
8. Couteau, C.; Larrue, A.; Wilhelm, C.; Soci, C. Nanowire lasers. *Nanophotonics* **2015**, *4*, 90–107. [CrossRef]
9. Zimmler, M.A.; Caspasso, F.; Müller, S.; Ronning, C. Optically pumped nanowire lasers: Invited review. *Semicond. Sci. Technol.* **2000**, *25*, 024001. [CrossRef]

10. Versteegh, M.A.M.; Vanmaekelbergh, D.; Dijkhuis, J.I. Room-temperature lasers emission of ZnO nanowires explained by many-body theory. *Phys. Rev. Lett.* **2012**, *108*, 157402. [CrossRef]
11. Gradečak, S.; Qian, F.; Li, Y.; Park, H.G.; Lieber, C.M. GaN nanowire lasers with low lasing thresholds. *Appl. Phys. Lett.* **2005**, *87*, 173111. [CrossRef]
12. Geburt, S.; Thielmann, A.; Röder, R.; Borschel, C.; McDonnell, A.; Kozlik, M.; Kühnel, J.; Sunter, K.A.; Capasso, F.; Ronning, C. Low threshold room-temperature lasing of CdS nanowires. *Nanotechnology* **2012**, *23*, 365204. [CrossRef] [PubMed]
13. Saxena, D.; Mokkapati, S.; Parkinson, P.; Jiang, N.; Gao, Q.; Tan, H.H.; Jagadish, C. Optically pumped room-temperature GaAs nanowire lasers. *Nat. Photon.* **2013**, *7*, 963–968. [CrossRef]
14. Wei, W.; Liu, Y.; Zhang, X.; Wang, Z.; Ren, X. Evanescent-wave pumped room-temperature single-mode GaAs/AlGaAs core-shell nanowire lasers. *Appl. Phys. Lett.* **2014**, *104*, 223103. [CrossRef]
15. Jiang, N.; Parkinson, P.; Gao, Q.; Breuer, S.; Tan, H.H.; Wong-Leung, J.; Jagadish, C. Long minority carrier lifetime in Au-catalyzed GaAs/$Al_xGa_{1-x}As$ core-shell nanowires. *Appl. Phys. Lett.* **2012**, *101*, 023111. [CrossRef]
16. Ding, Y.; Motohisa, J.; Hua, B.; Hara, S.; Fukui, T. Observation of microcavity modes and waveguides in InP nanowires fabricated by selective-are metalorganic vapor-phase epitaxy. *Nano Lett.* **2007**, *7*, 3598–3602. [CrossRef]
17. Oulton, R.F.; Sorger, V.J.; Zentgraf, T.; Ma, R.; Gladden, C.; Dai, L.; Bartal, G.; Zhang, X. Plasmon lasers at deep subwavelength scale. *Nature* **2009**, *461*, 629–632. [CrossRef]
18. Sidiropoulos, T.P.H.; Röder, R.; Geburt, S.; Hess, O.; Maier, S.A.; Ronning, C.; Oulton, R.F. Ultrafast plasmonic nanowire lasers near the surface plasmon frequency. *Nat. Phys.* **2014**, *10*, 870–876. [CrossRef]
19. Ho, J.; Tatebayashi, H.; Sergent, S.; Fong, C.F.; Iwamoto, S.; Arakawa, Y. Low-threshold near-infrared GaAs-AlGaAs core-shell nanowire plasmon laser. *ACS Photon.* **2015**, *2*, 165–171. [CrossRef]
20. Palik, E.D. *Handbook of Optical Constants of Solids*; Academic Press: San Diego, CA, USA, 1998.
21. Johnson, P.B.; Christy, R.W. Optical constants of the noble metals. *Phys. Rev. B* **1972**, *6*, 4370–4379. [CrossRef]
22. Luebbers, R.; Hunsberger, F. FDTD for Nth-Order Dispersive Media. *IEEE Tran. Antenn. Prop.* **1972**, *40*, 1297–1301. [CrossRef]
23. Vial, A.; Grimault, A.; Macías, D.; Barchiesi, D.; Chapelle, M. Improved analytical fit of gold dispersion: Applications to the modeling of extinction spectra with a finite-difference time-domain method. *Phys. Rev. B* **2005**, *71*, 085416. [CrossRef]
24. Guo, J.; Huang, H.; Ren, X.; Yan, X.; Cai, S.; Guo, X.; Huang, Y.; Wang, Q.; Zhang, X.; Wang, W. Grwoth of zinc blende GaAs/AlGaAs radial heterostructure nanowires by a two-temperature process. *Chin. Phys. Lett.* **2011**, *28*, 036101. [CrossRef]
25. Yan, X.; Wei, W.; Tang, F.; Wang, X.; Li, L.; Zhang, X. Low-threshold room-temperature AlGaAs/GaAs nanowire/single-quantum-well heterostructure laser. *Appl. Phys. Lett.* **2017**, *110*, 061104. [CrossRef]
26. Ashcroft, N.W.; Mermin, N.D. *Solid State Physics*; Saunders College Publishing: New York, NY, USA, 1976.
27. Yaris, A. *Quantum Electronics*; Wiley: New York, NY, USA, 1975.
28. Visser, T.D.; Blok, H.; Demeulenaere, B.; Lenstra, D. Confinement factors and gain in optical amplifiers. *IEEE J. Quantum Electron.* **1997**, *33*, 1763–1766. [CrossRef]
29. Maslov, A.V.; Ning, C.Z. Modal gain in a semiconductor nanowire laser with anisotropic bandstructure. *IEEE J. Quantum Electron.* **2004**, *40*, 1389–1397. [CrossRef]
30. Chang, S.W.; Chuang, S.L. Fundamental formulation for plasmonic nanolasers. *IEEE J. Quantum Electron.* **2009**, *45*, 1014–1023. [CrossRef]
31. Wei, W.; Yan, X.; Zhang, X. Ultrahigh Purcell factor in low-threshold nanolaser based on asymmetric hybrid plasmonic cavity. *Sci. Rep.* **2016**, *6*, 33063. [CrossRef]
32. Bermúdez-Ureña, E.; Tutuncuoglu, G.; Cuerda, J.; Smith, C.L.C.; Bravo-Abad, J.; Bozhevolnyi, S.I.; Morral, A.F.I.; García-Vidal, F.J.; Quidant, R. Plasmonic waveguide-integrated nanowire laser. *Nano Lett.* **2017**, *17*, 747–754. [CrossRef]

nanomaterials

MDPI

Article

Lightning-Rod Effect of Plasmonic Field Enhancement on Hydrogen-Absorbing Transition Metals

Norihiko Fukuoka and Katsuaki Tanabe *

Department of Chemical Engineering, Kyoto University, Kyoto 615-8510, Japan
* Correspondence: tanabe@cheme.kyoto-u.ac.jp

Received: 1 August 2019; Accepted: 27 August 2019; Published: 30 August 2019

Abstract: The plasmonic enhancement of electromagnetic field energy density at the sharp tips of nanoparticles or nanoscale surface roughnesses of hydrogen-absorbing transition metals, Pd, Ti, and Ni, is quantitatively investigated. A large degree of energy focusing is observed for these transition metals in the microwave region, even surpassing the enhancement for noble metals according to the conditions. Pd, for instance, exhibits peak field enhancement factors of 6000 and 2×10^8 in air for morphological aspect ratios of 10 and 100, respectively. Metal surfaces possibly contain such degrees of nano- or micro-scale native random roughnesses, and, therefore, the field enhancement effect may have been unknowingly produced in existing electrical and optical systems. In addition, for future devices under development, particularly in hydrogen-related applications, it is desirable to design and optimize the systems, including the choice of materials, structures, and operating conditions, by accounting for the plasmonic local energy enhancement effect around the metal surfaces.

Keywords: transition metal; surface plasmon; nanoparticle; nanophotonics; hydrogen storage; sensing; nuclear fusion; energy device

1. Introduction

Free electrons in metallic materials, particularly around metal surfaces or interfaces with dielectric media, exhibit a strong interaction with electromagnetic fields or light in the form of collective oscillation, named "surface plasmons" [1–3]. Surface-plasmon-induced electromagnetic field enhancement on metal surfaces [4–7] has been utilized for various applications, such as chemical/biomedical sensing [8–10], photodetectors [11–13], light-emitting diodes [14–16], nanolasers [17–19], solar cells [16,20–22], and optical cloaking [23–25]. Recently, we found through numerical analysis that a large degree of energy focusing, with enhancement factors over several hundred, is available for planar surfaces of hydrogen-absorbing transition metals, Pd, Ti, and Ni, in the microwave region, even surpassing the enhancement for noble metals [26]. We therein discussed the potential applications of the plasmonic field enhancement effect on hydrogen-absorbing transition metals, such as hydrogen storage [27–31], sensing [32–34], and nuclear fusion [35–38]. In contrast to the studied plasmonic field enhancement effect on planar metal surfaces, it is known that surfaces with sharp curvatures allow the electromagnetic field to concentrate further, called the "lightning-rod effect" [39–43]. In the present study, we numerically investigate the plasmonic field enhancement on sharp surfaces of hydrogen-absorbing transition metals, Pd, Ti, and Ni.

2. Theory and Calculation Methods

We calculate the field enhancement factors, which represent the intensity ratios for fields around the object (metals in this case) to those in the absence of the object, or the original incident fields, for prolate-spheroidal metal nanoparticles in air, H_2, or vacuum, and H_2O. We specifically calculate the field enhancement factors at the tips of the prolate-spheroidal metal nanoparticles, to represent sharp-curvature metal surfaces. These calculations, based on the classical electromagnetic field theory in the quasistatic limit [7,44], quantitatively show how much energy can be concentrated from the incident optical or electric power. The intensities of electromagnetic fields around subwavelength-size metal nanoparticles can be described by the formalism below in the quasistatic limit [44]. Consider a homogeneous, prolate spheroid with radii of the major and minor axes a and b, respectively, placed in a medium in which there exists a uniform static electric field $\vec{E}_0$ applied along the major axis of the spheroid, as schematically depicted in Figure 1. If the permittivities or dielectric constants of the spheroid and the medium are different, a charge will be induced on the surface of the spheroid. The initially uniform field will, therefore, be distorted by the introduction of the spheroid. Based on the calculation schemes described in earlier articles [7,42,44,45], in short, the electric field outside the spheroid and at the tip of the prolate spheroid is given by

$$\vec{E}_{tip} = \frac{\varepsilon_1(\lambda)}{\varepsilon_m(\lambda)+L_1\{\varepsilon_1(\lambda)-\varepsilon_m(\lambda)\}}\vec{E}_0, \tag{1}$$

where ε_1 and ε_m are the frequency-dependent complex permittivities or dielectric functions of the spheroid and the surrounding medium, respectively. L_1 is the geometrical factor for the major axis of the prolate spheroid, calculated as

$$L_1 = \frac{1-e^2}{e^2}\left(-1 + \frac{1}{2e}\ln\frac{1+e}{1-e}\right), \tag{2}$$

where e is the eccentricity of the particle shape:

$$e = \sqrt{1-\frac{b^2}{a^2}}. \tag{3}$$

The field enhancement factor is then calculated as

$$\eta \equiv \frac{\left|\vec{E}_{tip}\right|^2}{\left|\vec{E}_0\right|^2} = \left|\frac{\varepsilon_1(\lambda)}{\varepsilon_m(\lambda)+L_1\{\varepsilon_1(\lambda)-\varepsilon_m(\lambda)\}}\right|^2. \tag{4}$$

Note that this field enhancement factor is defined as the ratio of field intensities and not field magnitudes. Incidentally, for the spherical case, which provides $L_1 = 1/3$, η reduces to the equation of the field enhancement factor derived in Reference [7]. The empirical complex dielectric functions of metals and of the surroundings on the frequencies listed in References [7] and [26] are used for the computations in this study. We assume that $\varepsilon' = 1$ and $\varepsilon'' = 0$ throughout the entire frequencies for air, H_2, and vacuum. The electrostatic calculations carried out in this study are valid for particle sizes in the range of 10–100 nm, for which the phase retardation is negligible throughout the particle, the field enhancement will be largest, and metal nanoparticles and nanoscale roughnesses will, therefore, become most applicable, as discussed in the Results and Discussion section.

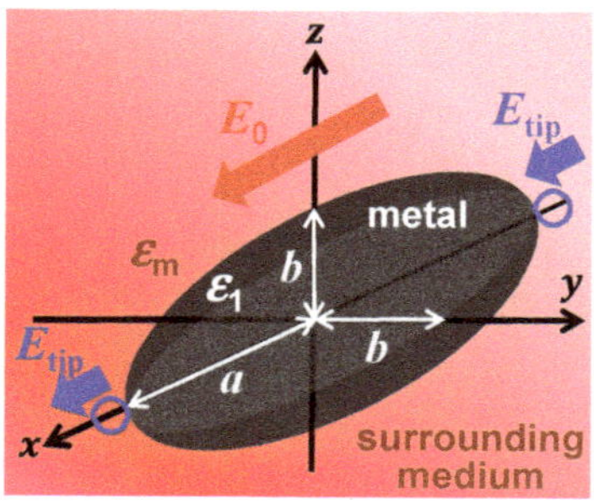

Figure 1. Schematic bird's-eye view of the system considered in this study for the calculations of field enhancement factors.

3. Results and Discussion

Firstly, as a reference, we present in Figure 2 the calculated electromagnetic field enhancement factors for the simple, spherical nanoparticle case, which corresponds to the spheroid's aspect ratio, *a*/*b*, of one. The peaks seen in these spectra are associated with the resonance or surface mode, characterized by internal electric fields with no radial nodes. A local energy enhancement around 10 times is decently observed for the hydrogen-absorbing transition metals, Pd, Ni, and Ti, in this spherical-shape case. Incidentally, the results in Figure 2 for shorter wavelengths are consistent with those reported in Reference [7]. It should be noted that the field enhancement factors, for our calculations, are independent of the particle size under the quasistatic approximation, and are valid for particle diameters around the range of 10–100 nm [7]. Figure 3 shows the field enhancement factors for spheroidal metal nanoparticles with an aspect ratio of three. It is observed that even for such a relatively small aspect ratio or morphological surface sharpness, enhancement factors over 100 are attainable for Pd, Ni, and Ti for a wide range of frequencies, through visible to infrared. Such nanoparticles or nanoscale surface roughness thus concentrate electromagnetic or optical energy in their vicinity like antennae. The artifact discontinuities for the curves for Cu and Ti around 1 and 4 µm, respectively, in Figures 3 and 4 are incidentally because of the discontinuities in the source empirical data of the dielectric functions. Remarkably, electromagnetic field enhancement factors of several thousand are observed for Pd, Ni, and Ti for the aspect ratio of 10 (Figure 4). The resonant peak enhancement factors for Au, Ag, and Cu are observed to be even larger, on the order of 10^5. Among the whole metal elements, Al and the noble metals Ag, Au, and Cu are known to exhibit distinctively higher field enhancement factors than other metals because of their high electrical conductivities [7,46]. Therefore, the combination of such noble-metal spheroidal nanoparticles and hydrogen-absorbed transition metals, available, for instance, by coating bulk metal surfaces by colloidal metal nanoparticles, may be another strategy for applications to harvest the photonic or electromagnetic energy focusing effect. Strikingly, as observed in Figure 5, the field enhancement factor of Pd for the sharp particle or surface morphology case of an aspect ratio of 100 reaches the order of 10^8 in the infrared region, even exceeding those for the noble metals. This consequence is consistent with the results reported in Reference [26] that the plasmonic field enhancement factors of the hydrogen-absorbing transition metals become higher than those of noble metals for planar metal surfaces. Incidentally, the peak or resonant wavelength of Ti may unfortunately locate outside of the range of frequencies handled in this study. Figures 6 and 7 summarize the dependence of the peak field enhancement factors and wavelengths, respectively, on the aspect ratio. It is observed that the electromagnetic field enhancement factors dramatically increase with the aspect ratio, namely, with the sharpness of the metal surfaces. As the sharpness increases, the resonant peak wavelength red-shifts. Large-aspect-ratio metal particles or high-curvature edges of surface irregularities exhibit high polarizabilities and, thus, large dipole moments, particularly at the resonance, to produce strong local field enhancement in the vicinity of such edges [44,47]. To discuss the electromagnetic similarity between isolated metal spheroidal particles and rough surfaces, it is worth mentioning that the detailed numerical calculation results reported in Reference [40], where rough metal surfaces were modeled as prolate hemispheroids protruding from a grounded flat plane, are quantitatively similar to our results in Figures 6 and 7, for the peak field enhancement factors and wavelengths.

The large field-enhancement effect on the hydrogen-absorbing transition metals, Pd, Ni, and Ti, observed in the series of calculations in this study can be used for various hydrogen-energy applications. As discussed in Reference [26], potential applications include hydrogen storage, sensing [48,49], laser fusion [46], and condensed-matter fusion. In addition, for the reported experiments so far, for instance, in the condensed-matter nuclear fusion field, it is highly possible that the deuterium-absorbed Pd, Ni, and Ti surfaces contained certain degrees of nano- or micro-scale native random roughnesses [39,40,50,51] corresponding to such morphological aspect ratios as those studied in this article. Therefore, some of the experimental material systems may have unknowingly benefited from the plasmonic field enhancement effect. The electrostatic calculation results shown in this paper are valid for particle sizes smaller than the fields' wavelengths at which the phase retardation is negligible throughout the particle object. In addition, the dielectric functions of materials used for our calculations are empirical values for bulk materials, whose validity is debatable when the particle sizes become smaller than 10 nm, because of the electron mean free path limitation or scattering of conduction electrons off particle surfaces [44,47,52,53]. The calculation results for optical wavelengths under the quasistatic approximation are, therefore, valid for metal particles with diameters in the range of 10–100 nm. Metal particles with sizes smaller and larger than these limits both exhibit broader plasmon resonances and smaller field enhancements, because of the surface scattering losses and the radiative losses or electrodynamic damping, respectively [47,52,53]. Therefore, the choice of particle sizes, 10–100 nm, for our calculations is most suitable for plasmon-enhanced electromagnetic and optical applications, because of the largest field enhancements. This size aspect should, therefore, also be accounted for in the optimized design of the material structures in potential applications. In addition, surface plasmons located in between multiple metallic objects with nanoscale separation distances, or so-called "gap plasmons" [54–56], would also provide large field enhancements on the conditions. The gap-plasmon effect for hydrogen-energy applications is important partially because gap plasmons are also commonly observed in real structures such as rough metal surfaces, and will be discussed in future work.

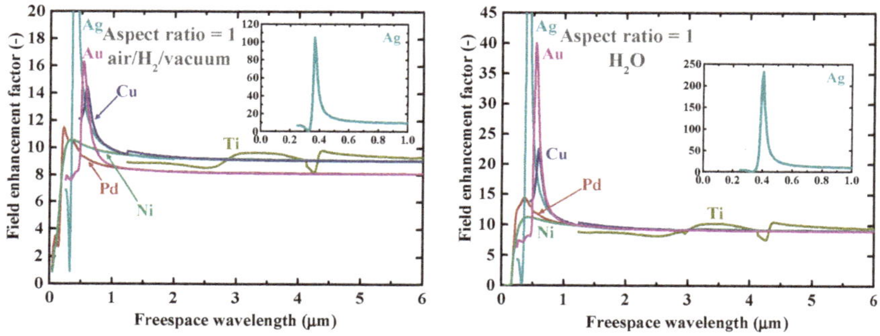

Figure 2. Calculated electromagnetic field enhancement factors around spherical nanoparticles of Au, Ag, Cu, Pd, Ti, and Ni in (**left**) air/H_2/vacuum and (**right**) H_2O. The insets are the clarified plots for Ag.

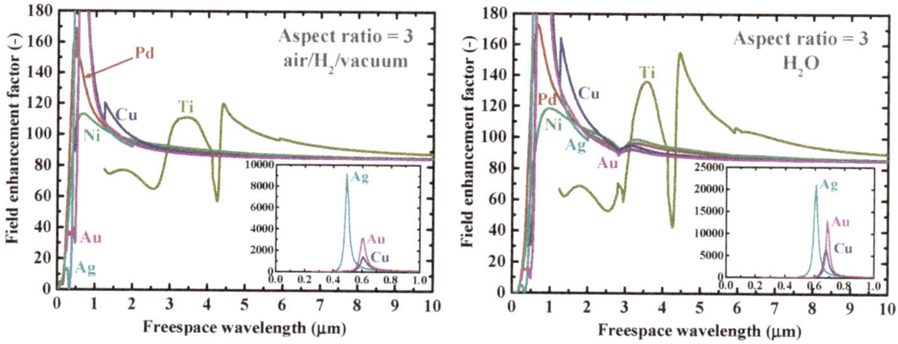

Figure 3. Calculated electromagnetic field enhancement factors at the tips of spheroidal nanoparticles of Au, Ag, Cu, Pd, Ti, and Ni with an aspect ratio of 3 in (**left**) air/H_2/vacuum and (**right**) H_2O. The insets are the clarified plots for Au, Ag, and Cu.

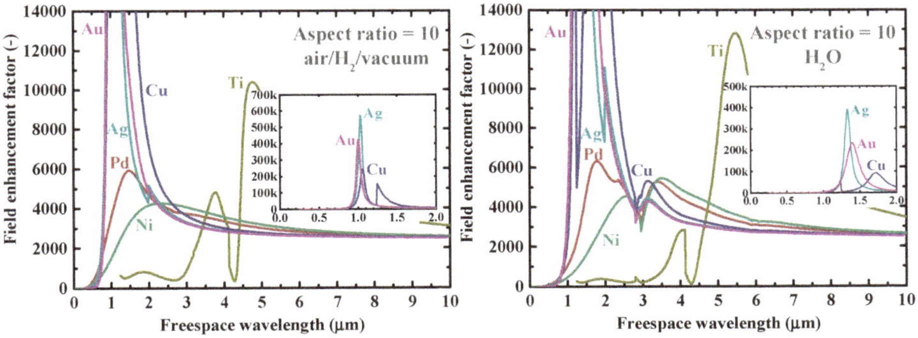

Figure 4. Calculated electromagnetic field enhancement factors at the tips of spheroidal nanoparticles of Au, Ag, Cu, Pd, Ti, and Ni with an aspect ratio of 10 in (**left**) air/H_2/vacuum and (**right**) H_2O. The insets are the clarified plots for Au, Ag, and Cu.

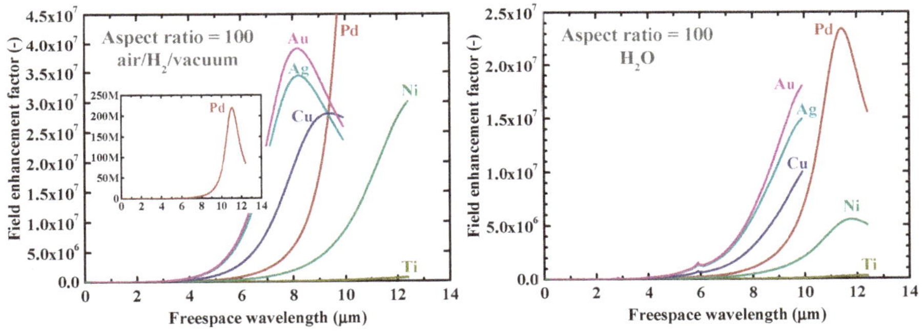

Figure 5. Calculated electromagnetic field enhancement factors at the tips of spheroidal nanoparticles of Au, Ag, Cu, Pd, Ti, and Ni with an aspect ratio of 100 in (**left**) air/H_2/vacuum and (**right**) H_2O. The inset of (**left**) is the clarified plot for Pd.

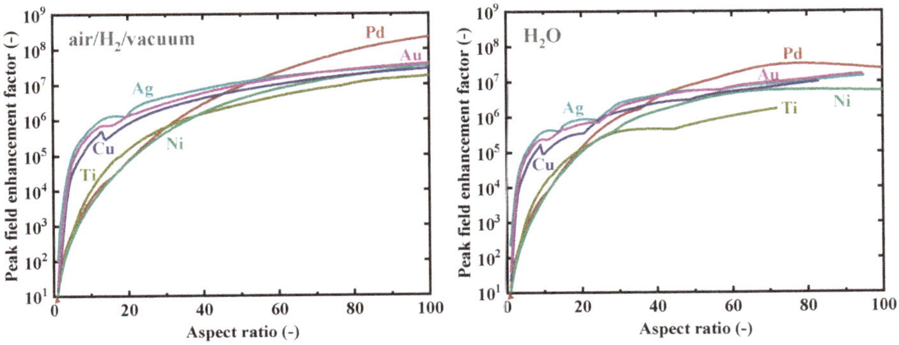

Figure 6. Dependence of the peak electromagnetic field enhancement factors at the tips of spheroidal nanoparticles of Au, Ag, Cu, Pd, Ti, and Ni on the aspect ratio in (**left**) air/H_2/vacuum and (**right**) H_2O.

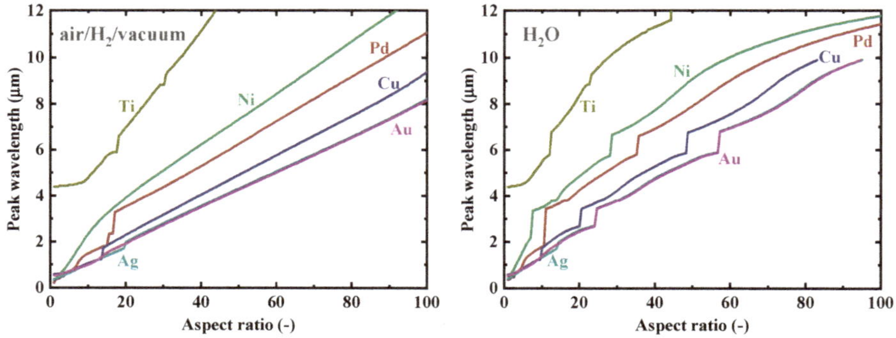

Figure 7. Dependence of the peak wavelength of the electromagnetic field enhancement factors at the tips of spheroidal nanoparticles of Au, Ag, Cu, Pd, Ti, and Ni on the aspect ratio in (**left**) air/H_2/vacuum and (**right**) H_2O.

4. Conclusions

In this work, we numerically investigated the lightning-rod effect of plasmonic field enhancement on hydrogen-absorbing transition metals. A large degree of energy focusing was observed for these transition metals in the microwave region, even surpassing the enhancement for noble metals according to the conditions. Pd, for instance, exhibited peak field enhancement factors of 6000 and 2×10^8 in air for morphological aspect ratios of 10 and 100, respectively. The metal surfaces possibly contained such degrees of nano- or micro-scale native random roughnesses, and, therefore, the field enhancement effect may have been unknowingly produced in existing electrical and optical systems. Active utilization of the plasmonic local energy enhancement effect around the metal surfaces by proper material and structure choices, such as the introduction of sharp nanoparticles or sharply roughened surfaces, can potentially improve hydrogen-related device performance.

Author Contributions: K.T. conceived the idea for the study. N.F. carried out the calculations and analyzed the data. Both authors contributed to discussion of the results and composed the manuscript.

Funding: This research was funded by the Thermal & Electric Energy Technology Foundation and the Research Foundation for Opto-Science and Technology.

Conflicts of Interest: The authors declare no conflict of interest. The funders had no role in the design of the study; in the collection, analyses, or interpretation of data; in the writing of the manuscript, or in the decision to publish the results.

References

1. Ritchie, R.H. Plasma losses by fast electrons in thin films. *Phys. Rev.* **1957**, *106*, 874–881. [CrossRef]
2. Barnes, W.L.; Dereux, A.; Ebbesen, T.W. Surface plasmon subwavelength optics. *Nature* **2003**, *424*, 824–830. [CrossRef] [PubMed]
3. Maier, S.A.; Atwater, H.A. Plasmonics: Localization and guiding of electromagnetic energy in metal/dielectric structures. *J. Appl. Phys.* **2005**, *98*, 011101. [CrossRef]
4. Ritchie, R.H.; Arakawa, E.T.; Cowan, J.J.; Hamm, R.N. Surface-plasmon resonance effect in grating diffraction. *Phys. Rev. Lett.* **1968**, *21*, 1530–1533. [CrossRef]
5. Weber, W.H.; McCarthy, S.L. Surface-plasmon resonance as a sensitive optical probe of metal-film properties. *Phys. Rev. B* **1975**, *12*, 5643–5650. [CrossRef]
6. Kim, S.; Jin, J.H.; Kim, Y.J.; Park, I.Y.; Kim, Y.; Kim, S.W. High-harmonic generation by resonant plasmon field enhancement. *Nature* **2008**, *453*, 757–760. [CrossRef] [PubMed]

7. Tanabe, K. Field enhancement around metal nanoparticles and nanoshells: A systematic investigation. *J. Phys. Chem. C* **2008**, *112*, 15721–15728. [CrossRef]
8. Nie, S.M.; Emery, S.R. Probing single molecules and single nanoparticles by surface-enhanced Raman scattering. *Science* **1997**, *275*, 1102–1106. [CrossRef]
9. Haes, A.J.; Van Duyne, R.P. A nanoscale optical blosensor: Sensitivity and selectivity of an approach based on the localized surface plasmon resonance spectroscopy of triangular silver nanoparticles. *J. Am. Chem. Soc.* **2002**, *124*, 10596–10604. [CrossRef]
10. Homola, J. Surface plasmon resonance sensors for detection of chemical and biological species. *Chem. Rev.* **2008**, *108*, 462–493. [CrossRef]
11. Shackleford, J.A.; Grote, R.; Currie, M.; Spanier, J.E.; Nabet, B. Integrated plasmonic lens photodetector. *Appl. Phys. Lett.* **2009**, *94*, 083501. [CrossRef]
12. Berini, P. Surface plasmon photodetectors and their applications. *Laser Photonics Rev.* **2014**, *8*, 197–220. [CrossRef]
13. Echtermeyer, T.J.; Milana, S.; Sassi, U.; Eiden, A.; Wu, M.; Lidorikis, E.; Ferrari, A.C. Surface plasmon polariton graphene photodetectors. *Nano Lett.* **2016**, *16*, 8–20. [CrossRef] [PubMed]
14. Vuckovic, J.; Loncar, M.; Scherer, A. Surface plasmon enhanced light-emitting diode. *IEEE J. Quantum Electron.* **2000**, *36*, 1131–1144. [CrossRef]
15. Hobson, P.A.; Wedge, S.; Wasey, J.A.E.; Sage, I.; Barnes, W.L. Surface plasmon mediated emission from organic light-emitting diodes. *Adv. Mater.* **2002**, *14*, 1393–1396. [CrossRef]
16. Pillai, S.; Catchpole, K.R.; Trupke, T.; Zhang, G.; Zhao, J.; Green, M.A. Enhanced emission from Si-based light-emitting diodes using surface plasmons. *Appl. Phys. Lett.* **2006**, *88*, 161102. [CrossRef]
17. Noginov, M.A.; Zhu, G.; Belgrave, A.M.; Bakker, R.; Shalaev, V.M.; Narimanov, E.E.; Stout, S.; Herz, E.; Suteewong, T.; Wiesner, U. Demonstration of a spaser-based nanolaser. *Nature* **2009**, *460*, 1110–1112. [CrossRef]
18. Oulton, R.F.; Sorger, V.J.; Zentgraf, T.; Ma, R.M.; Gladden, C.; Dai, L.; Bartal, G.; Zhang, X. Plasmon lasers at deep subwavelength scale. *Nature* **2009**, *461*, 629–632. [CrossRef]
19. Berini, P.; De Leon, I. Surface plasmon–polariton amplifiers and lasers. *Nat. Photonics* **2012**, *6*, 16–24. [CrossRef]
20. Hayashi, S.; Kozaru, K.; Yamamoto, K. Enhancement of photoelectric conversion efficiency by surface-plasmon excitation: A test with an organic solar-cell. *Solid State Commun.* **1991**, *79*, 763–767. [CrossRef]
21. Schaadt, D.M.; Feng, B.; Yu, E.T. Enhanced semiconductor optical absorption via surface plasmon excitation in metal nanoparticles. *Appl. Phys. Lett.* **2005**, *86*, 063106. [CrossRef]
22. Nakayama, K.; Tanabe, K.; Atwater, H.A. Plasmonic nanoparticle enhanced light absorption in GaAs solar cells. *Appl. Phys. Lett.* **2008**, *93*, 121904. [CrossRef]
23. Schurig, D.; Mock, J.J.; Justice, B.J.; Cummer, S.A.; Pendry, J.B.; Starr, A.F.; Smith, D.R. Metamaterial electromagnetic cloak at microwave frequencies. *Science* **2006**, *314*, 977–980. [CrossRef] [PubMed]
24. Tanaka, T. Plasmonic metamaterials. *IEICE Electron. Express* **2012**, *9*, 34–50. [CrossRef]
25. Amemiya, T.; Taki, M.; Kanazawa, T.; Hiratani, T.; Arai, S. Optical lattice model toward nonreciprocal invisibility cloaking. *IEEE J. Quantum Electron.* **2015**, *51*, 6100110. [CrossRef]
26. Fukuoka, N.; Tanabe, K. Large plasmonic field enhancement on hydrogen-absorbing transition metals at lower frequencies: Implications for hydrogen storage, sensing, and nuclear fusion. *J. Appl. Phys.* **2019**, *126*, 023102. [CrossRef]
27. Tripodi, P.; Armanet, N.; Asarisi, V.; Avveduto, A.; Marmigi, A.; Vinko, J.D.; Biberian, J.P. The effect of hydrogenation/dehydrogenation cycles on palladium physical properties. *Phys. Lett. A* **2009**, *373*, 3101–3108. [CrossRef]
28. Li, G.Q.; Kobayashi, H.; Taylor, J.M.; Ikeda, R.; Kubota, Y.; Kato, K.; Takata, M.; Yamamoto, T.; Toh, S.; Matsumura, S.; et al. Hydrogen storage in Pd nanocrystals covered with a metal-organic framework. *Nat. Mater.* **2014**, *13*, 802–806. [CrossRef]
29. Mohtadi, R.; Orimo, S. The renaissance of hydrides as energy materials. *Nat. Rev. Mater.* **2016**, *2*, 16091. [CrossRef]

30. Tanabe, K. Modeling of hydrogen/deuterium dynamics and heat generation on palladium nanoparticles for hydrogen storage and solid-state nuclear fusion. *Heliyon* **2016**, *2*, e00057. [CrossRef]
31. Kitagawa, Y.; Tanabe, K. Development of a kinetic model of hydrogen absorption and desorption in magnesium and analysis of the rate-determining step. *Chem. Phys. Lett.* **2018**, *699*, 132–138. [CrossRef]
32. Lundström, K.I.; Shivaraman, M.S.; Svensson, C.M. A hydrogen-sensitive Pd-gate MOS transistor. *J. Appl. Phys.* **1975**, *46*, 3876–3881. [CrossRef]
33. Favier, F.; Walter, E.C.; Zach, M.P.; Benter, T.; Penner, R.M. Hydrogen sensors and switches from electrodeposited palladium mesowire arrays. *Science* **2001**, *293*, 2227–2231. [CrossRef] [PubMed]
34. Hubert, T.; Boon-Brett, L.; Black, G.; Banach, U. Hydrogen sensors—A review. *Sens. Actuators B* **2011**, *157*, 329–352. [CrossRef]
35. Nuckolls, J.; Thiessen, A.; Wood, L.; Zimmerman, G. Laser compression of matter to super-high densities: Thermonuclear (CTR) applications. *Nature* **1972**, *239*, 139–142. [CrossRef]
36. Fleischmann, M.; Pons, S. Electrochemically induced nuclear-fusion of deuterium. *J. Electroanal. Chem.* **1989**, *261*, 301–308. [CrossRef]
37. Iwamura, Y.; Sakano, M.; Itoh, T. Elemental analysis of Pd complexes: Effects of D_2 gas permeation. *Jpn. J. Appl. Phys.* **2002**, *41*, 4642–4650. [CrossRef]
38. Hurricane, O.A.; Callahan, D.A.; Casey, D.T.; Celliers, P.M.; Cerjan, C.; Dewald, E.L.; Dittrich, T.R.; Doppner, T.; Hinkel, D.E.; Hopkins, L.F.B.; et al. Fuel gain exceeding unity in an inertially confined fusion implosion. *Nature* **2014**, *506*, 343–348. [CrossRef]
39. Gersten, J.I. The effect of surface roughness on surface enhanced Raman scattering. *J. Chem. Phys.* **1980**, *72*, 5779–5780. [CrossRef]
40. Gersten, J.; Nitzan, A. Electromagnetic theory of enhanced Raman scattering by molecules adsorbed on rough surfaces. *J. Chem. Phys.* **1980**, *73*, 3023–3037. [CrossRef]
41. Wang, D.S.; Kerker, M. Enhanced Raman scattering by molecules adsorbed at the surface of colloidal spheroids. *Phys. Rev. B* **1981**, *24*, 1777–1790. [CrossRef]
42. Liao, P.F.; Wokaun, A. Lightning rod effect in surface enhanced Raman scattering. *J. Chem. Phys.* **1982**, *76*, 751–752. [CrossRef]
43. Mohamed, M.B.; Volkov, V.; Link, S.; El-Sayed, M.A. The 'lightning' gold nanorods: Fluorescence enhancement of over a million compared to the gold metal. *Chem. Phys. Lett.* **2000**, *317*, 517–523. [CrossRef]
44. Bohren, C.F.; Huffman, D.R. *Absorption and Scattering of Light by Small Particles*, 1st ed.; Wiley: Weinheim, Germany, 1983; pp. 130–148. ISBN 978-0-471-29340-8.
45. Tanabe, K. A simple optical model well explains plasmonic-nanoparticle-enhanced spectral photocurrent in optically thin solar cells. *Nanoscale Res. Lett.* **2016**, *11*, 236. [CrossRef] [PubMed]
46. Tanabe, K. Plasmonic energy nanofocusing for high-efficiency laser fusion ignition. *Jpn. J. Appl. Phys.* **2016**, *55*, 08RG01. [CrossRef]
47. Moskovits, M. Surface-enhanced spectroscopy. *Rev. Mod. Phys.* **1985**, *57*, 783–826. [CrossRef]
48. Langhammer, C.; Zoric, I.; Kasemo, B. Hydrogen storage in Pd nanodisks characterized with a novel nanoplasmonic sensing scheme. *Nano Lett.* **2007**, *7*, 3122–3127. [CrossRef]
49. Baldi, A.; Narayan, T.C.; Koh, A.L.; Dionne, J.A. In situ detection of hydrogen-induced phase transitions in individual palladium nanocrystals. *Nat. Mater.* **2014**, *13*, 1143–1148. [CrossRef]
50. Boyd, G.T.; Rasing, T.; Leite, J.R.R.; Shen, Y.R. Local-field enhancement on rough surfaces of metals, semimetals, and semiconductors with the use of optical second-harmonic generation. *Phys. Rev. B* **1984**, *30*, 519–526. [CrossRef]
51. Boyd, G.T.; Yu, Z.H.; Shen, Y.R. Photoinduced luminescence from the noble metals and its enhancement on roughened surfaces. *Phys. Rev. B* **1986**, *33*, 7923–7936. [CrossRef]
52. Zeman, E.J.; Schatz, G.C. An accurate electromagnetic theory study of surface enhancement factors for Ag, Au, Cu, Li, Na, Al, Ga, In, Zn, and Cd. *J. Phys. Chem.* **1987**, *91*, 634–643. [CrossRef]
53. Hao, E.; Schatz, G.C. Electromagnetic fields around silver nanoparticles and dimers. *J. Chem. Phys.* **2004**, *120*, 357–366. [CrossRef] [PubMed]
54. Inoue, M.; Ohtaka, K. Surface enhanced Raman scattering by metal spheres. I. Cluster effect. *J. Phys. Soc. Jpn.* **1983**, *52*, 3853–3864. [CrossRef]

55. Tanaka, K.; Tanaka, M. Simulations of nanometric optical circuits based on surface plasmon polariton gap waveguide. *Appl. Phys. Lett.* **2003**, *82*, 1158–1160. [CrossRef]
56. Ward, D.R.; Huser, F.; Pauly, F.; Cuevas, J.C.; Natelson, D. Optical rectification and field enhancement in a plasmonic nanogap. *Nat. Nanotechnol.* **2010**, *5*, 732–736. [CrossRef] [PubMed]

nanomaterials

MDPI

Review

Surface/Interface Engineering for Constructing Advanced Nanostructured Photodetectors with Improved Performance: A Brief Review

Meng Ding [1,*], Zhen Guo [2,3,*], Xuehang Chen [1], Xiaoran Ma [1] and Lianqun Zhou [2,4,*]

1 School of Physics and Technology, University of Jinan, 336 Nanxinzhuang West Road, Jinan 250022, China; chenxuehang666@163.com (X.C.); mammxxxr@163.com (X.M.)
2 Key Lab of Bio-Medical Diagnostics, Suzhou Institute of Biomedical Engineering and Technology, Chinese Academy of Sciences, Suzhou 215163, China
3 Zhongke Mass Spectrometry (Tianjin) Medical Technology Co., Ltd., Tianjin 300399, China
4 Jihua Institute of Biomedical Engineering Technology, Jihua Laboratory, Foshan 528251, China
* Correspondence: dingmeng0207@163.com (M.D.); guozhen@sibet.ac.cn (Z.G.); zhoulq@sibet.ac.cn (L.Z.)

Received: 10 January 2020; Accepted: 13 February 2020; Published: 19 February 2020

Abstract: Semiconductor-based photodetectors (PDs) convert light signals into electrical signals via a photon–matter interaction process, which involves surface/interface carrier generation, separation, and transportation of the photo-induced charge media in the active media, as well as the extraction of these charge carriers to external circuits of the constructed nanostructured photodetector devices. Because of the specific electronic and optoelectronic properties in the low-dimensional devices built with nanomaterial, surface/interface engineering is broadly studied with widespread research on constructing advanced devices with excellent performance. However, there still exist some challenges for the researchers to explore corresponding mechanisms in depth, and the detection sensitivity, response speed, spectral selectivity, signal-to-noise ratio, and stability are much more important factors to judge the performance of PDs. Hence, researchers have proposed several strategies, including modification of light absorption, design of novel PD heterostructures, construction of specific geometries, and adoption of specific electrode configurations to modulate the charge-carrier behaviors and improve the photoelectric performance of related PDs. Here, in this brief review, we would like to introduce and summarize the latest research on enhancing the photoelectric performance of PDs based on the designed structures by considering their surface/interface engineering and how to obtain advanced nanostructured photo-detectors with improved performance, which could be applied to design and fabricate novel low-dimensional PDs with ideal properties in the near future.

Keywords: photodetectors; interface/interface engineering; nanostructures; charge carriers

1. Introduction

In recent years, following the developmental step forward of microelectronics technology, optoelectronic technology, as one of burgeoning advanced technologies, has developed rapidly and will have far-reaching effects on the lives of human beings. It involves optical display, optical storage, lasers, and other fields and is the core technology of the future information industry. It also has significant strategic importance for the state economy, technology, and defense. Various materials with different structures have attracted much more attention for their potential applications in integrated nano- or micro-optoelectronics, including light emitting diodes (LEDs) [1], solar cells [2,3], sensors [4,5], field-effect transistors (FETs) [6,7], photodetectors [8–10], and so on.

At present, smart photoelectric devices with excellent performance play important roles in our daily life, scientific research frontiers, high school education, and biomedical fields, and so on.

The photodetector is actually a device that converts light signals into electrical signals. Compared with the classical photodetector devices with bulk materials, improved photoelectric conversion performance could be realized in constructed nanostructured photodetectors though from light absorption, photon charge carrier generation, and collection of contributing electronic signals, the field of which researchers have paid more and more attention.

Specially, a photodetector is an indispensable and important device in photoelectronic systems for military detection, aerospace satellites, and so on and helps scouts or researchers to collect key information for formulating strategies, which can also be widely used in national safety fields including flame monitoring and personal property safety [11], and so on. For an ideal photodetector, the main characteristics could be high sensitivity, fast response time, wavelength selectivity, and so on. Photodetectors could be classified in several ways: wavelength spectra (IR to UV detectors), information sampling (smoke, temperature, gas), structure (p–n junction; PN), metal-semiconductors (MS), and metal-semiconductor-metal (MSM). Photodetectors can be classified according to the dimension of the photodetector (PD): zero-dimensional (0D), one-dimensional (1D, such as nanowire, nanorod, nanotube, and so on), two-dimensional (2D, such as graphene (Gr) and transition metal dichalcogenides (TMDCs), three-dimensional (3D), or bulk. Semiconductor PDs have the unique advantages of high quantum efficiency, small size, low consumption energy, and high stability; therefore, PDs have attracted wide attention and interest of researchers.

So far, PDs based on single crystal, film, and nanostructured materials have been applied for building designed PDs to detect ultraviolet [12–14], visible [15–18], or infrared [19–25] photons with actual needs. For example, highly narrow band (bandwidth of 10 nm) solar-blind photodetectors of β-Ga_2O_3 single crystals with a peak responsivity of 0.23 A/W at 262 nm and an EQE of 110% were reported [26]. Metal-oxide-semiconductor ultraviolet PDs based on Au/MgO/MgZnO with high internal gain have been constructed [14]. A novel hybrid visible PD was realized using a planar p-type inorganic NiO layer in a junction with an organic electron acceptor layer. A hexagonal boron nitride (hBN)/b-$As_{0.83}P_{0.17}$/hBN sandwiched structured mid-infrared photodetector was constructed with responsivity of 190, 16, and 1.2 mA/W at 3.4, 5.0, and 7.7 μm at room temperature, respectively [20]. Conventional silicon (Si)-based PDs usually show peak photoresponse between 700 and 900 nm with rather low dark current because of the high crystalline quality and excellent passivation properties of Si [27]. Moreover, much research on flexible/stretchable image sensors based on thin-film Si PDs has been studied in materials science and engineering [28–31]. Especially, thin film silicon devices with nanophotonic structures could effectively improve the absorption efficiency of incident light [32,33], which was one of the directions to improve the performance of PDs. From the conclusion of the literature review, it could be illustrated that compared with film and bulk materials, low-dimensional nanostructures have the advantages of unique conductivity caused by high quality of crystal and carrier mobility and a confined carrier transport channel; thus, they are much more suitable candidate materials in assembling high-performance PDs on a large scale [34]. It could also be found that at this stage, various nanostructures of different materials have been utilized to fabricate PDs, including quantum dots (QDs) [24], nanoparticles (NPs) [35,36], nanowires [37–39], nanotubes [40], nanosheets [41,42], nanoribbons [43], nanobelts [44], and so on.

From the point view of dimension and easy assembly, nanowires are one of the best choices to be applied for building different kinds of photoelectric devices. The large surface-to-volume ratio and the presence of deep level surface trap states in NWs greatly prolong the photocarrier lifetime; the reduced dimensionality of the active area in NW devices shortens the carrier transit time. Indeed, the combination of long lifetime and short transit time of charge carriers can result in substantial photoconductive gain [45]. Especially, quasi-one-dimensional nanowires with semiconducting properties have been widely investigated as active materials for high-performance photodetectors [23]. It could be also explored from the published literature related to PDs that, generation, separation, transportation, and collection of photo-induced charge carrier are the key parameters for improving performance of the designed devices [12].

In general, the photoelectric conversion process from optical signals to electric signals in PDs mainly involves three steps:

(1) Generation of the photo-induced carriers in the case of external light radiation;
(2) Separation, transportation, and multiplication of photo-induced charge carriers derived by the applied electric field or built-in electric field formed at the interface of the heterojunction;
(3) Collection of the photocurrent generated by photo-excited carriers at both ends of electrodes, thus realizing the detection of external light radiation.

It can be observed that charge carrier generation, diffusion, and recombination modulation are all very important considerations in the construction of high-efficiency PDs [46]. The sensitivity, response speed, spectral selectivity, signal-to-noise ratio, and stability of detection are much more important factors to judge the performance of PDs [34].

To rationally implement surface and interface engineering in the PDs' design, an understanding of the effects of surface and interface on reactions is required. In this section, we highlight a few of the key factors that need to be taken into account when designing nanostructures and their hybrid nanostructures. The carrier behaviors such as generation, recombination, separation, and collection are closely related to the surface/interface of devices, especially for the nanostructured devices; thus, surface/interface engineering of nanostructured materials is critically important for influencing the performance of the fabricated devices, and how to suitably apply surface/interface properties of nanostructured PDs for building advanced devices needs continuous exploration. In this review, how the surface/interface engineering improves the performance of the constructed nanostructured PDs is demonstrated theoretically through a review of the literature.

2. Surface and Interface States

A surface could be defined as atomic layers that do not have three-dimensional continuous environment of bulk materials, three-dimensional continuous environment, or the periodicity of the infinite lattice that is destroyed by the existence of nanostructured surface. For crystal structures, in the direction of the vertical surface, the potential energy of lattice atoms could not have corresponding symmetry, and some new eigenvalues could be obtained in the Hamiltonian characteristic value when the Schrodinger equation is applied through the theory of quantum mechanics [47]. A new energy level could appear and be defined as a surface state, as shown in Figure 1. There are actually two kinds of surface states: intrinsic and external surface states. The intrinsic one is the surface state without foreign impurities, and the other is due to the existent of impurities from adsorbed atoms or other imperfections on the surface.

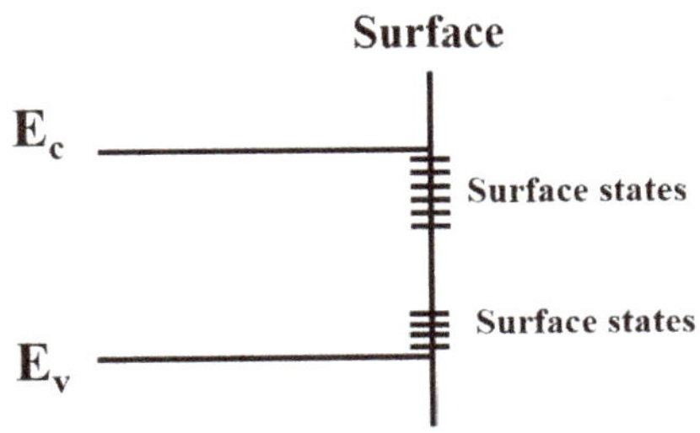

Figure 1. Schematic diagram of surface level.

Semiconductor surface research is mainly concerned with various electrical properties of surface phenomena. These phenomena involve the free carriers in the space charge region, surface states, and their mutual interactions. Although the free carrier transport in the space charge layer should be paid considerable attention, the most concerted effort in the study of the surface electrical properties has been directed towards the surface states [48]. Much research on semiconductor surfaces caused by cleaving in ultrahigh vacuum suggested that the formation of surface states is attributed to dangling bonds at the surface [49]. According to a previous report [47], the schematic diagram of the surface level is shown in Figure 1. The surface states could also be divided into donor states or acceptor states based on the behavior of the electrons. Surface states could capture, scatter generated carriers, and influence electric field effect, and the results could be effective photon capture, carrier generation, and combination centers, which controls surface properties of the absorbed photons and generated carriers. Scattered carriers influence surface mobility, thus affecting surface conductance.

An interface could be defined as the interface between heterostructure phases, similar to surface states. Interface states that are introduced at the interface state occur at the interface where two heterostructure media exist and contact. Thus, the continuous environment of atoms or lattice structures is destroyed by the existence of other substances. Hence, the interface states could be introduced by other atoms, lattice mismatch, interface roughness, and thermal expansion of two materials' interface. Generally speaking, there are two interface states: donor and recipient. Considering energy levels of the built heterostructures, the existing interface energy levels greatly influence carrier transport behaviors, which controls performance of the devices. The performance of the photodetector devices could be considerably improved by considering interface carrier transport properties by designing optimal heterostructures.

Commonly, interface engineering includes surface charge transfer, charge injection, and collection on the metal electrode/semiconductor interface and carrier bound state on the dielectric/semiconductor interface. On the other hand, there are a large number of defects in lattice structure, including vacancies, adsorbed atoms, grain boundaries, and impurities. Interfaces and defects have a critical influence on the properties and operational stability of metal halide perovskite optoelectronic devices. Therefore, interface and defect engineering is crucial to control the behavior of the charge carriers and to grow high quality, defect-free perovskite crystals. Yang et al. summarized the strategies of interface and defect engineering in perovskite solar cells and light-emitting diodes [50]. Due to the atomic thickness and super high specific surface area of two-dimensional TMDCs, the interface of TMDCs plays a decisive role in the device performance. Chen and other researchers summarized and highlighted advances in TMDC interfaces and defect engineering and applications in electronic devices. Various appropriate interfaces and defect engineering that effectively adjust the electrical and optical properties of the TMDCs were presented in order to achieve the ultimate goal of improving device performance [51]. By adopting appropriate defect engineering strategy, the defect fix, controlling the type and concentration of carriers, and the reduction of contact resistance could be achieved to realize high-performance electronic devices. On the other hand, controlling the defect state could increase the sensitivity of the photoconductive device or improve the response speed of the device. A heterojunction can be constructed to produce a photovoltaic effect through defect doping [52], and also the quantum electroluminescence effect can be produced using defects reasonably [53].

3. Surface/Interface Engineering for Improvement of Photodetector Properties

For the design and fabrication of optoelectronic devices based on nanostructures, the surface-to-volume ratios, interface areas, and interaction between surface and environment are increased dramatically, which has great impact on the performance of a PD device. For example, trapping at surface states of ZnO nanowire obviously influenced the transport and photoconduction properties of nanowires. The high surface-to-volume ratios and the existence of deep level surface trap states in nanostructures greatly extend the lifetime of photogenerated carriers. Moreover, the low dimensionality of the active area in a nanostructure device can reduce the transit time of carriers [45]. Zhou and colleagues fabricated UV photodetectors based on single In_2O_3 nanowire, and the conductance was significantly increased by four orders of magnitude [54]. Fang et al. constructed individual SnO_2 nanowire UV photodetectors, which showed excellent optical selectivity and ultrahigh external quantum efficiency [55]. Yang et al. reported the ZnO nanowire UV PD with high internal gain, which was attributed to the presence of oxygen-related hole-trap states at the surface of the nanowire. At the same time, the slow dynamics of the surface oxygen molecules' adsorption and desorption processes could result in a long response time in both the rise and decay process of a PD [45,56]. That is, the major shortcoming for ZnO based on PD applications is the strong persistent photo-induced conductivity after light illumination, which inhibits a fast recovery of the dark current [57]. Moreover, the persistent photoconductivity phenomenon was observed in PDs based on MoS_2 [58,59]. The performance of PD devices can be improved by the following ways: (a) enhancing the interaction between light and matter; (b) decreasing the adverse effects of defects; and (c) adjustment of electronic characteristics [60]. Therefore, various approaches, such as doping [61], surface functionalization [56,62–68], surface carrier transport modulation [34], interface carrier-trapping/transport control [46], piezo-phototronic effects [69–74], and so on, have been utilized to improve the photoresponse performance of low-dimension nanostructure PDs for use in practical applications. The following section lists several ways to deal with surface/interface issues for improving performance of the photodetectors.

3.1. Surface-State Passivation for Terminating Dangling Bonds

As nanostructured photodetector devices have a much larger surface to volume ratio, and surface states play an important role in controlling their performance, surface-state passivation was one of the most useful solutions to deal with the mentioned issues. Surface-state passivation has been considered as one of the most effective and advanced methods to improve the performance of PDs through terminating their dangling bonds to decrease their surface states' influences. Fang et al. [64] modified the ZnO nanowalls by CdS nanoparticles and investigated the effect of CdS nanoparticles on the optical and photoelectrical properties of a PD. The surface states of ZnO nanowalls was suppressed due to the introduction of CdS nanoparticles' passivation layer; thus, the deep-level emission was reduced, and the recombination of carriers was prevented, and then the photoconductivity of ZnO nanowalls was improved obviously. Ren et al. [38] constructed nanowire-based photodetectors at mid-wavelength infrared composed of vertical selective-area n-InAsSb nanowire photoabsorber arrays on large bandgap p-InP substrate. In order to effectively inhibit the nonradiative recombination at the surface of InAsSb nanowire, the Al_2O_3 passivation shells were introduced, as displayed in Figure 2a,b. Furthermore, it was demonstrated that the photoluminescence (PL) emission intensity of InAsSb nanowire arrays with Al_2O_3 passivation layer increased 10- to 50-fold at 77 K. The spectral response of n-InAsSb/p-InP PDs with Al_2O_3 passivation at a reverse bias of 0.5 V is shown in Figure 2c. Two detection peaks located at about 2.0 and 3.4μm were observed at room temperature.

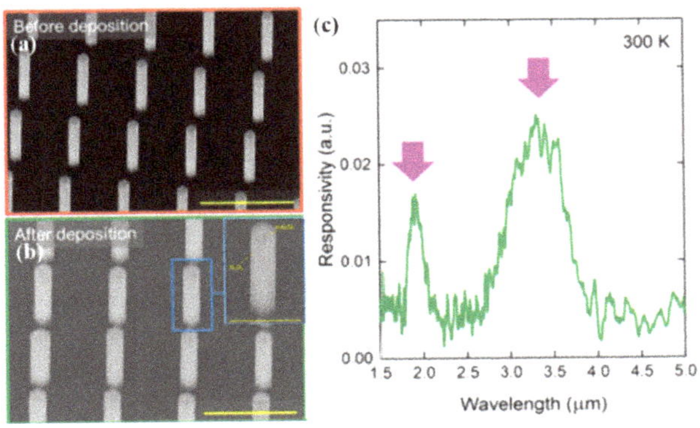

Figure 2. SEM images of (**a**) InAsSb nanowires and (**b**) InAsSb nanowires with ~60 nm Al_2O_3 passivation, the scale bar is 2 μm. The inset is single nanowire coated with Al_2O_3 passivation layer, the scale bar is 1 μm. (**c**) Spectral response at a reverse bias of 0.5 V at room temperature [38].

Yang et al. reported the single ZnO nanowire PDs with large photoresponse and high internal gain; the hole-trapping mechanism was also proposed. The holes could be trapped through oxygen adsorption and desorption at surface states caused by the dangling bonds, further prolonging the lifetime of photogenerated carriers, and multiple electrons could transit through the nanowire. Finally, the photoconductive gain was realized [45]. However, the surface adsorption process could make the response time (rise and decay) of PDs become large, which is unfavorable for the performance of PDs. Chen et al. [56] constructed photoconductive ZnO nanowire/copper phthalocyanine (CuPc) hybrid PDs. A p–n heterojunction was formed at the interface of CuPc film and ZnO nanowires, leading to reduced conductivity and inhibited dark current in the ZnO NW. Under UV illumination, electron–hole pairs were generated and separated by a built-in internal electric field, and photoinduced potential was built. Meanwhile, the width of the surface depletion layer reduced, resulting in higher conductivity and increased photocurrent. The transition was rather fast and did not consider the adsorption or desorption process of molecules; thus, the rise and decay time of ZnO nanowire/CuPc photoresponse was shorter than that of the ZnO nanowire, which is shown in Figure 3. Therefore, the main reason for the improved photoresponse speed was the passivation of ZnO nanowires' surface states by CuPc film.

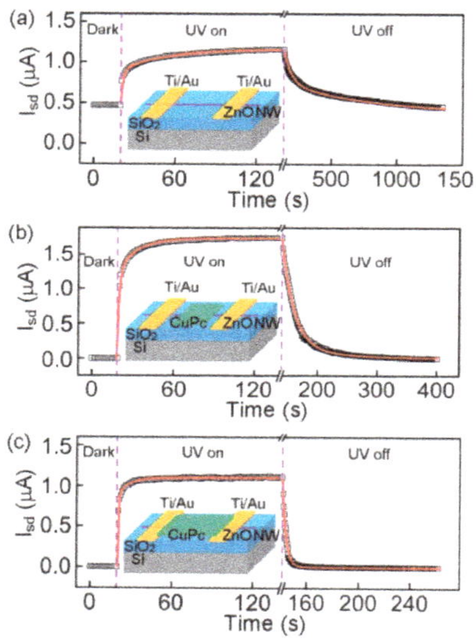

Figure 3. (**a**) The photoresponse curve of the bare ZnO nanowire device, (**b**) the partial coverage device, (**c**) the full coverage device. The insets are schematic diagrams of corresponding devices, respectively. The photoresponses of three devices with different CuPc coverage were measured at bias voltage of 0.5 V, and the UV illumination source was 350 nm light with intensity of 100 uW·cm^{-2}. [56].

For two-dimensional nanomaterials, many reports about transition metal dichalcogenides (TMDCs) with layer structure material and interacting layers held together with weak van der Waals interaction, including MoS_2 [75], $MoSe_2$ [76], WSe_2 [77], WS_2 [78], and ReS_2 [79], have been published. Because of their large surface volume ratio, the performance of these detectors is more sensitive to the environment. Studies have shown that physically adsorbed gas molecules, such as O_2 and H_2O, could deplete MoS_2 and $MoSe_2$ by removing electrons from the channels [80,81]. Konstantatos et al. constructed a high-stability and excellent-performance photodetector with an atomic layer deposited hafnium oxide (HfO_2) encapsulated monolayer and bilayer MoS_2, and the schematic of the PD device is shown in Figure 4a [82]. The oxide could suppress strong current drifting and degradation caused by environmental effects and effectively remove the surface atmospheric adsorbates. Unprotected MoS_2 photodetector devices usually show slow response speed, which is displayed in Figure 4b.

After encapsulation with HfO_2, the photocurrent of the PD device was increased by about 40 times, while the decay time was reduced by more than one order of magnitude, as shown in Figure 4c. Moreover, the responsivity of MoS_2 PDs was promoted by more than one order of magnitude after HfO_2 encapsulation, as shown in Figure 4d. Therefore, the investigation of effective packaging technology and the prior complete removal of surface binding adsorbates is an effective way to achieve high speed and sensitive light detection.

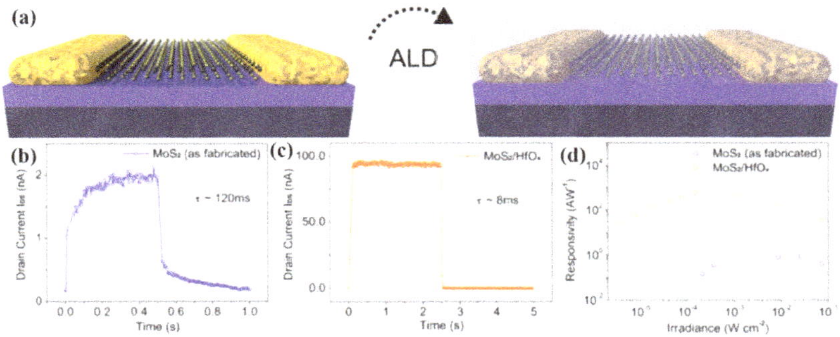

Figure 4. (**a**) Device schematic of MoS_2 and MoS_2/HfO_2 photodetectors (PDs). The photocurrent of MoS_2 before (**b**) and after (**c**) HfO_2 encapsulation. (**d**) The power-dependent responsivity of a bilayer MoS_2 device before and after HfO_2 encapsulation [82].

Certainly, other research groups have also used surface passivation to improve the performance of different detectors. Ni et al. investigated the influence of the MgO surface modification on the photocurrent and responsivity performance of TiO_2@MgO core-shell nanowire array self-powered UV PDs [66]. The energy level structure between TiO_2 and MgO is beneficial to the process of the separation of photogenerated electron–hole pairs and thus restrains the recombination of photogenerated carriers. Chen's group [61] have utilized a surface transfer doping method to adjust the electronic properties and work function of graphene films and then tune the Schottky barrier of Gr/Si junctions. The transition metal oxide molybdenum trioxide (MoO_3) possesses high work function [83] and has displayed obvious surface transfer hole doping effects in graphene, MoS_2, and other two-dimensional materials [67]. Therefore, they fabricated a Gr/Si self-powered photodetector using in situ surface functionalization of the MoO_3 layer on graphene. It was demonstrated that the photocurrent responsivity of MoO_3 doped Gr/Si PDs was significantly improved for a wide spectrum ranging from ultraviolet to near infrared. In addition, the external quantum efficiency, photocurrent responsivity, photovoltage responsivity, and the specific detectivity (D*) of Gr/Si devices were significantly promoted after MoO_3 decoration, which was attributed to the more efficient photocarrier separation and collection process caused by the increased height of the Schottky barrier at the Gr/Si interface and the reduced series resistance of the Gr/Si device after MoO_3 modification [61]. Sangyeon Pak [84] demonstrated the effects of surface functionalization on charge carrier density and photoresponse performance of the MoS_2 photodetector.

3.2. *Surface Plasmonic Resonance for Strong Scattering and Absorption of Incident Light*

In recent years, many research groups have found that metal-surface plasma could be used to enhance the performance of detectors, which was attributed to the localized surface plasmon resonance (LSPR) effect, resulting in strong scattering and absorption of incident light, effectively separating photocarriers and transporting carriers at the interfaces of metal/semiconductor. Metal surface plasma is a hybrid electromagnetic wave mode formed by coupling of free vibrating electrons and incident photons on the metal surface. Both conductive surface plasmas at metal/dielectric interfaces and localized surface plasmas on metal nanoparticles have unique optical properties: high spatial local properties and local field enhancement properties. The unique photonic properties of surface plasmas make them widely used in subwavelength sensors, detectors, and modulators. To date, various plasmonic structures have been explored to improve the performance of photodetectors or phototransistors. In general, noble metals, for example Au [85–89], Ag [90–92], and Pt [93–95], are used as plasmonic metals for their useful range in the visible wavelengths, due to the excellent optical property as a result of SPR. In the development of plasmonic photodetectors using Au@MoS_2 heterostructures, the photoresponsivity of the Au@ MoS_2 device was about 10 times larger than that of planar MoS_2 devices. Another type of Si-supported Au@MoS_2 p–n junction photodiode demonstrated photoresponsivity of 10–30 A/W, higher than the values reported for similar MoS_2 gateless photodevices, and the response time was less than 20 ms [88]. Luo et al. demonstrated that the photodetector device modifying plasmonic Au nanostructures onto the surface of CdTe NW exhibited the obvious photoresponse to 510 nm light illumination, with high response speed and fast recovery time [87]. A high-performance near-infrared (NIR) PD was constructed by coating single layer graphene (SLG)/InP Schottky junction diode with plasmonic SiO_2@AuNRs; the schematic diagram of the PD and the TEM image of the SiO_2@AuNRs on SLG film are shown in Figure 5a,b [96]. The decoration of SiO_2@AuNRs could slightly improve the Schottky junction barrier, resulting in the increased built-in electric field, and then the separation efficiency of carriers was improved. The light trapping effect of the SiO_2@AuNRs-SLG/InP device nearly promoted the absorption of incident light; as a result, many more holes and electrons were generated, and therefore photocurrent and responsivity were increased obviously (displayed in Figure 5c,d). Furthermore, the response rate of the SiO_2@AuNRs-SLG/InP device was rather fast, which was able to monitor switching optical signals with a frequency up to 1 MHz, indicating its potential application in sensing high-frequency optical signals.

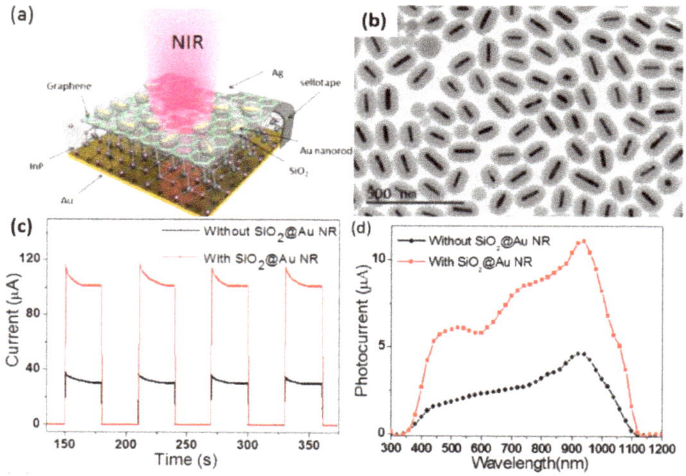

Figure 5. (**a**) The schematic diagram of the SiO_2@AuNRs-SLG/InP PD device. (**b**) The TEM image of the SiO_2@AuNRs on single layer graphene film. (**c**) Photoresponse of two different devices with and without SiO_2@AuNRs modification under 980 nm light illumination of 6.7 mW cm^{-2} without applied bias. (**d**) Photoresponsivity of the NIR PDs with and without SiO_2@Au decoration [96].

It was demonstrated that by coating Ag metal nanoparticles onto ZnO nanowires, the UV photodetectors exhibited higher sensitivity by four orders of magnitude with rather fast and stable response speed [90]. Arquer and colleagues reported that by introducing Ag metal nanoparticles into the PbS colloidal quantum dot photoconductive photodetectors, the absorption ability was increased due to a plasmonic scattering layer of Ag metal nanoparticles. As a result, the responsivity enhanced about 2.4-fold in the near infrared with the absorption band edge of ~1 μm [91]. Apart from noble metals, aluminum (Al) has been reported as a viable plasmonic metal for its useful range in the UV wavelengths [97,98]. Unlike the noble metals, the localized plasmon resonances can extend from the visible spectrum to the ultraviolet because the d-band of Al lies above its Fermi energy [99]. The Al element is a kind of abundant and low-cost metal found in the world, which is a better plasmonic material than Au and Ag in the UV range due to the negative real part and relatively low imaginary part of its dielectric function [100]. Xu et al. [97] improved the response characteristic of ZnO nanorod array UV PDs using surface plasmonic resonance by Al nanoparticles. Compared with the pure ZnO nanorod array PDs, the responsivity of the ZnO nanorod photodetector modified with Al nanoparticles was improved from 0.12 A/W to 1.59 A/W.

Recently, it has been observed that boron (B)-doped and phosphorus (P)-doped silicon (Si) nanocrystals enable the localized surface plasmon resonance of Si nanocrystals in the mid-infrared (MIR) region [101]. Especially, the heavily B-doped silicon nanocrystals induced band-tail states that could expand the optical absorption of Si from the UV-visible region into the near-infrared region. Furthermore, the localized surface plasmon resonance of heavily B-doped silicon nanocrystals could be tunable, which is attributed to free holes above the Fermi level from the B-induced impurity band [102]. Yang et al. [103] constructed quantum dots (QDs)/graphene hybrid MIR PDs by using plasmonic B-doped Si QDs; the schematic diagram of B-doped Si-QDs/graphene PDs is illustrated in Figure 6a. In the process of operating PDs, two different optical phenomena of silicon quantum dots can be adopted, as shown in Figure 6b. In the MIR region, the LSPR of silicon quantum dots generated a strong electric field that increased the direct excitation of the underlying graphene. Therefore, the phototransistor could effectively respond to the MIR light. The UV-to-near-infrared absorption of B-doped Si QDs resulted in the generation of electrons and holes in quantum dots. The photoresponse of the PD device was obtained by transferring one type of carrier from the quantum dot to graphene and capturing another carrier in the quantum dot. Figure 6c displays that the responsivity of Si-QD/graphene PDs was reduced with the increase of the irradiance power. The responsivity was in the range of 1.2×10^8 to 2.2×10^8 A/W with the UV-to-NIR light illumination at the lowest irradiance of 0.2 μW/cm^2. As can be seen from Figure 6d, the noise equivalent power (NEP) values of the PDs in the MIR region and the UV-to-NIR region were about 10^{-10} and 10^{-18} W/Hz$^{0.5}$, respectively. The fairly small NEP value suggests that the Si-QD/graphene photodetector can be applied to low light detection. The values of D* were about 10^5 and 10^{13} Jones for the MIR and the UV-to-NIR photodetection, respectively.

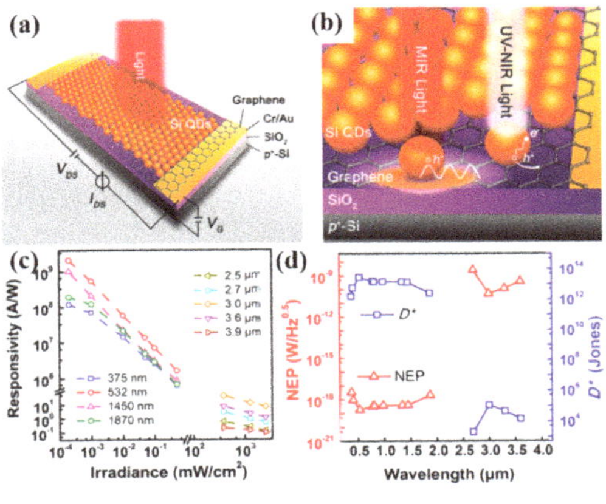

Figure 6. (**a**) Schematic diagram of B-doped Si-quantum dots (QDs)/graphene PDs. (**b**) Two different optical phenomena of B-doped Si QDs. (**c**) The irradiance power-dependent responsivity of the Si-QD/graphene PDs with different laser wavelengths at $V_G = 0$ V and $V_{DS} = 1$ V. (**d**) The spectral dependence of the noise equivalent power (NEP) and D* of the PDs. The measurements were performed at room temperature and 77 K, respectively [103].

3.3. Interface Carrier-Trapping/Transport Modulation

Once the design principles are defined, important parameters of interface engineering can be directly identified, containing interface compositions, areas, defects, faces, electronic coupling, and band bending. Before the interface engineering, the roles of the interface in the design structure should be fully understood according to the specific charge dynamics models [104]. For example, defects at the interface tend to be the recombination centers of electrons and holes, thus hindering the charge transfer across the interfaces. Therefore, eliminating the interfacial defects can improve the performance of the PDs in most situations. It has been proved that improving the quality of crystals, optimizing the arrangement of energy levels of layers, and constructing p–n or Schottky heterojunctions could be effective interfacial engineering methods to improve the performance of photodetectors [105].

It is demonstrated that surface and interface engineering play a key role in improving the performances of PDs. The generation, diffusion, and recombination/transport processes of charge carriers can influence the performance of PDs; thus, these factors should be taken into account when the PDs are built. Huang et al. [106] have constructed a new type of nanocomposite ultraviolet photodetector; the schematic drawing of the PD is shown in Figure 7a. The Frenkel excitons generated in ZnO nanoparticles and polymers could diffuse to the polymer/nanoparticles interface. Under reverse bias voltage, holes were transported in the semiconducting polymer, but the electrons were still trapped in the nanoparticles, attributed to the short percolation network for electrons and the strong quantum confinement effect of ZnO nanoparticles. Because of the strong electron trapping effect, the PDs had Schottky contact in the dark and an ohmic contact under illumination, confirmed by the dark current and photocurrent. That is, the PD device transitioned from a photodiode (in dark) to photoconductor (under illumination) via interfacial trap-controlled charge injection. The measured total noise current in the frequency range 1 to 5 kHz was mainly the shot noise, as shown in Figure 7b. The specific detectivity of hybrid PDs (shown in Figure 7c) was tens- to hundreds-fold better than that of inorganic semiconductor photodetectors. Furthermore, the response speed of the PD was rather high compared to other photodetectors based on any nanoparticles or colloidal quantum dots. It was indicated that two channels for the recombination of holes existed through the decay of the photocurrent experiment (displayed in Figure 7d). The decay process was combined with a fast component of 142 ms and a slow component of 558 ms.

Semiconductor heterojunctions with different bandgaps could exhibit different interface properties. By adjusting the arrangement of interfacial energy bands, the heterojunction structures may possess unique photoresponse characteristics. Moreover, when constructing the heterojunction for multicolor photodetection, it is necessary to have a large band deviation at the interface of the heterojunction in order to effectively extract carriers. There are larger band offsets between the Si/TiO_2 heterojunction; according to the calculation, the conduction band offset (ΔE_C) and the valence band offset (ΔE_V) were 0.75 eV and 2.66 eV, respectively, which would be conducive to control the transport of the carriers. Hu et al. [107] fabricated a multicolor photodetector based on the n-Si(111)/TiO_2 nanorod array heterojunction ranging from ultraviolet to visible light by controlling the applied voltage. In the case of forward bias, the band offsets would not hinder the diffusion of carriers; thus, there was no difference between the photocurrent and the dark current. The PD device exhibited multicolor detection capability under reverse bias because the motion of these carriers was controlled by the applied bias and the barrier of conduction band offset and the valence band offset.

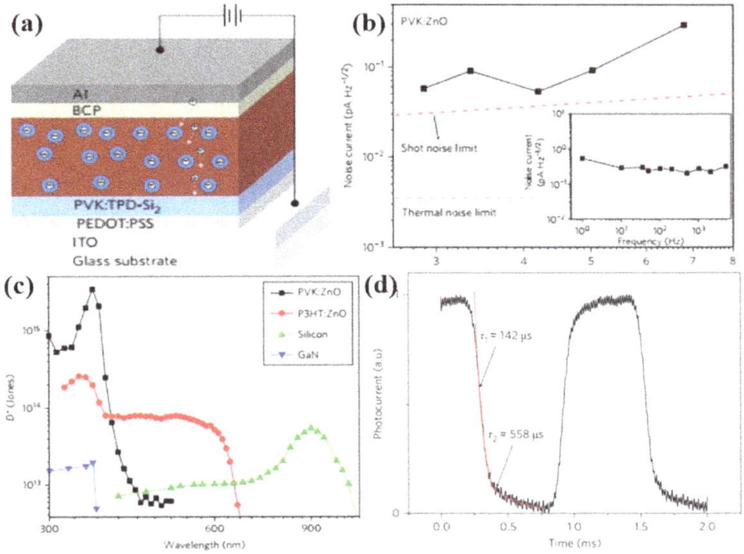

Figure 7. (**a**) The schematic diagram of the PD device. (**b**) Noise characteristics of the PVK/ZnO PDs under different currents, the inset is the frequency-dependent noise current with −9 V bias. (**c**) Specific detectivities of the photodetector at different wavelengths. (**d**) Transient photocurrent of the P_3HT/ZnO device [106].

Nowadays, Perovskite material has become one of the superstar media for constructing different kinds of photoelectric devices; especially, all-inorganic cesium lead halides ($CsPbX_3$, X = Cl, Br, I) represent an emerging class of materials owing to their high carrier mobility, long carrier diffusion length, and excellent visible light absorption. The high quantum efficiency (over 90%), narrow line width, and high stability make these all-inorganic perovskites suitable for application in novel optoelectronics [108]. Song et al. [108] utilized the advantages of large absorption coefficient and high quantum efficiency of the perovskites by fabricating the superior performance hybrid $CsPbBr_3$ perovskite/MoS_2 PD with a high photoresponsivity of 4.4 A/W and an external quantum efficiency of 302%; the schematic diagram of the hybrid PDs is displayed in Figure 8a,b. Especially, the photogenerated electrons in the perovskite were effectively transferred and injected to MoS_2. MoS_2 monolayer acted as an electron-collecting layer for perovskite photodetectors to improve the photoelectrical response, as shown in Figure 8c–e, which could decrease the probability of carrier recombination; therefore, the photoresponsivity of the hybrid $CsPbBr_3$ perovskite/MoS_2 PDs improved by three orders of magnitude compared with the pure MoS_2 PD. The dark current was also obviously reduced. Furthermore, the MoS_2 layer could induce the trap passivation on the substrate and benefit the carrier transport and then improve the separation efficiency of carriers; thus, the rising time of the hybrid PD was also lowered from 65.2 to 0.72 ms after combining with MoS_2 layers, as displayed in Figure 8f. It was demonstrated that 2D nanomaterials combined with the perovskite layers could be considered as a superior candidate to realize the next-generation high-performance photodetectors.

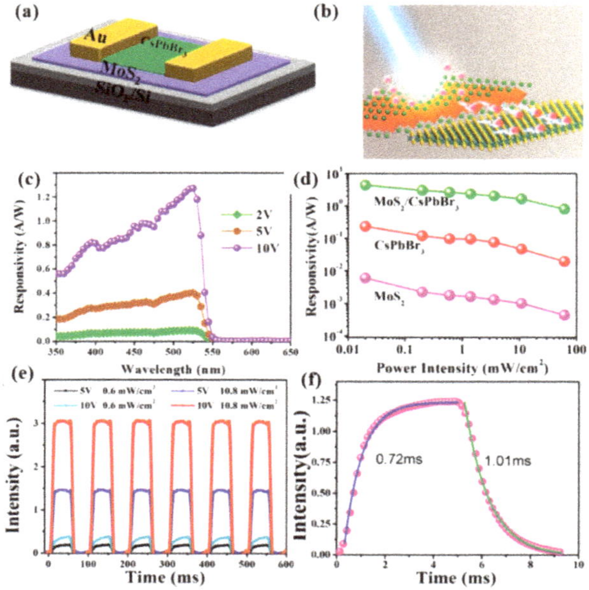

Figure 8. (**a**) The schematic diagram of the hybrid $MoS_2/CsPbBr_3$ PDs. (**b**) The schematic diagram of the charge transfer from $CsPbBr_3$ to MoS_2 under illumination. (**c**) Photoresponsivity spectra of the hybrid $MoS_2/CsPbBr_3$ PDs with different applied bias of 2, 5, and 10 V. (**d**) Photoresponsivity of the three PDs as a function of the power intensity. (**e**) Cyclic response of the hybrid $MoS_2/CsPbBr_3$ PDs at different voltages and incident optical power intensity under 442 nm laser illumination. (**f**) The temporal photocurrent response of the hybrid PDs with rising time (0.72 ms) and decay time (1.01 ms), respectively [108].

For fiber-shaped devices, the interface problems of the fiber device were more serious, which was attributed to the curved and rough surface, many more defects, and poor contacts. Hence, smoothing the rough surface and achieving better contact between all layers was particularly important for improving the performance of fiber-shaped devices. Zeng et al. [105] built and constructed an inorganic-organic-graphene hybrid fiber-shaped PD via "soft" interfaces of all layers; the schematic diagram illustrating the fabrication procedure of ZnO nanorods/PVK/graphene hybrid fiber-shaped PDs is displayed in Figure 9a. In this hybrid fiber-shaped PD, the organic semiconductor completely covered the inorganic functional layer, forming "soft" contact and smoothing its rough surface. In addition, the ultra-soft graphene could tightly wrap around the surface, creating another "soft" interface, as shown in Figure 9b. These "soft" interfaces between the various functional layers even at curved interfaces have established the tight contact; thus, the contact resistance could be significantly reduced. The ZnO nanorods array with good crystal quality was uniformly grown on the Zn wire. As seen in Figure 9d,e, the rise time of the PD with structure of ZnO/PVK/graphene was about 280 ms with fast response speed. Meanwhile, by using hard metal Ag wire instead of soft graphene as surface electrode, the response speed slowed down and the rise time of PDs increased to 6 s, as shown in Figure 9f,g. Therefore, the tight contact and reasonable interfacial energy level alignment (shown in Figure 9c) were beneficial to the separation and transportation of photogenerated electrons and holes. The results showed that the performance of the fiber-shaped PDs could be significantly improved by effective interface optimization.

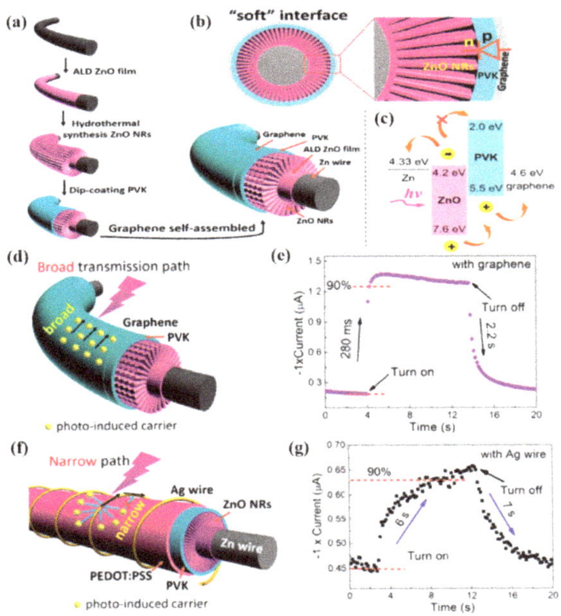

Figure 9. (**a**) Schematic diagram of the preparation process procedure of hybrid fiber-shaped photodetector based on ZnO NRs array/PVK/graphene. (**b**) The cross-sectional diagram of the device that revealed the p–n heterojunction and all tight soft interfaces in the PDs. (**c**) Schematic diagram of the corresponding energy levels of the material involved in the PD device, as well as the transportation of electrons and holes caused by the arrangement of interfacial energy levels. (**d**) Schematic representation of the transport property of the photoinduced carriers in the PD with structure of ZnO/PVK/graphene. Photoresponse characteristics of (**e**) corresponding current–time curves of ZnO/PVK/graphene PDs. (**f**) Schematic representation of the transport property of the photoinduced carriers in the PD with structure along Ag wire. (**g**) Corresponding current–time curves [105].

Furthermore, interface nanojunction engineering of the electron depletion effect could be utilized to improve the performance of PDs. For instance, Liu et al. [109] constructed nanojunction-interlinked ZnO nanoparticle networks without changing the dimension and morphology of the nanoparticles using an ultrafast thermal annealing (UTA) method. Figure 10 indicates band bending, electric field, and the low conductivity surface depletion layer present due to the adsorption of oxygen molecules on the ZnO nanoparticle surfaces. Contaminants (orange shell in schematic) and surface defects could reduce the electron depletion effect, resulting in high dark current and low photocurrent, as shown in Figure 10a,b. Surface modification using UTA can eliminate surface defects and contaminants and thus restore the electron depletion effect. Finally, this caused a decrease in the dark current and an increase in photocurrent, as shown in Figure 10c,d. In addition, ZnO NP–NP interface nanojunctions could be fused into interlinked ZnO-NP networks, which could significantly reduce the potential barrier of inter-NP and thus suppress the recombination of carriers in the transport process. From Figure 10e it is evident that the photoresponse of PDs increased dramatically by about four orders of magnitude with the decreased wavelength across the band edge after UTA treatment at 800 °C for 2 s. Figure 10f displays the relationship between applied bias and detectivity for a set of samples treated with UTA at different temperatures in the range of 500 to 900 °C. The detectivity value increased at first and then descended with the increasing UTA temperature and obtained the maximum value of 1.4×10^{13} Jones for the sample treated at 800 °C. Especially, D* after UTA treatment at 800 °C was enhanced by approximately three orders of magnitude compared with that of untreated, demonstrating the significance of removing surface defects and contaminants and improving the interconnections of NP. Therefore, it is very important to achieve high-performance detection making full use of the electron depletion effect with the minimum charge combination.

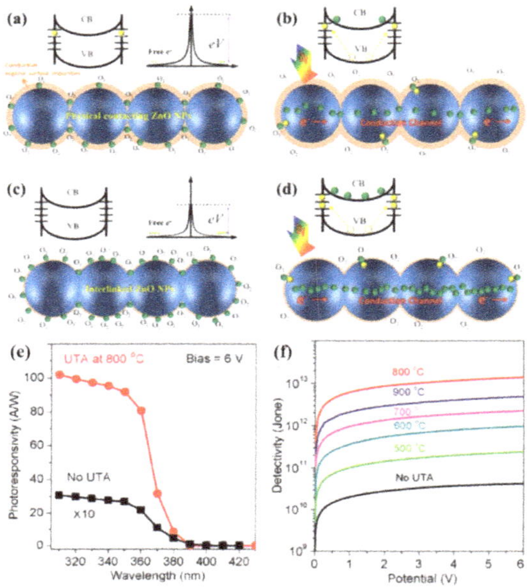

Figure 10. (**a**,**b**) The ZnO-NP before ultrafast thermal annealing (UTA). Schematic diagram of the energy band (top) and the photoresponse mechanism (bottom). (**c**,**d**) the nanojunction-interlinked ZnO-NP networks after the UTA treatment. The orange shell on ZnO-NPs represents the surface defects and contaminants. Green and yellow dots denote electrons and holes, respectively. (**e**) Spectral responsivity of the same PDs before and after UTA treatment at 800 °C for 2 s under the 340 nm UV illumination. (**f**) The relationship between applied bias and detectivity for a set of samples treated with UTA at different temperatures ranging from 500 to 900 °C [109].

3.4. Piezo-Phototronic Effects for Modulating Carrier Transport Behavior

In fact, piezoelectric phenomena exist in materials such as semiconductors, polymers, ceramics, and even biological media. Among them, the well-known piezoelectric materials are quartz and Pb(Zr, Ti)O_3, but the non-semiconductor and insulating properties limit their applications in photonic and electronic fields [110,111]. In theory, any semiconductor material with non-central symmetry could produce piezoelectric effect. Generally speaking, ZnO with wurtzite structure is the most well-known material in the study of the piezoelectric effect; it not only has rich nanostructural morphology but also possesses much higher piezoelectric coefficients than other II–VI family materials. The transmission characteristics of carriers can be dynamically adjusted by using the effect of piezoelectric potential, so that they have potential application prospects in mechanical electronic devices, including sensors, nanorobots, and so on. When considering the interaction with light, by adjusting the magnitude of the strain and the intensity of light, the transmission properties of carriers in the device can be effectively modulated. Light-emitting diodes [112,113], solar cells [114–116], and photodetectors [74,117–120] associated with the piezo-phototronic effect have attracted wide attention because of their potential application prospects. The band structure and carrier transport behavior can be dynamically modulated by taking advantage of the piezo-phototronic effect in the photodetectors.

The piezo-phototronic effect could be applied in different ways with designed structure [111,114]. Wang et al. [121] utilized the piezo-phototronic effect to adjust the height of the Schottky barrier at local contact; the performances of the Schottky contact metal-semiconductor-metal photodetector based on the GaN nanobelt was enhanced. Generally, the response ability of the PDs was obviously improved due to the piezotronic effect when a strain was applied on devices. In particular, the responsivity of PDs increased by 18% with a compressive strain, and their sensitivity increased from 22% to 31% [121]. The piezo-phototronic effect can not only improve the responsivity of the PDs but can also adjust the dark current of the detector. Wang et al. [122] constructed MSM ZnO micro-/nanowire PDs; the responsivity of the PDs was increased by 530% under −0.36% compressive strain, while the dark current of PDs was tuned with the applied piezo-phototronic effect.

Inspired by the enhanced performance caused by the piezo-phototronic effect, a hybrid PD based on heterojunction structure was constructed [72,73]. Zhu and colleagues [72] fabricated broad-band UV/visible/near-infrared self-powered PDs based on p-P3HT and n-CdS. When a 0.67% tensile strain was applied in the [001] direction of CdS contacting with P3HT, the photocurrent of the PDs could be increased to over 330% under UV illumination. This improvement by means of the piezo-phototronic effect could be ascribed to the increased built-in electric field in the interface of the p–n junction, which was more conducive to the separation of photogenerated carriers in CdS and P3HT. It has been reported that the piezoelectric effect that exists in MoS_2 with an odd number of atomic layers could tune the height of the Schottky barrier in metal-semiconductor contact and/or drive the nanometer device due to its non-central symmetric structure [123,124]. Wang et al. [125] fabricated a flexible photodetector based on a p-CuO film/n-MoS_2 monolayers heterojunction; the structure diagram and the optical image of PDs are shown in Figure 11a and the inset of Figure 11b, respectively. Figure 11b shows the I-V characteristics of p-CuO/n-MoS_2 heterojunction had good rectification characteristics in the dark without strain condition. Even at a forward bias voltage of 10 V, it had an ultra-low dark current of 0.039 nA, which was conducive to obtain high-performance optical sensing. Figure 11c exhibits the relationship between current and voltage of the heterojunction with different tensile strains in the dark. The dark current increased from 0.039 nA to 0.12 nA with the increase of tensile strain from 0% to 0.65% at 10 V bias, which resulted from the piezotronic effect of monolayer MoS_2 (the positive piezopotential caused by tensile strain can modulate the bandgap at the interface of the CuO and MoS_2 heterojunction). The relation between photocurrent and different tensile strains under different power density of 532 nm laser illumination is shown in Figure 11d; the photocurrent obviously increased with the increased strain due to the piezo-phototronic effect. The strain dependence of R/R_0 and detectivity of heterojunction PDs with different power densities at a bias of 10 V are displayed in Figure 11e,f, respectively. The photoresponsivity and detectivity of the PDs increased with increasing strain because of the piezo-phototronic effect.

A pure semiconductor absorbs photons with an energy larger than the band gap, which limits the wide spectral sensitivity in PD devices. In order to remove the restriction, the core/shell system is adopted for adequate light absorption. The core-shell architecture not only effectively utilizes the inherent material properties of individual components but also provides direct electrical conducting channels for the transport of photocarriers and results in reducing recombination losses; thus, it is a promising choice for improving the performance of photodetectors [118,119,126]. Wang et al. [118] investigated the influence of the piezo-phototronic effect on the photosensitivity of visible and UV PDs based on ZnO/CdS core-shell micro/nanowire, which is shown in Figure 12a. The Schottky barrier heights (SBHs) at the source and drain contacts could be adjusted by the strain induced piezopotential in the ZnO core, and the experimental results in Figure 12b demonstrated that SBHs at the source and drain contacts decreased with the increase of compressive strains. Moreover, the responsivity of ZnO/CdS nanowire-based PDs was improved dramatically by more than ten times with −0.31% compressive strain, as shown in Figure 12c. On this basis, the research group fabricated the branched ZnO/CdS double-shell nanowires array on carbon fiber (CF/ZnO-CdS wire) [117], as illustrated in Figure 12d,e. As shown in Figure 4f, the responsivity of this single CF/ZnO-CdS wire PD was enhanced about 40%–60% under compressive strain of −0.38% and decreased about 8%–20% under a 0.31% tensile strain. In addition, Zhou et al. [126] fabricated broad band PDs based on a type-II CdSe/ZnTe heterojunction core/shell nanowire array using a 100 nm thin Ag layer as top electrode, as illustrated in Figure 12g. When the core/shell nanowires array was bent under the action of external forces, the positive piezopotential on the stretched surface lowered the Schottky barrier height at the Ag-ZnTe junction (Figure 12h), which resulted in the increase of minority carrier diffusion and more carriers flowing through the interface. Combined with the type-II band alignment at the CdSe–ZnTe interfaces and the small lattice mismatch between the CdSe and ZnTe, the responsivity of this PD was enhanced by four orders of magnitude with an applied load of 0.25 kgf under 1.8 V bias, as shown in Figure 12i. Therefore, it is feasible to optimize the performance of PDs based on a multi-function NW array by the piezoelectric photoelectric effect.

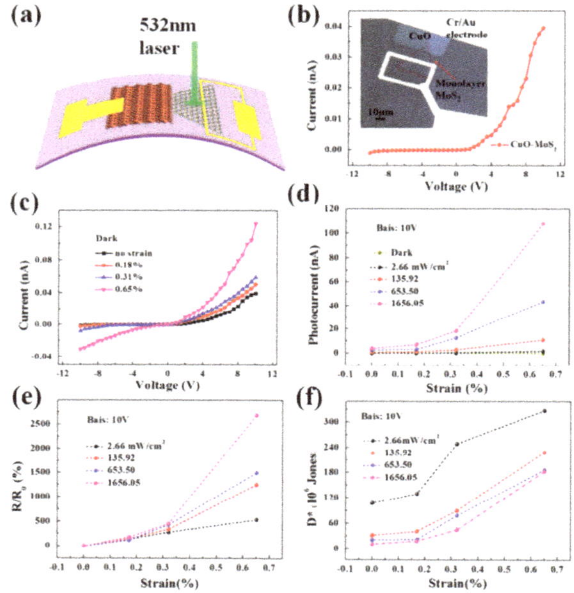

Figure 11. (**a**) The structure diagram of p-CuO/n-MoS_2 heterojunction PDs on the PET substrate. (**b**) I-V curve of p-CuO/n-MoS_2 heterojunction without strain in the dark. The inset is the optical image of the PD device. (**c**) I-V curves of the heterojunction with different tensile strains in the dark. (**d**) The strain dependence of the photocurrent. (**e**) The strain dependence of R/R_0. (**f**) The strain dependence of detectivity with different power densities at a bias of 10 V [125].

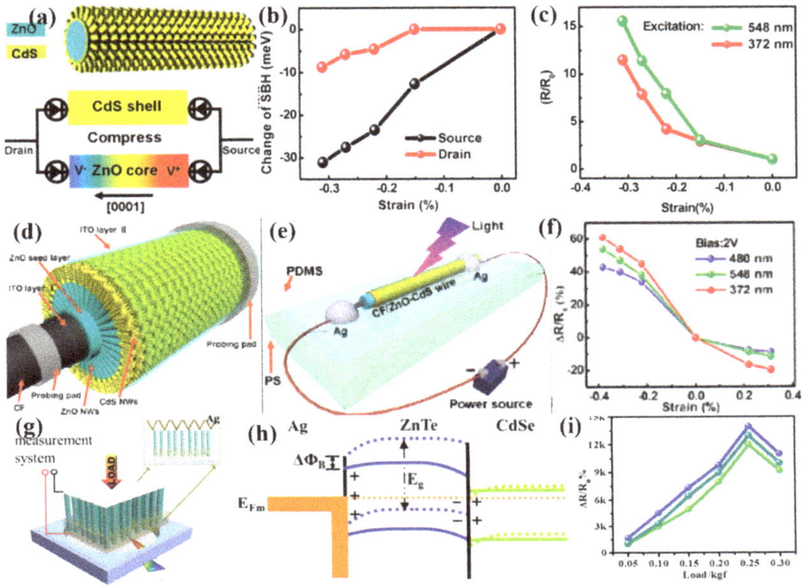

Figure 12. (**a**) Schematic diagram of ZnO/CdS core-shell structure (top), the sandwich model of the PD device, and simulation of the piezopotential distribution in the ZnO nanowire under compressive strain (bottom). (**b**) Relationship between compressive strain and changing Schottky barrier heights. Black curve and red curve represent the SBHs change with strains at source-drain bias of V = 2 and −2 V, respectively. (**c**) The derived change of responsivity in SBH as a function of compressive strains under illumination of green light (548 nm) and UV light (372 nm); R_0 represents the responsivity of the PD device without compressive strains. (**d**) Schematic representation of CF/ZnO-CdS wire structure. (**e**) Schematic diagram of the PDs based on a single CF/ZnO-CdS wire. (**f**) The strain dependence of $\Delta R/R_0$ excited by blue light (480 nm), green light (548 nm), and UV light (372 nm); R_0 is the responsivity without strain, and ΔR was set as the variation of responsivity. (**g**) Schematic diagram of CdSe/ZnTe core/shell nanowire array PDs on PVC with illumination source underneath; the inset is a cross-sectional schematic diagram of the assembled PDs. (**h**) Schematic energy band alignment of the Ag/ZnTe/CdSe structure under compressive load and illumination. (**i**) The strain dependence of $\Delta R/R_0$ under constant illumination [105,106,114].

4. Conclusions

In conclusion, through surface/interface engineering of nanostructured photodetector devices, advanced nanostructured PDs could be constructed by engineering through modulating photon absorption, charge carrier generation, transport, and collection behaviors. Based on the review, it could be well observed that surface/interface engineering of nanostructured PDs is crucially important for affecting PD performance. More importantly, as well illustrated in the paper, through surface or interface engineering such as surface-state passivation for terminating dangling bonds, surface plasmonic resonance for strong scattering and absorption of incident light, interface carrier-trapping/transport modulation, and piezo-phototronic effects for modulating carrier transport behavior in constructing PD devices, the constructed photodetectors' performance could be considerably improved as supported by reviewing published literature. Even though this brief review paper lists several key factors in the improvement of the constructed photodetector devices, it could be pointed out that several challenges still exist in the way of commercialization of nanostructured photodetectors as shown in the following:

1. Well controlled fabrication of nanostructured materials with uniform morphology and size for constructing designed photodetectors;
2. Well controlled growth of nanostructures with strong absorption of incident photons;
3. Well controlled surface/interface modification for enhancing carrier transportation;
4. Well applied material properties (such as piezoelectric effects) for modulating carrier behavior.

Finally, we believe that advanced nanostructured photodetectors with excellent performance could be designed and fabricated for practical applications through studying the mentioned issues in depth.

Author Contributions: The manuscript was written through contributions of all authors. All authors have read and agreed to the published version of the manuscript.

Funding: This research was funded by the Natural Science Foundation of China (Grant nos: 51675517, and 61874133), STS Projects of the Chinese Academy of Sciences (nos. KFJ-STS-SCYD-217, and KFJ-STS-ZDTP-061), Youth Innovation Promotion Association CAS (No.2019322, 2014280, No. 2018360), Instrument Developing Project of the Chinese Academy of Sciences (no. YJKYYQ20190057), Key Research and Development Program of Jiangsu Province (no. BE2018080), Tianjin Science and Technology Project (no. 19YFYSQY00040), and Jihua Laboratory (no. X190181TD190).

Conflicts of Interest: The authors declare that they have no competing interests.

References

1. Zhang, C.; Zhu, F.; Xu, H.; Liu, W.; Yang, L.; Wang, Z.; Ma, J.; Kang, Z.; Liu, Y. Significant improvement of near-UV electroluminescence from ZnO quantum dot LEDs via coupling with carbon nanodot surface plasmons. *Nanoscale* **2017**, *9*, 14592–14601. [CrossRef] [PubMed]
2. Oener, S.Z.; van de Groep, J.; Macco, B.; Bronsveld, P.C.P.; Kessels, W.M.M.; Polman, A.; Garnett, E.C. Metal–Insulator–Semiconductor Nanowire Network Solar Cells. *Nano Lett.* **2016**, *16*, 3689–3695. [CrossRef] [PubMed]
3. Gonzalez-Pedro, V.; Juarez-Perez, E.J.; Arsyad, W.S.; Barea, E.M.; Fabregat-Santiago, F.; Mora-Sero, I.; Bisquert, J. General Working Principles of $CH_3NH_3PbX_3$ Perovskite Solar Cells. *Nano Lett.* **2014**, *14*, 888–893. [CrossRef] [PubMed]
4. Katoch, A.; Abideen, Z.U.; Kim, H.W.; Kim, S.S. Grain-Size-Tuned Highly H_2-Selective Chemiresistive Sensors Based on ZnO–SnO_2 Composite Nanofibers. *ACS Appl. Mater. Interfaces* **2016**, *8*, 2486–2494. [CrossRef]
5. Huang, L.; Zhang, Z.; Li, Z.; Chen, B.; Ma, X.; Dong, L.; Peng, L.M. Multifunctional Graphene Sensors for Magnetic and Hydrogen Detection. *ACS Appl. Mater. Interfaces* **2015**, *7*, 9581–9588. [CrossRef]
6. Wu, P.; Ameen, T.; Zhang, H.; Bendersky, L.A.; Ilatikhameneh, H.; Klimeck, G.; Rahman, R.; Davydov, A.V.; Appenzeller, J. Complementary Black Phosphorus Tunneling Field-Effect Transistors. *ACS Nano* **2019**, *13*, 377–385. [CrossRef]
7. Pradhan, N.R.; Rhodes, D.; Feng, S.; Xin, Y.; Memaran, S.; Moon, B.H.; Terrones, H.; Terrones, M.; Balicas, L. Field-Effect Transistors Based on Few-Layered α-$MoTe_2$. *ACS Nano* **2014**, *8*, 5911–5920. [CrossRef] [PubMed]
8. Dou, L.; Yang, Y.; You, J.; Hong, Z.; Chang, W.H.; Li, G.; Yang, Y. Solution-processed hybrid perovskite photodetectors with high detectivity. *Nat. Commun.* **2014**, *5*, 5404. [CrossRef]
9. Liang, H.L.; Mei, Z.X.; Zhang, Q.H.; Gu, L.; Liang, S.; Hou, Y.N.; Ye, D.Q.; Gu, C.Z.; Yu, R.C.; Du, X.L. Interface engineering of high-Mg-content MgZnO/BeO/Si for p-n heterojunction solar-blind ultraviolet photodetectors. *Appl. Phys. Lett.* **2011**, *98*, 221902. [CrossRef]
10. Liu, C.H.; Chang, Y.C.; Norris, T.B.; Zhong, Z. Graphene photodetectors with ultra-broadband and high responsivity at room temperature. *Nat. Nanotechnol.* **2014**, *9*, 273. [CrossRef]
11. Xiong, D.; Deng, W.; Tian, G.; Gao, Y.; Chu, X.; Yan, C.; Jin, L.; Su, Y.; Yan, W.; Yang, W. A piezo-phototronic enhanced serrate-structured ZnO-based heterojunction photodetector for optical communication. *Nanoscale* **2019**, *11*, 3021–3027. [CrossRef] [PubMed]
12. Oh, S.; Kim, C.K.; Kim, J. High Responsivity β-Ga_2O_3 Metal–Semiconductor–Metal Solar-Blind Photodetectors with Ultraviolet Transparent Graphene Electrodes. *ACS Photonics* **2018**, *5*, 1123–1128. [CrossRef]

13. Lin, R.; Zheng, W.; Zhang, D.; Zhang, Z.; Liao, Q.; Yang, L.; Huang, F. High-Performance Graphene/β-Ga_2O_3 Heterojunction Deep-Ultraviolet Photodetector with Hot-Electron Excited Carrier Multiplication. *ACS Appl. Mater. Interfaces* **2018**, *10*, 22419–22426. [CrossRef] [PubMed]
14. Zhu, H.; Shan, C.X.; Wang, L.K.; Zheng, J.; Zhang, J.Y.; Yao, B.; Shen, D.Z. Metal–Oxide–Semiconductor-Structured MgZnO Ultraviolet Photodetector with High Internal Gain. *J. Phys. Chem. C* **2010**, *114*, 7169–7172. [CrossRef]
15. Shalev, E.; Oksenberg, E.; Rechav, K.; Popovitz-Biro, R.; Joselevich, E. Guided CdSe Nanowires Parallelly Integrated into Fast Visible-Range Photodetectors. *ACS Nano* **2017**, *11*, 213–220. [CrossRef]
16. Wei, Q.; Chen, J.; Ding, P.; Shen, B.; Yin, J.; Xu, F.; Xia, Y.; Liu, Z. Synthesis of Easily Transferred 2D Layered BiI3 Nanoplates for Flexible Visible-Light Photodetectors. *ACS Appl. Mater. Interfaces* **2018**, *10*, 21527–21533. [CrossRef]
17. Zhai, T.; Fang, X.; Liao, M.; Xu, X.; Li, L.; Liu, B.; Koide, Y.; Ma, Y.; Yao, J.; Bando, Y.; et al. Fabrication of High-Quality In_2Se_3 Nanowire Arrays toward High-Performance Visible-Light Photodetectors. *ACS Nano* **2010**, *4*, 1596–1602. [CrossRef]
18. Mallows, J.; Planells, M.; Thakare, V.; Bhosale, R.; Ogale, S.; Robertson, N. p-Type NiO Hybrid Visible Photodetector. *ACS Appl. Mater. Interfaces* **2015**, *7*, 27597–27601. [CrossRef]
19. Dias, S.; Kumawat, K.; Biswas, S.; Krupanidhi, S.B. Solvothermal Synthesis of Cu_2SnS_3 Quantum Dots and Their Application in Near-Infrared Photodetectors. *Inorg. Chem.* **2017**, *56*, 2198–2203. [CrossRef]
20. Yuan, S.; Shen, C.; Deng, B.; Chen, X.; Guo, Q.; Ma, Y.; Abbas, A.; Liu, B.; Haiges, R.; Ott, C.; et al. Air-Stable Room-Temperature Mid-Infrared Photodetectors Based on hBN/Black Arsenic Phosphorus/hBN Heterostructures. *Nano Lett.* **2018**, *18*, 3172–3179. [CrossRef]
21. Sarasqueta, G.; Choudhury, K.R.; So, F. Effect of Solvent Treatment on Solution-Processed Colloidal PbSe Nanocrystal Infrared Photodetectors. *Chem. Mater.* **2010**, *22*, 3496–3501. [CrossRef]
22. Ackerman, M.M.; Tang, X.; Guyot-Sionnest, P. Fast and Sensitive Colloidal Quantum Dot Mid-Wave Infrared Photodetectors. *ACS Nano* **2018**, *12*, 7264–7271. [CrossRef]
23. Zheng, D.; Fang, H.; Long, M.; Wu, F.; Wang, P.; Gong, F.; Wu, X.; Ho, J.C.; Liao, L.; Hu, W. High-Performance Near-Infrared Photodetectors Based on p-Type SnX (X = S, Se) Nanowires Grown via Chemical Vapor Deposition. *ACS Nano* **2018**, *12*, 7239–7245. [CrossRef]
24. Sun, Z.; Liu, Z.; Li, J.; Tai, G.; Lau, S.P.; Yan, F. Infrared Photodetectors Based on CVD-Grown Graphene and PbS Quantum Dots with Ultrahigh Responsivity. *Adv. Mater.* **2012**, *24*, 5878–5883. [CrossRef] [PubMed]
25. Ma, L.; Hu, W.; Zhang, Q.; Ren, P.; Zhuang, X.; Zhou, H.; Xu, J.; Li, H.; Shan, Z.; Wang, X.; et al. Room-Temperature Near-Infrared Photodetectors Based on Single Heterojunction Nanowires. *Nano Lett.* **2014**, *14*, 694–698. [CrossRef] [PubMed]
26. Chen, X.; Mu, W.; Xu, Y.; Fu, B.; Jia, Z.; Ren, F.F.; Gu, S.; Zhang, R.; Zheng, Y.; Tao, X.; et al. Highly Narrow-Band Polarization-Sensitive Solar-Blind Photodetectors Based on β-Ga_2O_3 Single Crystals. *ACS Appl. Mater. Interfaces* **2019**, *11*, 7131–7137. [CrossRef]
27. Berencén, Y.; Prucnal, S.; Liu, F.; Skorupa, I.; Hübner, R.; Rebohle, L.; Zhou, S.; Schneider, H.; Helm, M.; Skorupa, W. Room-temperature short-wavelength infrared Si photodetector. *Sci. Rep.* **2017**, *7*, 43688. [CrossRef]
28. Song, Y.M.; Xie, Y.; Malyarchuk, V.; Xiao, J.; Jung, I.; Choi, K.J.; Liu, Z.; Park, H.; Lu, C.; Kim, R.H.; et al. Digital cameras with designs inspired by the arthropod eye. *Nature* **2013**, *497*, 95–99. [CrossRef]
29. Kim, T.I.; McCall, J.G.; Jung, Y.H.; Huang, X.; Siuda, E.R.; Li, Y.; Song, J.; Song, Y.M.; Pao, H.A.; Kim, R.H.; et al. Injectable, Cellular-Scale Optoelectronics with Applications for Wireless Optogenetics. *Science* **2013**, *340*, 211. [CrossRef]
30. Kim, M.S.; Lee, G.J.; Kim, H.M.; Song, Y.M. Parametric Optimization of Lateral NIPIN Phototransistors for Flexible Image Sensors. *Sensors* **2017**, *17*, 1774. [CrossRef]
31. Li, J.; Li, R.; Chiang, C.H.; Zhong, Y.; Shen, H.; Song, E.; Hill, M.; Won, S.M.; Yu, K.J.; Baek, J.M.; et al. Ultrathin, High Capacitance Capping Layers for Silicon Electronics with Conductive Interconnects in Flexible, Long-Lived Bioimplants. *Adv. Mater. Technol.* **2020**, *5*, 1900800. [CrossRef]
32. Yu, K.J.; Gao, L.; Park, J.S.; Lee, Y.R.; Corcoran, C.J.; Nuzzo, R.G.; Chanda, D.; Rogers, J.A. Light Trapping in Ultrathin Monocrystalline Silicon Solar Cells. *Adv. Energy Mater* **2013**, *3*, 1401–1406. [CrossRef]
33. Nam, W.I.; Yoo, Y.J.; Song, Y.M. Geometrical shape design of nanophotonic surfaces for thin film solar cells. *Opt. Express* **2016**, *24*, A1033–A1044. [CrossRef] [PubMed]

34. Ouyang, W.; Teng, F.; He, J.H.; Fang, X. Enhancing the Photoelectric Performance of Photodetectors Based on Metal Oxide Semiconductors by Charge-Carrier Engineering. *Adv. Funct. Mater.* **2019**, *29*, 1807672. [CrossRef]
35. Mondal, S.; Dutta, K.; Dutta, S.; Jana, D.; Kelly, A.G.; De, S. Efficient Flexible White-Light Photodetectors Based on $BiFeO_3$ Nanoparticles. *ACS Appl. Nano Mater.* **2018**, *1*, 625–631. [CrossRef]
36. Jin, Y.; Wang, J.; Sun, B.; Blakesley, J.C.; Greenham, N.C. Solution-Processed Ultraviolet Photodetectors Based on Colloidal ZnO Nanoparticles. *Nano Lett.* **2008**, *8*, 1649–1653. [CrossRef]
37. Fang, H.; Hu, W.; Wang, P.; Guo, N.; Luo, W.; Zheng, D.; Gong, F.; Luo, M.; Tian, H.; Zhang, X.; et al. Visible Light-Assisted High-Performance Mid-Infrared Photodetectors Based on Single InAs Nanowire. *Nano Lett.* **2016**, *16*, 6416–6424. [CrossRef]
38. Ren, D.; Azizur-Rahman, K.M.; Rong, Z.; Juang, B.C.; Somasundaram, S.; Shahili, M.; Farrell, A.C.; Williams, B.S.; Huffaker, D.L. Room-Temperature Midwavelength Infrared InAsSb Nanowire Photodetector Arrays with Al_2O_3 Passivation. *Nano Lett.* **2019**, *19*, 2793–2802. [CrossRef]
39. Ahmadi, M.; Wu, T.; Hu, B. A Review on Organic–Inorganic Halide Perovskite Photodetectors: Device Engineering and Fundamental Physics. *Adv. Mater.* **2017**, *29*, 1605242. [CrossRef]
40. Barone, P.W.; Baik, S.; Heller, D.A.; Strano, M.S. Near-infrared optical sensors based on single-walled carbon nanotubes. *Nat. Mater.* **2005**, *4*, 86–92. [CrossRef]
41. Liu, H.; Yu, M.; Qin, F.; Feng, W.; Hu, P. Two-Dimensional Nonlayered $CuInSe_2$ Nanosheets for High-Performance Photodetectors. *ACS Appl. Nano Mater.* **2018**, *1*, 5414–5418. [CrossRef]
42. Lin, S.Y.; Haider, G.; Liao, Y.M.; Chang, C.H.; Lin, W.J.; Su, C.Y.; Liou, Y.R.; Huang, Y.F.; Lin, H.I.; Chung, T.C.; et al. Transient and Flexible Photodetectors. *ACS Appl. Nano Mater.* **2018**, *1*, 5092–5100. [CrossRef]
43. Deng, Z.; Cao, D.; He, J.; Lin, S.; Lindsay, S.M.; Liu, Y. Solution Synthesis of Ultrathin Single-Crystalline SnS Nanoribbons for Photodetectors via Phase Transition and Surface Processing. *ACS Nano* **2012**, *6*, 6197–6207. [CrossRef] [PubMed]
44. Hu, L.; Yan, J.; Liao, M.; Xiang, H.; Gong, X.; Zhang, L.; Fang, X. An Optimized Ultraviolet-A Light Photodetector with Wide-Range Photoresponse Based on ZnS/ZnO Biaxial Nanobelt. *Adv. Mater.* **2012**, *24*, 2305–2309. [CrossRef]
45. Soci, C.; Zhang, A.; Xiang, B.; Dayeh, S.A.; Aplin, D.P.R.; Park, J.; Bao, X.Y.; Lo, Y.H.; Wang, D. ZnO Nanowire UV Photodetectors with High Internal Gain. *Nano Lett.* **2007**, *7*, 1003–1009. [CrossRef]
46. Guo, Z.; Zhou, L.; Tang, Y.; Li, L.; Zhang, Z.; Yang, H.; Ma, H.; Nathan, A.; Zhao, D. Surface/Interface Carrier-Transport Modulation for Constructing Photon-Alternative Ultraviolet Detectors Based on Self-Bending-Assembled ZnO Nanowires. *ACS Appl. Mater. Interfaces* **2017**, *9*, 31042–31053. [CrossRef]
47. Cao, L.; Liu, X.; Guo, Z.; Zhou, L. Surface/Interface Engineering for Constructing Advanced Nanostructured Light-Emitting Diodes with Improved Performance: A Brief Review. *Micromachines* **2019**, *10*, 821. [CrossRef]
48. Many, A. Relation between physical and chemical processes on semiconductor surfaces. *CRC Crit. Rev. Solid State Sci.* **1973**, *4*, 515–539. [CrossRef]
49. Nozik, A.J.; Memming, R. Physical Chemistry of Semiconductor−Liquid Interfaces. *J. Phys. Chem.* **1996**, *100*, 13061–13078. [CrossRef]
50. Han, T.H.; Tan, S.; Xue, J.; Meng, L.; Lee, J.W.; Yang, Y. Interface and Defect Engineering for Metal Halide Perovskite Optoelectronic Devices. *Adv. Mater.* **2019**, *31*, 1803515. [CrossRef]
51. Hu, Z.; Wu, Z.; Han, C.; He, J.; Ni, Z.; Chen, W. Two-dimensional transition metal dichalcogenides: Interface and defect engineering. *Chem. Soc. Rev.* **2018**, *47*, 3100–3128. [CrossRef]
52. Proskuryakov, Y.Y.; Durose, K.; Major, J.D.; Al Turkestani, M.K.; Barrioz, V.; Irvine, S.J.C.; Jones, E.W. Doping levels, trap density of states and the performance of co-doped CdTe(As,Cl) photovoltaic devices. *Solar Energy Mater. Solar Cells* **2009**, *93*, 1572–1581. [CrossRef]
53. Perlin, P.; Iota, V.; Weinstein, B.A.; Wiśniewski, P.; Suski, T.; Eliseev, P.G.; Osiński, M. Influence of pressure on photoluminescence and electroluminescence in GaN/InGaN/AlGaN quantum wells. *Appl. Phys. Lett.* **1997**, *70*, 2993–2995. [CrossRef]
54. Zhang, D.; Li, C.; Han, S.; Liu, X.; Tang, T.; Jin, W.; Zhou, C. Ultraviolet photodetection properties of indium oxide nanowires. *Appl. Phys. A* **2003**, *77*, 163–166. [CrossRef]
55. Hu, L.; Yan, J.; Liao, M.; Wu, L.; Fang, X. Ultrahigh External Quantum Efficiency from Thin SnO_2 Nanowire Ultraviolet Photodetectors. *Small* **2011**, *7*, 1012–1017. [CrossRef] [PubMed]

56. Chen, Q.; Ding, H.; Wu, Y.; Sui, M.; Lu, W.; Wang, B.; Su, W.; Cui, Z.; Chen, L. Passivation of surface states in the ZnO nanowire with thermally evaporated copper phthalocyanine for hybrid photodetectors. *Nanoscale* **2013**, *5*, 4162–4165. [CrossRef]
57. Cammi, D.; Ronning, C. Persistent Photoconductivity in ZnO Nanowires in Different Atmospheres. *Adv. Condens. Matter Phys.* **2014**, *2014*, 1–5. [CrossRef]
58. Pezeshki, A.; Hosseini Shokouh, S.H.; Raza, S.R.A.; Kim, J.S.; Min, S.W.; Shackery, I.; Jun, S.C.; Im, S. Top and back gate molybdenum disulfide transistors coupled for logic and photo-inverter operation. *J. Mater. Chem. C* **2014**, *2*, 8023–8028. [CrossRef]
59. Hosseini Shokouh, S.H.; Pezeshki, A.; Raza, S.R.A.; Choi, K.; Min, S.W.; Jeon, P.J.; Lee, H.S.; Im, S. Molybdenum Disulfide Nanoflake–Zinc Oxide Nanowire Hybrid Photoinverter. *ACS Nano* **2014**, *8*, 5174–5181. [CrossRef]
60. Hassan, M.S.; Bera, S.; Gupta, D.; Ray, S.K.; Sapra, S. $MoSe_2$–Cu_2S Vertical p–n Nanoheterostructures for High-Performance Photodetectors. *ACS Appl. Mater. Interfaces* **2019**, *11*, 4074–4083. [CrossRef]
61. Xiang, D.; Han, C.; Hu, Z.; Lei, B.; Liu, Y.; Wang, L.; Hu, W.P.; Chen, W. Surface Transfer Doping-Induced, High-Performance Graphene/Silicon Schottky Junction-Based, Self-Powered Photodetector. *Small* **2015**, *11*, 4829–4836. [CrossRef] [PubMed]
62. Zhou, J.; Gu, Y.; Hu, Y.; Mai, W.; Yeh, P.H.; Bao, G.; Sood, A.K.; Polla, D.L.; Wang, Z.L. Gigantic enhancement in response and reset time of ZnO UV nanosensor by utilizing Schottky contact and surface functionalization. *Appl. Phys. Lett.* **2009**, *94*, 191103. [CrossRef] [PubMed]
63. Shen, T.; Li, B.; Zheng, K.; Pullerits, T.; Cao, G.; Tian, J. Surface Engineering of Quantum Dots for Remarkably High Detectivity Photodetectors. *J. Phys. Chem. Lett.* **2018**, *9*, 3285–3294. [CrossRef] [PubMed]
64. Fang, F.; Zhao, D.X.; Li, B.H.; Zhang, Z.Z.; Zhang, J.Y.; Shen, D.Z. The enhancement of ZnO nanowalls photoconductivity induced by CdS nanoparticle modification. *Appl. Phys. Lett.* **2008**, *93*, 233115. [CrossRef]
65. Garg, M.; Tak, B.R.; Rao, V.R.; Singh, R. Giant UV Photoresponse of GaN-Based Photodetectors by Surface Modification Using Phenol-Functionalized Porphyrin Organic Molecules. *ACS Appl. Mater. Interfaces* **2019**, *11*, 12017–12026. [CrossRef] [PubMed]
66. Ni, S.; Guo, F.; Wang, D.; Liu, G.; Xu, Z.; Kong, L.; Wang, J.; Jiao, S.; Zhang, Y.; Yu, Q.; et al. Effect of MgO Surface Modification on the TiO_2 Nanowires Electrode for Self-Powered UV Photodetectors. *ACS Sustain. Chem. Eng.* **2018**, *6*, 7265–7272. [CrossRef]
67. Lin, J.; Zhong, J.; Zhong, S.; Li, H.; Zhang, H.; Chen, W. Modulating electronic transport properties of MoS_2 field effect transistor by surface overlayers. *Appl. Phys. Lett.* **2013**, *103*, 063109. [CrossRef]
68. Ai, N.; Zhou, Y.; Zheng, Y.; Chen, H.; Wang, J.; Pei, J.; Cao, Y. Achieving high sensitivity in single organic submicrometer ribbon based photodetector through surface engineering. *Org. Electron.* **2013**, *14*, 1103–1108. [CrossRef]
69. Shen, Y.; Yan, X.; Si, H.; Lin, P.; Liu, Y.; Sun, Y.; Zhang, Y. Improved Photoresponse Performance of Self-Powered ZnO/Spiro-MeOTAD Heterojunction Ultraviolet Photodetector by Piezo-Phototronic Effect. *ACS Appl. Mater. Interfaces* **2016**, *8*, 6137–6143. [CrossRef]
70. Lu, S.; Qi, J.; Liu, S.; Zhang, Z.; Wang, Z.; Lin, P.; Liao, Q.; Liang, Q.; Zhang, Y. Piezotronic Interface Engineering on ZnO/Au-Based Schottky Junction for Enhanced Photoresponse of a Flexible Self-Powered UV Detector. *ACS Appl. Mater. Interfaces* **2014**, *6*, 14116–14122. [CrossRef]
71. Zhang, Z.; Liao, Q.; Yu, Y.; Wang, X.; Zhang, Y. Enhanced photoresponse of ZnO nanorods-based self-powered photodetector by piezotronic interface engineering. *Nano Energy* **2014**, *9*, 237–244. [CrossRef]
72. Yu, X.X.; Yin, H.; Li, H.X.; Zhang, W.; Zhao, H.; Li, C.; Zhu, M.Q. Piezo-phototronic effect modulated self-powered UV/visible/near-infrared photodetectors based on CdS:P3HT microwires. *Nano Energy* **2017**, *34*, 155–163. [CrossRef]
73. Peng, W.; Wang, X.; Yu, R.; Dai, Y.; Zou, H.; Wang, A.C.; He, Y.; Wang, Z.L. Enhanced Performance of a Self-Powered Organic/Inorganic Photodetector by Pyro-Phototronic and Piezo-Phototronic Effects. *Adv. Mater.* **2017**, *29*, 1606698. [CrossRef] [PubMed]
74. Liu, X.; Yang, X.; Gao, G.; Yang, Z.; Liu, H.; Li, Q.; Lou, Z.; Shen, G.; Liao, L.; Pan, C.; et al. Enhancing Photoresponsivity of Self-Aligned MoS_2 Field-Effect Transistors by Piezo-Phototronic Effect from GaN Nanowires. *ACS Nano* **2016**, *10*, 7451–7457. [CrossRef]
75. Lopez-Sanchez, O.; Lembke, D.; Kayci, M.; Radenovic, A.; Kis, A. Ultrasensitive photodetectors based on monolayer MoS_2. *Nat. Nanotechnol.* **2013**, *8*, 497–501. [CrossRef]

76. Chang, Y.H.; Zhang, W.; Zhu, Y.; Han, Y.; Pu, J.; Chang, J.K.; Hsu, W.T.; Huang, J.K.; Hsu, C.L.; Chiu, M.H.; et al. Monolayer $MoSe_2$ Grown by Chemical Vapor Deposition for Fast Photodetection. *ACS Nano* **2014**, *8*, 8582–8590. [CrossRef]
77. Zhang, W.; Chiu, M.H.; Chen, C.H.; Chen, W.; Li, L.J.; Wee, A.T.S. Role of Metal Contacts in High-Performance Phototransistors Based on WSe_2 Monolayers. *ACS Nano* **2014**, *8*, 8653–8661. [CrossRef]
78. Tan, H.; Fan, Y.; Zhou, Y.; Chen, Q.; Xu, W.; Warner, J.H. Ultrathin 2D Photodetectors Utilizing Chemical Vapor Deposition Grown WS_2 With Graphene Electrodes. *ACS Nano* **2016**, *10*, 7866–7873. [CrossRef]
79. Thakar, K.; Mukherjee, B.; Grover, S.; Kaushik, N.; Deshmukh, M.; Lodha, S. Multilayer ReS_2 Photodetectors with Gate Tunability for High Responsivity and High-Speed Applications. *ACS Appl. Mater. Interfaces* **2018**, *10*, 36512–36522. [CrossRef]
80. Tongay, S.; Zhou, J.; Ataca, C.; Liu, J.; Kang, J.S.; Matthews, T.S.; You, L.; Li, J.; Grossman, J.C.; Wu, J. Broad-Range Modulation of Light Emission in Two-Dimensional Semiconductors by Molecular Physisorption Gating. *Nano Lett.* **2013**, *13*, 2831–2836. [CrossRef]
81. Baugher, B.W.H.; Churchill, H.O.H.; Yang, Y.; Jarillo-Herrero, P. Intrinsic Electronic Transport Properties of High-Quality Monolayer and Bilayer MoS_2. *Nano Lett.* **2013**, *13*, 4212–4216. [CrossRef] [PubMed]
82. Kufer, D.; Konstantatos, G. Highly Sensitive, Encapsulated MoS_2 Photodetector with Gate Controllable Gain and Speed. *Nano Lett.* **2015**, *15*, 7307–7313. [CrossRef] [PubMed]
83. Meyer, J.; Khalandovsky, R.; Görrn, P.; Kahn, A. MoO_3 Films Spin-Coated from a Nanoparticle Suspension for Efficient Hole-Injection in Organic Electronics. *Adv. Mater.* **2011**, *23*, 70–73. [CrossRef] [PubMed]
84. Pak, S.; Jang, A.R.; Lee, J.; Hong, J.; Giraud, P.; Lee, S.; Cho, Y.; An, G.H.; Lee, Y.W.; Shin, H.S.; et al. Surface functionalization-induced photoresponse characteristics of monolayer MoS_2 for fast flexible photodetectors. *Nanoscale* **2019**, *11*, 4726–4734. [CrossRef]
85. Sun, Z.; Aigouy, L.; Chen, Z. Plasmonic-enhanced perovskite–graphene hybrid photodetectors. *Nanoscale* **2016**, *8*, 7377–7383. [CrossRef]
86. Chen, Z.; Li, X.; Wang, J.; Tao, L.; Long, M.; Liang, S.J.; Ang, L.K.; Shu, C.; Tsang, H.K.; Xu, J.B. Synergistic Effects of Plasmonics and Electron Trapping in Graphene Short-Wave Infrared Photodetectors with Ultrahigh Responsivity. *ACS Nano* **2017**, *11*, 430–437. [CrossRef]
87. Luo, L.B.; Huang, X.L.; Wang, M.Z.; Xie, C.; Wu, C.Y.; Hu, J.G.; Wang, L.; Huang, J.A. The Effect of Plasmonic Nanoparticles on the Optoelectronic Characteristics of CdTe Nanowires. *Small* **2014**, *10*, 2645–2652. [CrossRef]
88. Li, Y.; DiStefano, J.G.; Murthy, A.A.; Cain, J.D.; Hanson, E.D.; Li, Q.; Castro, F.C.; Chen, X.; Dravid, V.P. Superior Plasmonic Photodetectors Based on Au@MoS_2 Core–Shell Heterostructures. *ACS Nano* **2017**, *11*, 10321–10329. [CrossRef]
89. Miao, J.; Hu, W.; Jing, Y.; Luo, W.; Liao, L.; Pan, A.; Wu, S.; Cheng, J.; Chen, X.; Lu, W. Surface Plasmon-Enhanced Photodetection in Few Layer MoS_2 Phototransistors with Au Nanostructure Arrays. *Small* **2015**, *11*, 2392–2398. [CrossRef]
90. Lin, D.; Wu, H.; Zhang, W.; Li, H.; Pan, W. Enhanced UV photoresponse from heterostructured Ag–ZnO nanowires. *Appl. Phys. Lett.* **2009**, *94*, 172103. [CrossRef]
91. Arquer, F.P.G.d.; Beck, F.J.; Bernechea, M.; Konstantatos, G. Plasmonic light trapping leads to responsivity increase in colloidal quantum dot photodetectors. *Appl. Phys. Lett.* **2012**, *100*, 043101. [CrossRef]
92. Wang, X.; Liu, K.; Chen, X.; Li, B.; Jiang, M.; Zhang, Z.; Zhao, H.; Shen, D. Highly Wavelength-Selective Enhancement of Responsivity in Ag Nanoparticle-Modified ZnO UV Photodetector. *ACS Appl. Mater. Interfaces* **2017**, *9*, 5574–5579. [CrossRef] [PubMed]
93. Hu, K.; Chen, H.; Jiang, M.; Teng, F.; Zheng, L.; Fang, X. Broadband Photoresponse Enhancement of a High-Performance t-Se Microtube Photodetector by Plasmonic Metallic Nanoparticles. *Adv. Funct. Mater.* **2016**, *26*, 6641–6648. [CrossRef]
94. Kumar, R.; Sharma, A.; Kaur, M.; Husale, S. Pt-Nanostrip-Enabled Plasmonically Enhanced Broad Spectral Photodetection in Bilayer MoS_2. *Adv. Opt. Mater.* **2017**, *5*, 1700009. [CrossRef]
95. Zhang, X.; Liu, Q.; Liu, B.; Yang, W.; Li, J.; Niu, P.; Jiang, X. Giant UV photoresponse of a GaN nanowire photodetector through effective Pt nanoparticle coupling. *J. Mater. Chem. C* **2017**, *5*, 4319–4326. [CrossRef]
96. Luo, L.B.; Zou, Y.F.; Ge, C.W.; Zheng, K.; Wang, D.D.; Lu, R.; Zhang, T.F.; Yu, Y.Q.; Guo, Z.Y. A Surface Plasmon Enhanced Near-Infrared Nanophotodetector. *Adv. Opt. Mater.* **2016**, *4*, 763–771. [CrossRef]
97. Lu, J.; Xu, C.; Dai, J.; Li, J.; Wang, Y.; Lin, Y.; Li, P. Improved UV photoresponse of ZnO nanorod arrays by resonant coupling with surface plasmons of Al nanoparticles. *Nanoscale* **2015**, *7*, 3396–3403. [CrossRef]

98. Zhang, W.; Xu, J.; Ye, W.; Li, Y.; Qi, Z.; Dai, J.; Wu, Z.; Chen, C.; Yin, J.; Li, J.; et al. High-performance AlGaN metal–semiconductor–metal solar-blind ultraviolet photodetectors by localized surface plasmon enhancement. *Appl. Phys. Lett.* **2015**, *106*, 021112. [CrossRef]
99. Knight, M.W.; Liu, L.; Wang, Y.; Brown, L.; Mukherjee, S.; King, N.S.; Everitt, H.O.; Nordlander, P.; Halas, N.J. Aluminum Plasmonic Nanoantennas. *Nano Lett.* **2012**, *12*, 6000–6004. [CrossRef]
100. West, P.R.; Ishii, S.; Naik, G.V.; Emani, N.K.; Shalaev, V.M.; Boltasseva, A. Searching for better plasmonic materials. *Laser Photonics Rev.* **2010**, *4*, 795–808. [CrossRef]
101. Kramer, N.J.; Schramke, K.S.; Kortshagen, U.R. Plasmonic Properties of Silicon Nanocrystals Doped with Boron and Phosphorus. *Nano Lett.* **2015**, *15*, 5597–5603. [CrossRef]
102. Zhou, S.; Ni, Z.; Ding, Y.; Sugaya, M.; Pi, X.; Nozaki, T. Ligand-Free, Colloidal, and Plasmonic Silicon Nanocrystals Heavily Doped with Boron. *ACS Photonics* **2016**, *3*, 415–422. [CrossRef]
103. Ni, Z.; Ma, L.; Du, S.; Xu, Y.; Yuan, M.; Fang, H.; Wang, Z.; Xu, M.; Li, D.; Yang, J.; et al. Plasmonic Silicon Quantum Dots Enabled High-Sensitivity Ultrabroadband Photodetection of Graphene-Based Hybrid Phototransistors. *ACS Nano* **2017**, *11*, 9854–9862. [CrossRef] [PubMed]
104. Bai, S.; Jiang, W.; Li, Z.; Xiong, Y. Surface and Interface Engineering in Photocatalysis. *ChemNanoMat* **2015**, *1*, 223–239. [CrossRef]
105. Zhu, Z.; Ju, D.; Zou, Y.; Dong, Y.; Luo, L.; Zhang, T.; Shan, D.; Zeng, H. Boosting Fiber-Shaped Photodetectors via "Soft" Interfaces. *ACS Appl. Mater. Interfaces* **2017**, *9*, 12092–12099. [CrossRef]
106. Guo, F.; Yang, B.; Yuan, Y.; Xiao, Z.; Dong, Q.; Bi, Y.; Huang, J. A nanocomposite ultraviolet photodetector based on interfacial trap-controlled charge injection. *Nat. Nanotechnol.* **2012**, *7*, 798. [CrossRef]
107. Ji, T.; Liu, Q.; Zou, R.; Sun, Y.; Xu, K.; Sang, L.; Liao, M.; Koide, Y.; Yu, L.; Hu, J. An Interface Engineered Multicolor Photodetector Based on n-Si (111)/TiO_2 Nanorod Array Heterojunction. *Adv. Funct. Mater.* **2016**, *26*, 1400–1410. [CrossRef]
108. Song, X.; Liu, X.; Yu, D.; Huo, C.; Ji, J.; Li, X.; Zhang, S.; Zou, Y.; Zhu, G.; Wang, Y.; et al. Boosting Two-Dimensional MoS_2/$CsPbBr_3$ Photodetectors via Enhanced Light Absorbance and Interfacial Carrier Separation. *ACS Appl. Mater. Interfaces* **2018**, *10*, 2801–2809. [CrossRef]
109. Liu, Q.; Gong, M.; Cook, B.; Ewing, D.; Casper, M.; Stramel, A.; Wu, J. Fused Nanojunctions of Electron-Depleted ZnO Nanoparticles for Extraordinary Performance in Ultraviolet Detection. *Adv. Mater. Interfaces* **2017**, *4*, 1601064. [CrossRef]
110. Wang, Z.L.; Wu, W. Piezotronics and piezo-phototronics: Fundamentals and applications. *Natl. Sci. Rev.* **2013**, *1*, 62–90. [CrossRef]
111. Guo, Z.; Pan, H.; Li, C.; Zhang, L.; Yan, S.; Zhang, W.; Yao, J.; Tang, Y.; Yang, H.; Wu, Y.; et al. Dynamic carrier transport modulation for constructing advanced devices with improved performance by piezotronic and piezo-phototronic effects: A brief review. *Semicond. Sci. Technol.* **2017**, *32*, 083001. [CrossRef]
112. Yang, Q.; Liu, Y.; Pan, C.; Chen, J.; Wen, X.; Wang, Z.L. Largely Enhanced Efficiency in ZnO Nanowire/p-Polymer Hybridized Inorganic/Organic Ultraviolet Light-Emitting Diode by Piezo-Phototronic Effect. *Nano Lett.* **2013**, *13*, 607–613. [CrossRef]
113. Wang, C.; Bao, R.; Zhao, K.; Zhang, T.; Dong, L.; Pan, C. Enhanced emission intensity of vertical aligned flexible ZnO nanowire/p-polymer hybridized LED array by piezo-phototronic effect. *Nano Energy* **2015**, *14*, 364–371. [CrossRef]
114. Pan, C.; Niu, S.; Ding, Y.; Dong, L.; Yu, R.; Liu, Y.; Zhu, G.; Wang, Z.L. Enhanced Cu_2S/CdS Coaxial Nanowire Solar Cells by Piezo-Phototronic Effect. *Nano Lett.* **2012**, *12*, 3302–3307. [CrossRef] [PubMed]
115. Hu, G.; Guo, W.; Yu, R.; Yang, X.; Zhou, R.; Pan, C.; Wang, Z.L. Enhanced performances of flexible ZnO/perovskite solar cells by piezo-phototronic effect. *Nano Energy* **2016**, *23*, 27–33. [CrossRef]
116. Zhu, L.; Wang, L.; Xue, F.; Chen, L.; Fu, J.; Feng, X.; Li, T.; Wang, Z.L. Piezo-Phototronic Effect Enhanced Flexible Solar Cells Based on n-ZnO/p-SnS Core–Shell Nanowire Array. *Adv. Sci.* **2017**, *4*, 1600185. [CrossRef]
117. Zhang, F.; Niu, S.; Guo, W.; Zhu, G.; Liu, Y.; Zhang, X.; Wang, Z.L. Piezo-phototronic Effect Enhanced Visible/UV Photodetector of a Carbon-Fiber/ZnO-CdS Double-Shell Microwire. *ACS Nano* **2013**, *7*, 4537–4544. [CrossRef]
118. Zhang, F.; Ding, Y.; Zhang, Y.; Zhang, X.; Wang, Z.L. Piezo-phototronic Effect Enhanced Visible and Ultraviolet Photodetection Using a ZnO–CdS Core–Shell Micro/nanowire. *ACS Nano* **2012**, *6*, 9229–9236. [CrossRef]

119. Rai, S.C.; Wang, K.; Ding, Y.; Marmon, J.K.; Bhatt, M.; Zhang, Y.; Zhou, W.; Wang, Z.L. Piezo-phototronic Effect Enhanced UV/Visible Photodetector Based on Fully Wide Band Gap Type-II ZnO/ZnS Core/Shell Nanowire Array. *ACS Nano* **2015**, *9*, 6419–6427. [CrossRef]
120. Wang, Z.; Yu, R.; Wen, X.; Liu, Y.; Pan, C.; Wu, W.; Wang, Z.L. Optimizing Performance of Silicon-Based p–n Junction Photodetectors by the Piezo-Phototronic Effect. *ACS Nano* **2014**, *8*, 12866–12873. [CrossRef]
121. Yu, R.; Pan, C.; Hu, Y.; Li, L.; Liu, H.; Liu, W.; Chua, S.; Chi, D.; Wang, Z.L. Enhanced performance of GaN nanobelt-based photodetectors by means of piezotronic effects. *Nano Res.* **2013**, *6*, 758–766. [CrossRef]
122. Yang, Q.; Guo, X.; Wang, W.; Zhang, Y.; Xu, S.; Lien, D.H.; Wang, Z.L. Enhancing Sensitivity of a Single ZnO Micro-/Nanowire Photodetector by Piezo-phototronic Effect. *ACS Nano* **2010**, *4*, 6285–6291. [CrossRef] [PubMed]
123. Wu, W.; Wang, L.; Li, Y.; Zhang, F.; Lin, L.; Niu, S.; Chenet, D.; Zhang, X.; Hao, Y.; Heinz, T.F.; et al. Piezoelectricity of single-atomic-layer MoS_2 for energy conversion and piezotronics. *Nature* **2014**, *514*, 470–474. [CrossRef] [PubMed]
124. Zhu, H.; Wang, Y.; Xiao, J.; Liu, M.; Xiong, S.; Wong, Z.J.; Ye, Z.; Ye, Y.; Yin, X.; Zhang, X. Observation of piezoelectricity in free-standing monolayer MoS_2. *Nat. Nanotechnol.* **2015**, *10*, 151–155. [CrossRef]
125. Zhang, K.; Peng, M.; Wu, W.; Guo, J.; Gao, G.; Liu, Y.; Kou, J.; Wen, R.; Lei, Y.; Yu, A.; et al. A flexible p-CuO/n-MoS_2 heterojunction photodetector with enhanced photoresponse by the piezo-phototronic effect. *Mater. Horiz.* **2017**, *4*, 274–280. [CrossRef]
126. Rai, S.C.; Wang, K.; Chen, J.; Marmon, J.K.; Bhatt, M.; Wozny, S.; Zhang, Y.; Zhou, W. Enhanced Broad Band Photodetection through Piezo-Phototronic Effect in CdSe/ZnTe Core/Shell Nanowire Array. *Adv. Electron. Mater.* **2015**, *1*, 1400050. [CrossRef]

nanomaterials

MDPI

Review

Inorganic Boron-Based Nanostructures: Synthesis, Optoelectronic Properties, and Prospective Applications

Yan Tian, Zekun Guo, Tong Zhang, Haojian Lin, Zijuan Li, Jun Chen, Shaozhi Deng and Fei Liu *

State Key Laboratory of Optoelectronic Materials and Technologies, Guangdong Province Key Laboratory of Display Material and Technology, School of Electronics and Information Technology, Sun Yat-sen University, Guangzhou 510275, China; tiany29@mail2.sysu.edu.cn (Y.T.); guozk@mail2.sysu.edu.cn (Z.G.); zt423109972@163.com (T.Z.); linhjian@foxmail.com (H.L.); lizj35@mail2.sysu.edu.cn (Z.L.); stscjun@mail.sysu.edu.cn (J.C.); stsdsz@mail.sysu.edu.cn (S.D.)

* Correspondence: liufei@mail.sysu.edu.cn

Received: 15 February 2019; Accepted: 21 March 2019; Published: 3 April 2019

Abstract: Inorganic boron-based nanostructures have great potential for field emission (FE), flexible displays, superconductors, and energy storage because of their high melting point, low density, extreme hardness, and good chemical stability. Until now, most researchers have been focused on one-dimensional (1D) boron-based nanostructures (rare-earth boride (REB_6) nanowires, boron nanowires, and nanotubes). Currently, two-dimensional (2D) borophene attracts most of the attention, due to its unique physical and chemical properties, which make it quite different from its corresponding bulk counterpart. Here, we offer a comprehensive review on the synthesis methods and optoelectronics properties of inorganic boron-based nanostructures, which are mainly concentrated on 1D rare-earth boride nanowires, boron monoelement nanowires, and nanotubes, as well as 2D borophene and borophane. This review paper is organized as follows. In Section I, the synthesis methods of inorganic boron-based nanostructures are systematically introduced. In Section II, we classify their optical and electrical transport properties (field emission, optical absorption, and photoconductive properties). In the last section, we evaluate the optoelectronic behaviors of the known inorganic boron-based nanostructures and propose their future applications.

Keywords: inorganic boron-based nanostructures; boron monoelement nanowire and nanotube; borophene; rare-earth boride (REB_6); optoelectronic properties

---

1. Introduction

Boron is very special, because it is the only nonmetallic element in group III, the lightest nonmetallic element in the periodic table, and excellent properties similar to carbon. Moreover, boron possesses unique physical and chemical properties due to the B_{12} icosahedra structural unit consisting of a particular three-center two-electron bond, such as high melting point, low density, extreme hardness, and nice chemical stability [1–4]. Over the past few decades, boron has attracted much attention from researchers, and boron-based inorganic compounds are popular lightweight structural materials and thermionic cathode materials [5,6]. As one of the greatest scientists of the 20th century, Dr. Lipscomb was respectively awarded the 1976 and 1979 Nobel Prizes in Chemistry for his outstanding contribution to the configuration study of borane and excellent research on metal borides [7]. The occurrence of these two important achievements heralded the first hot wave research of boron. However, over the next 20 years, the research of the boron-based materials gradually tended to stabilize. As the building blocks for the nanodevices, nanomaterials (including two-dimensional (2D) layered structures, one-dimensional (1D) nanowires and nanotubes, and zero-dimensional (0D) nanoparticles) have attracted more and more attention [8,9]. Compared with their corresponding bulk counterparts, inorganic boron-based nanomaterials exhibit more fascinating optical and electrical transport behaviors due to their higher aspect ratio, larger specific surface area, and smaller size. Among them, rare-earth boride (REB_6) nanomaterials (LaB_6, CeB_6 etc.) are excellent cold cathode candidates, since they have low work function, ultrahigh hardness, and a high melting point, as well as excellent electrical and thermal conductivity performances [10–16]. Being typical Kondo topological insulators, SmB_6, CeB_6, and YbB_6 nanomaterials have attracted the researchers since the 2016 Nobel Prize in Physics was awarded to David J. Thouless, F. Duncan M. Haldane, and J. Michael Kosterlitz for their theoretical discoveries on the topological phase transitions and topological phases of matter [17–22]. At the end of 2015, the 2D layered structure of the boron single element was first synthesized experimentally, and was named borophene [23–25]. Borophene has been theoretically proposed to have numerous excellent and unique properties, which may have a promising future in batteries, flexible devices and high-speed electronic devices [26–28]. As a result, the relevant studies on inorganic boron-based nanomaterials resonate in the material sciences and condense physics fields again.

In this paper, we review the current developments of inorganic boron-based nanostructures (boron monoelement and rare-earth borides). First, we introduce their synthesis methods. Second, the optoelectronic properties of the boron-based nanostructures are classified. Finally, we give the evaluations on the optoelectronic performances of inorganic boron-based nanostructures and propose their applications based on our research results.

2. Synthesis Methods of Inorganic Boron-Based Nanostructures

2.1. One-Dimensional Boron-Based Nanomaterials

One-dimensional (1D) nanostructures offer ideal platforms to study the dependence of electrical transport, thermal conductivity, and mechanical strength on the dimensionality or size effects [29]. In general, 1D nanostructures are widely used in field-effect transistors, light-emitting diodes (LEDs), photodetectors, solar cells, field emission displays, and sensors [29–36]. In 2010, we introduced 1D boron single element nanostructures [37]. However, to date, although the synthesis methods of 1D inorganic boron-based nanostructures (nanowires, nanobelts, and nanotubes) have been well described, they need to be further renewed to meet the requirements of their rapid developments.

2.1.1. Rare-Earth Boride Nanostructures

There are usually six kinds of stoichiometry for rare earth borides, namely REB_2, REB_4, RE_2B_5, REB_6, REB_{12}, and REB_{66}, in which RE represents rare-earth element (La, Sm, Ce, Eu, Gd, etc.). Among them, REB_6 is very popular, and has been widely used in commercial cathode materials. The typical lattice structure of REB_6 is given in Figure 1. REB_6 belongs to the cubic lattice, in which boron octahedrons consisting of six boron atoms are located at the vertex positions of the cubic lattice, and rare-earth atoms occupying the central position of the cubic lattice form the other body-centered CsCl-type cubic lattice.

In actual field emission (FE) applications, higher crystallinity, lower work function, and larger aspect ratio are highly demanded for REB_6 nanomaterials, which are believed to be the crucial criterions to evaluate the quality of cold cathode nanomaterials. The synthesis methods of REB_6 nanostructures can be classified into three types (Table 1): chemical vapor deposition (CVD), hydrothermal reaction, and electrochemical etching methods. Chi and Fan et al. [38–43] respectively applied a no-catalyst CVD method to synthesize different REB_6 (PrB_6, NdB_6, CeB_6, GdB_6, and EuB_6) nanowires at 1000–1150 °C by choosing rare-earth metal powders and B-contained gas (BCl_3 or B_2H_6) as source materials. As seen in Figure 2a,c, single crystalline PrB_6 and NdB_6 nanowires have diameters of several tens of nanometers and lengths extending to several micrometers as well as uniform diameter along the growth axis. Using the CVD method, Xu et al. [35] reported the successful growth of 1D LaB_6 nanoneedles with a diameter decreasing from the bottom to the top and nanowires with uniform diameter by changing the distance between the precursors and substrate. By the hydrothermal reaction method, Han et al. [39] obtained single crystalline SmB_6 nanowires growing along the [001] direction, whose averaged diameter and length were respectively 80 nm and 4 μm. To prepare REB_6 nanomaterials, nearly all of the researchers used toxic, inflammable, or expensive gas as the source materials, which inevitably bring some threat to our lives. To solve the above problem, our group developed a simple solid-source CVD method to synthesize single crystalline SmB_6 and LaB_6 nanowire arrays using Ni catalysts. In our method, the low-cost, nontoxic solid B, B_2O_3, and $LaCl_3$ powders or Sm film were the source materials, which can avoid the potential pollution of the nanomaterial production [44,45]. In comparison with other two ways, the CVD method is more convenient for the growth of high-quality REB_6 nanomaterials, because the control of their morphology, crystal structure, and stoichiometry are relatively easier.

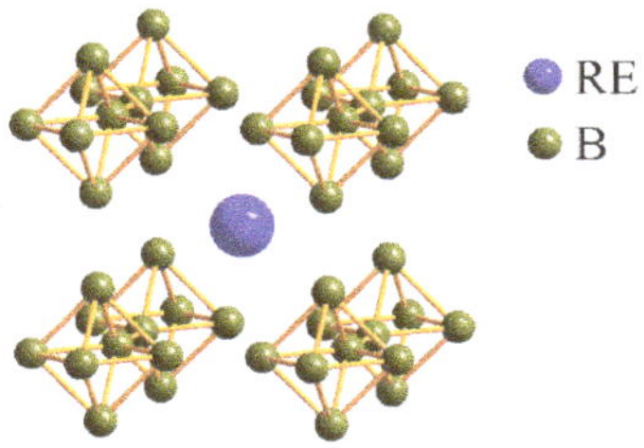

Figure 1. Typical lattice structure of rare-earth boride (REB_6).

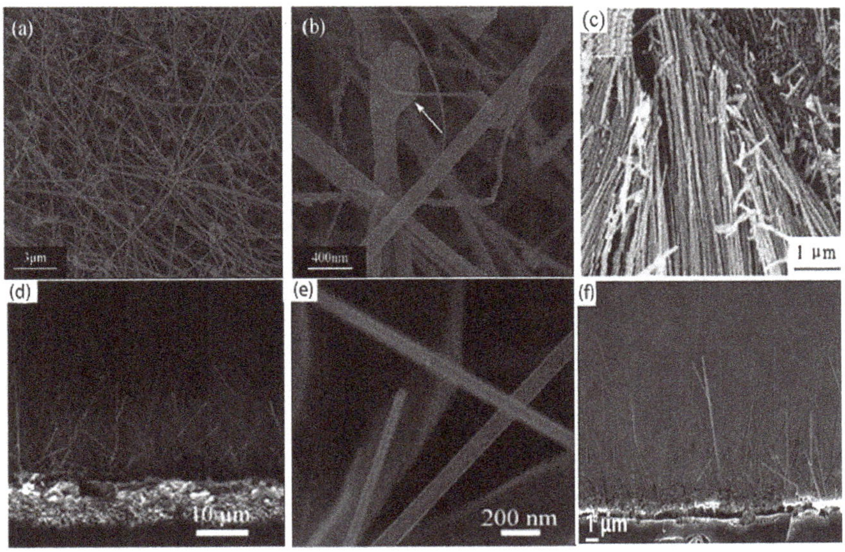

Figure 2. (**a**,**b**) Typical low-magnification and high-magnification SEM images of the PrB_6 nanowires [39]. Copyright 2014, Elsevier. (**c**) Low-magnification SEM images of the NdB_6 nanowires [41]. Copyright 2013, Elsevier. (**d**,**e**) Representative morphologies of the LaB_6 nanowires [44]. Copyright 2017, Royal Society of Chemistry. (**f**) Side-view SEM image of the SmB_6 nanowires [45]. Copyright 2017, IOP Publishing.

Table 1. The summary of the synthesis methods of different rare-earth boride nanostructures. CVD: chemical vapor deposition.

Materials	Source Materials	Method	T/°C	Catalyst	Ref.
LaB_6 nanowire	$LaCl_3·7H_2O$, B_2H_6	CVD	930	free	[35]
LaB_6 nanoneedle	$LaCl_3·7H_2O$, B_2H_6	CVD	970	free	[35]
PrB_6 nanowire	Pr powders, BCl_3 gas	CVD	1000-1150	free	[38]
NdB_6 nanowire	Nd powders, BCl_3 gas	CVD	1150	free	[40]
LaB_6 nanowire	B, B_2O_3,$LaCl_3$ powders	CVD	1100	Ni-assisted	[44]
SmB_6 nanowire	B, B_2O_3 powders, Sm film	CVD	1100	Ni-assisted	[45]
SmB_6 nanowire	Sm, H_3BO_3, Mg and I_2 powders	Hydrothermal reaction	220-240	I_2-assisted	[39]
LaB_6 nanowire	H_3PO_4, C_2H_5OH,LaB_6 target	Electrochemical etching	2	free	[43]

2.1.2. Boron Single Element Nanostructures

Boron nanostructures have potential applications in ideal cold-cathode electron sources, high-temperature semiconductor devices, and field-effect transistors, which have attracted much interest from researchers.

Boron Nanowires

Magnetron sputtering, laser ablation, thermal carbon reduction, thermal evaporation, and chemical vapor methods have been utilized to prepare amorphous or crystalline boron nanowires (BNWs) [46–50]. The morphology and crystallinity of the as-synthesized BNWs can differ with the growth methods. By the combination of CVD and ultraviolet photoresist technology, large-scale patterned boron nanowire (BNW) arrays were synthesized on Si substrate [51]. As found in Figure 3a,b, all of the nanowire patterns exhibit a regular square with a size of 25 μm × 60 μm, and the BNWs in each patterns have a uniform morphology, which may be more beneficial for future integrated circuit device applications. Using magnetron sputtering, Cao et al. [46] successfully prepared amorphous boron nanowire aligned arrays on silicon substrates. The as-grown BNWs have diameters from 20 to 80 nm and lengths up to several tens of micrometers, which are indexed as amorphous structures (Figure 3c,d).

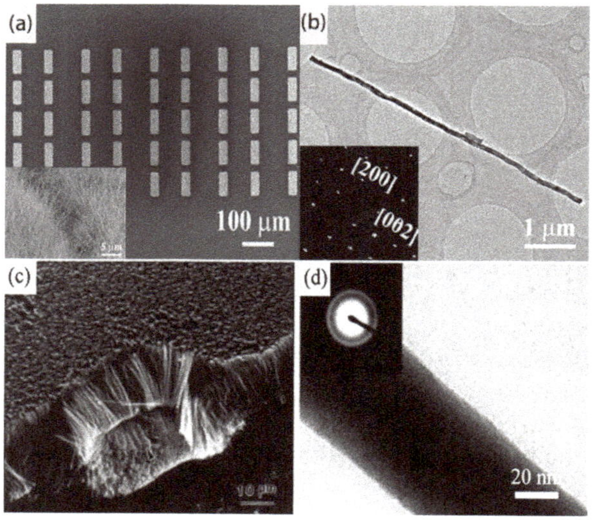

Figure 3. (**a**) Low-magnification SEM images of large area boron nanowire (BNW) arrays, and the inset gives their high-resolution image. (**b**) Typical HR-TEM image and SAED pattern of the single-crystalline BNW [51]. Copyright 2014, Wiley. (**c**,**d**) SEM and TEM image of the amorphous BNWs. Inset, SAED pattern taken from the nanowire showing some amorphous halo rings [46]. Copyright 2001, Wiley.

Boron Nanotube

Boron nanotubes (BNTs) have been theoretically proposed to have metallic properties, whether their structure configurations are armchair or zigzag. In 2004, the magnesium-substituted mesoporous silica template (Mg–MCM-41) method was applied to prepare pure boron single-wall nanotubes at 870 °C according to the method by Ciuparu et al. [52], using the mixture of BCl_3 and H_2 as gas sources. The authors attributed the Raman peaks at about 210 cm^{-1} (peak a) and 430 cm^{-1} (peak b) to typical tubular structures (Figure 4a), corresponding to the characteristic radial breathing mode. Using boron (99.99%) and boron oxide powders (99.99%) as source materials, our group reported the first large-scale fabrication of single crystalline multilayered BNTs in 2010 [53]. As seen in Figure 4b–e, the as-synthesized BNTs have lengths of several micrometers and diameters of about 30 nm. Moreover, these nanotubes are indexed as multilayered single crystalline boron nanotubes with the interlayer spacing of about 3.2 Å. Most of all, these nanotubes were experimentally proven to have metallic properties as theoretical predictions, which didn't vary with their chirality.

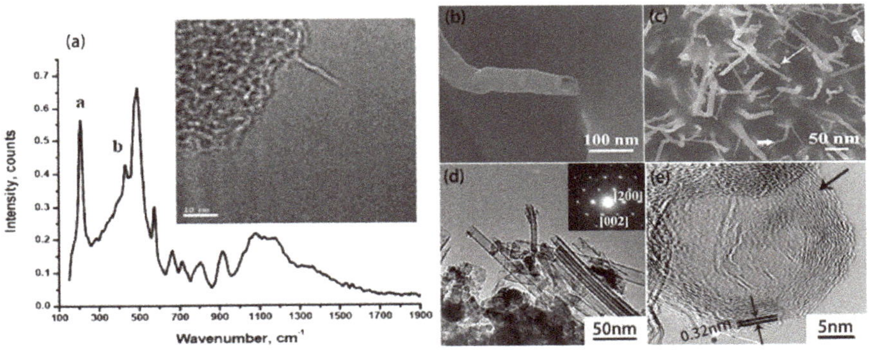

Figure 4. (**a**) Raman spectrum and TEM image (inset) of a single-wall boron nanotube [52]. Copyright 2004, American Chemical Society. (**b**,**c**) Low-resolution and high-resolution SEM images of the boron nanotubes (BNTs) (white arrows) at the initial growth stage. (**d**) TEM image of the BNTs. The inset is the corresponding SAED pattern. (**e**) The HRTEM image of the top of the BNT [53]. Copyright 2010, Royal Society of Chemistry.

At the viewpoint of future cold cathode applications of 1D boron nanostructures, the CVD method has advantages over other synthesis methods owing to its better morphology, aspect ratio, and crystallinity controllability. In addition, large-scale, high-quality 1D boron-based nanostructure arrays are much easier to obtain on rigid or flexible substrates by the CVD method, which is very beneficial for their cell or photosensitive device applications.

2.2. Two-Dimensional Boron-Based Nanomaterials

Inspired by the stunning achievements of 2D materials [54], much effort has been devoted to exploring 2D boron-based nanostructures in both theoretical calculations and experiments [55–57]. To date, there are three synthesis methods of 2D boron-based nanostructures: molecular beam epitaxy (MBE), liquid-phase exfoliation, and CVD methods. The researchers are devoted to modulating the morphology, layer number, and structure configuration of 2D boron-based nanostructures due to the effects on their physical properties.

2.2.1. Two-Dimensional Boron Monoelement Nanostructures

By MBE, Mannix and Wu [24,25] respectively reported the first growth of atomically thin, crystalline 2D boron sheets on silver surfaces under an ultrahigh-vacuum system. As shown in Figure 5a–c, Mannix et al. [24] obtained the atomic image of borophene, and ascertained that it has the same metallic and highly anisotropic characteristics as theoretically predicted. Wu et al. [25] uncovered two allotropes of boron sheets on Ag (111) substrate, which are respectively β_{12} and χ_3 sheets. Both have a similar triangular lattice but different arrangements of periodic holes, as revealed by STS spectra (Figure 5d–f). Bedzyk's research further proved that atomically thin borophene sheets on the Ag (111) surface have a Van der Waals-like structure [58]. By X-ray standing wave-excited X-ray photoelectron spectroscopy, the lattice positions of boron atoms with multiple chemical states were confirmed, revealing that the thickness of a single layer borophene is 2.4 Å on an unreconstructed Ag surface. Afterwards, Wu et al. [59] synthesized a pure honeycomb, graphene-like borophene on the Al (111) substrate by MBE. Moreover, their theoretical calculations discovered that the honeycomb borophene is stable on the Al (111) surface, because there exists one electron transfer from the Al (111) substrate to each boron atom, which stabilizes the structure. The atomic resolution STM images of the honeycomb borophene are shown in Figure 6, in which the lattice constant is about 0.29 nm. Very recently, single crystalline borophene with an area up to 100 μm^2 has been fabricated on the Cu (111) substrate via the MBE method by Gozar et al. [60]. The as-grown borophene was composed of novel triangular networks with a concentration of hexagonal vacancies of $\eta = 1/5$, belonging to β_{12} and χ_3 phases.

By CVD, Tai et al. also prepared 2D γ-boron films on copper foil using boron and boron oxide powders as the source materials [23], as seen in Table 2. The as-prepared γ-B_{28} film was identified as a semiconductor with a direct bandgap of around 2.25 eV, which had a strong photoluminescence emission band at 626 nm (Figure 7). Besides, Tsai et al. [61] applied a B plasma-assisted technique to deposit a multilayer β-borophene on an insulating SiN_x film on Si substrate.

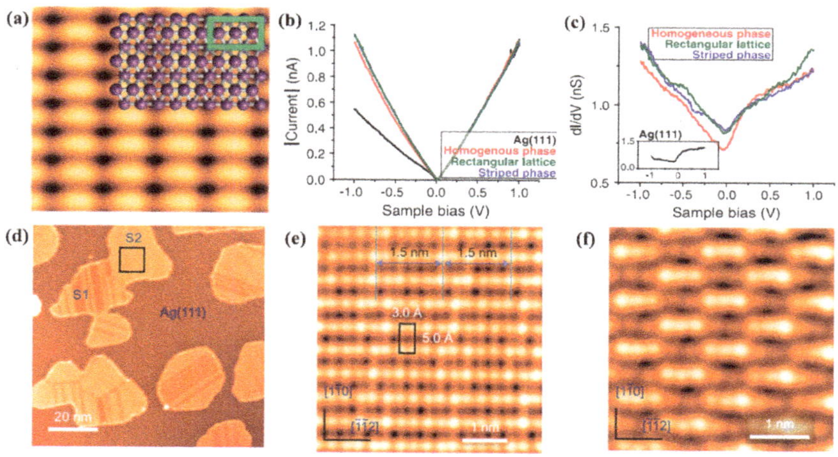

Figure 5. (**a**) Theoretical model and experimental STM images of borophene (V_{sample} = 0.1 V, I_{tunnel} = 1.0 nA). (**b**,**c**) I–V curves and dI/dV spectra of borophene [24]. Copyright 2015, Science. (**d**) STM image of boron sheets after 650 K annealing, in which two different phases are respectively labeled 'S1' and 'S2'. (**e**) The S1 unit cell and the 1.5-nm stripes are respectively marked by a black rectangle and the solid lines. (**f**) High-resolution STM images of the S2 phase. Bias voltages of STM images: −4.0 V (**d**), 1.0 V (**e**,**f**) [25]. Copyright 2016, Springer Nature.

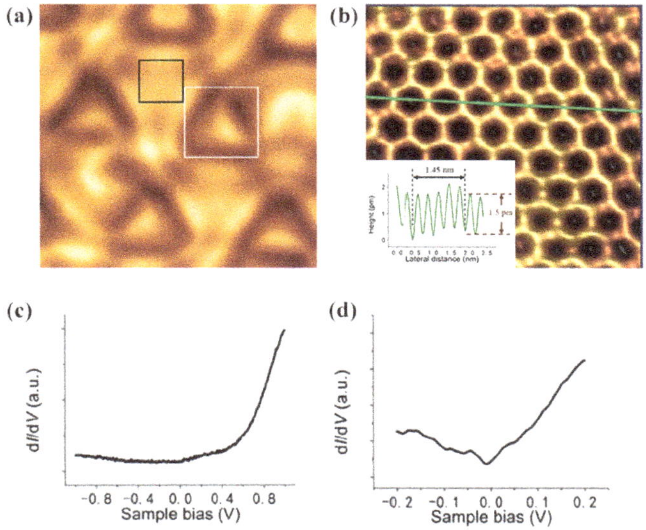

Figure 6. (**a**) STM image (15 nm × 15 nm) revealing the long-period configuration with triangular corrugation. (**b**) A high-resolution STM image (2.4 nm × 2.4 nm) of the area marked by the black rectangle in (**a**), showing a flat honeycomb lattice. Inset is a line profile along the green line. (**c**,**d**) The dI/dV curves taken on a borophene surface with a different bias voltage range. The scanning parameters for (**a**,**b**) are: sample bias −11 mV, I = 130 pA [59]. Copyright 2018, Elsevier.

Using the sonication-assisted liquid-phase exfoliation method, Teo et al. [62] obtained high-yield few-layer B sheets. As given in Figure 8, the as-synthesized few-layer B sheets have a clear diffraction pattern with a d spacing of 0.504 nm, conforming to the (104) plane of β-rhombohedral B crystal.

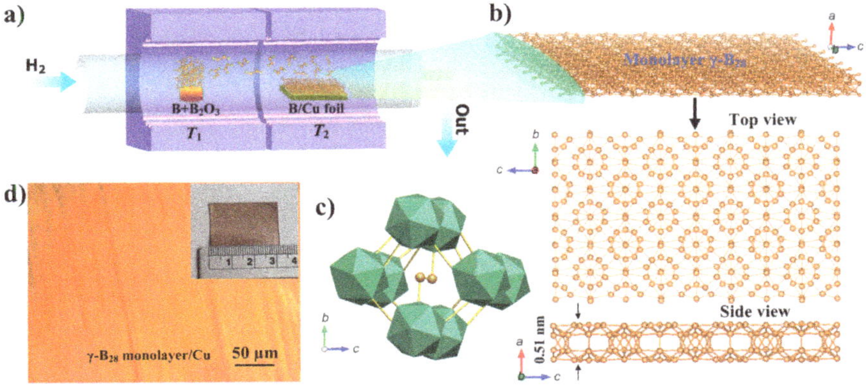

Figure 7. (**a**) Schematic diagram of the two-zone CVD furnace for two-dimensional (2D) γ-B_{28} film. (**b**) Top and side views of the borophene monolayer. (**c**) The corresponding polyhedral structure of the basic unit cell in bc projection, in which boron atoms (orange spheres) form the dumbbells. (**d**) Optical image of a monolayer on Cu foil [23]. Copyright 2015, Wiley.

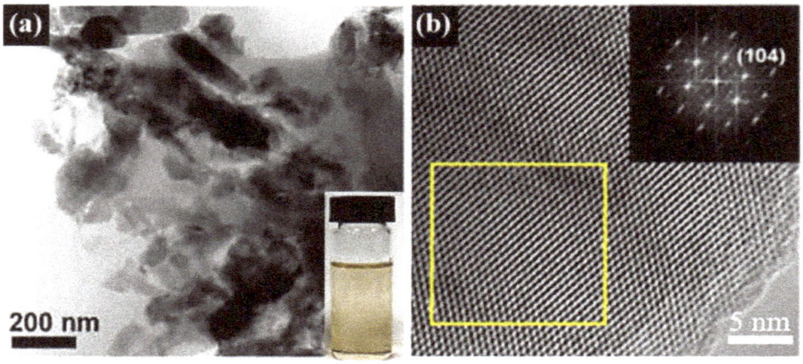

Figure 8. (**a**) Typical TEM images of the few-layer B sheets prepared by tip sonication. The inset is a photograph of B-sheet dispersions and sonication after 4 h. (**b**) Typical TEM image of the few-layer B sheets. The inset shows the corresponding FFT pattern [62]. Copyright 2018, American Chemical Society.

2.2.2. Two-Dimensional Boron-Based Nanostructures

Due to the structural instability of borophene, the fully-hydrogenated borophene (i.e., borophane) with enough stability has aroused much interest, which was theoretically predicted by Xu et al. [63]. By the first-principle calculations, borophane is anticipated to have a stable surface configuration due to electron transfer after full hydrogenation, as presented in Figure 9. Moreover, the Fermi velocity of borophane is as high as 3.5×10^6 m s^{-1} under HSE06 level, which is the highest among the known 2D materials, and is four times larger than that of graphene. Although there are many theoretical calculations and predictions on borophane [64,65], it has not been experimentally synthesized. On the other hand, the 2D boron oxide (BO) structure is also proposed to be stable after experiencing a suitable stoichiometric degree of oxidation. Yang et al. [66] studied the structural configurations of the 2D boron oxide sheet by using the first-principle global optimization method, suggesting that they have potential applications in high-speed nanoelectronic devices. Also, Zhang et al. [67] investigated the physical properties of the hexagonal borophene oxide sheets in Wu's work [59], revealing they should possess remarkable mechanical and electronic properties. Furthermore, 2D borophene oxide nanostructures with completely flat configuration may serve as a promising platform for studying 2D topological phases, because its energy band hosts a nodal loop centered around the Y point in the Brillouin zone, and exhibits different topological indices before and after transition [67].

Table 2. The summary table of the synthesis methods of borophene. MBE: molecular beam epitaxy.

Method	Temperature	Substrate	Structural Configuration	Source Materials	Sample Area
CVD [23]	1100 °C	Cu foil	γ-B_{28}	B, B_2O_3 powders	Nanometer
MBE [24,25,58]	300–750 °C	Ag (111)	β_{12} and χ_3	B powder	Nanometer
MBE [59]	230 °C	Al (111)	honeycomb	B powder	Nanometer
MBE [60]	300/490 °C	Cu(111)	β_{12} and χ_3	B powder	Micrometer
Plasma-assisted ion implantation [61]	800 °C	Si (001)	β	B powder	Nanometer
Liquid-phase exfoliation [62]	N/A	N/A	β-rhombohedral	B powder	Nanometer

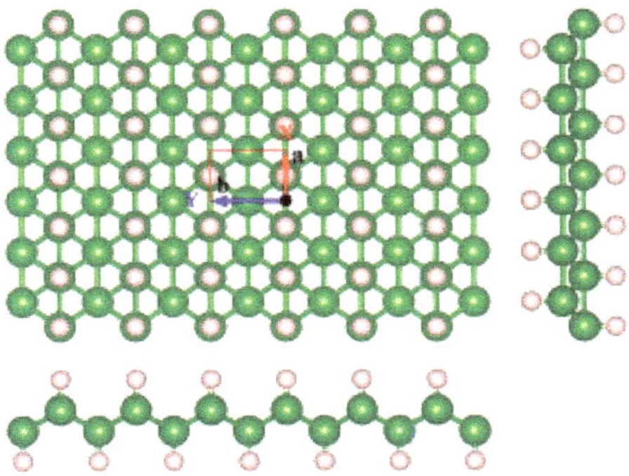

Figure 9. Top and side views of the optimized configuration of borophane. The unit cell is marked by a red box, in which the green and white balls respectively represent B and H atoms [64]. Copyright 2016, Royal Society of Chemistry.

As mentioned above, the MBE way is more suitable for the fabrication of a monolayer or few-layer high-crystallinity borophene, while CVD method has advantages on the layer or structure configuration control of large-area borophene with different standing directions to the substrate. For high-yield borophene, we suggest adopting the sonication-assisted liquid-phase exfoliation method, which has a more promising future in industry. Until now, only a few phases of borophene have been successfully prepared, which is far less than the several dozen phases in theoretical prediction. As a result, it still remains a big challenge for the researchers.

3. Optoelectronic Properties

Inorganic boron-based nanostructures have potential applications in optoelectronic nanodevices because of their particular physical and chemical properties. In this section, we review the electrical and optical properties measured on inorganic boron-based nanostructures in the past two decades.

3.1. Electrical Properties

3.1.1. Field Emission (FE) Properties

Inorganic boron-based nanostructures have been proposed as ideal cathode choices because they have small surface electron affinity, high electrical and thermal conductivity, a large aspect ratio, a high melting point, and large current endurance. Xu et al. [36] reported that the 1D LaB_6 nanostructures produced by CVD exhibited good FE characteristics, which had a turn-on field of 1.82 V μm^{-1} and a threshold field of 2.48 V μm^{-1}. In our recent works, the LaB_6 nanowire arrays had a low turn-on (2.2 V μm^{-1}) and threshold field (2.9 V μm^{-1}) as well as nice field emission (FE) stability with a current fluctuation of only 1.7% [44]. Further research showed that individual LaB_6 nanowires can bear the maximum current of 96 μA and the maximum emission current of 0.3 cm^2 LaB_6 nanowire film can reach as high as 5.6 mA. Most of all, the LaB_6 nanowire film exhibited the recoverable emission performance after O_2 absorption or desorption, making them suitable for applications in air. Recently, Zhang et al. [68] measured the working performance of cold field emission SEM equipped with a single LaB_6 nanowire emitter (Figure 10). The typical morphology of the LaB_6 nanowires (NWs) was presented in Figure 10a. Figure 10b schematically illustrates the fabrication of the LaB_6 emitter. An etched W tip is manipulated to approach individual LaB_6 nanowires, and subsequently, it is handled by an ion beam to form a tight contact with the LaB_6 nanowire via Van der Waals force. After the LaB_6 nanowire breaks off from the substrate and adheres onto the tip, the nanowire is fixed on a carbon pad by the electron beam-induced deposition (EBID) technique and the fabrication of the LaB_6 emitter is accomplished. In these FE experiments in the modified SEM system [68], the electron gun equipped with the LaB_6 NW displays an ultrahigh angular current density of ~2.4×10^5 μA sr^{-1}, which is 1000 times larger than the electron gun with the W (310) emitter (Figure 10c). Additionally, the current fluctuation (0.32%) for the LaB_6 emitter is obviously lower than that for the W (310) emitter (7.2%) under the same experimental conditions. The signal noise comparison for imaging the acceleration electrode using the LaB_6 NW and W (310) emitters are given in Figure 10d, in which the resolution and image quality of the LaB_6 emitter are obviously higher than those of the W emitter. Considering that the acquisition image time of the W emitter is twice as long as that of the LaB_6 nanowire emitter, the LaB_6 nanowire emitter should have more advantages than the commonly-used W (310) emitter. Also, the FE properties of the SmB_6 nanowires were compared with those of the SmB_6 nanopencils, which have a thick bottom and a sharp tip, to evaluate their application in cold cathodes [45]. The SmB_6 nanowires had a lower turn-on field of 6.5 V μm^{-1} in comparison with the SmB_6 nanopencils (6.9 V μm^{-1}), and their maximum emission current density can reach several hundred μA cm^{-2} (Figure 11).

Table 3. Comparison table of the FE properties of some nanomaterials with excellent emission performances.

Nanomaterials	Turn-on Field Vμm^{-1}	Threshold Field Vμm^{-1}	Emission Current Fluctuation	Ref.
Mo nanoscrew	1.65	2.4	0.46%, 1 h, 50 mA/cm^2	[69]
Carbon nanotube	3.2	5.8	25%, 20 h, 260 mA/cm^2	[70]
ZnO nanobelt	6.6	8.5	14%, 16 h, 7.4 mA/cm^2	[71]
SiC nanowire	0.9	1.7	3%, 24 h, 5 mA/cm^2	[72]
LaB_6 nanowire	1.9	2.9	1.2%, 0.5 h, 2.6 3mA/cm^2	[44]
Boron nanowire	4.3	10.4	5.6%, 8 h, 1 mA/cm^2	[51]

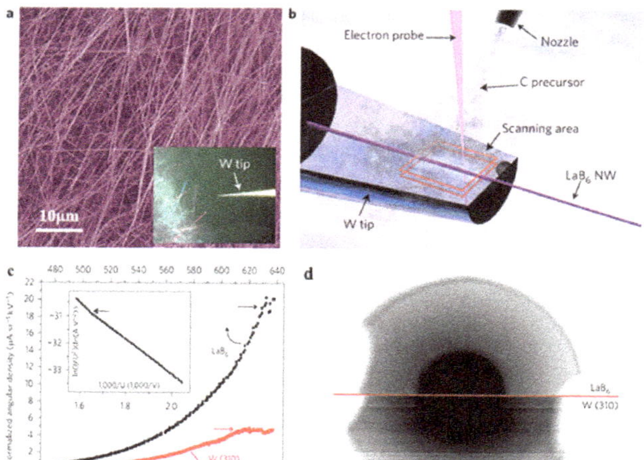

Figure 10. (**a**) SEM image of LaB_6 nanowires (NWs) on a monocrystalline LaB_6 (100) substrate. The inset shows the optical image in which a LaB_6 nanowire is picked up by a W tip. (**b**) Schematic diagram showing the electron beam deposition process of fixing a LaB_6 NW onto a W needle. (**c**) The dependence of normalized angular current density on extraction voltage with maximum value limited by emission instability. The inset gives their corresponding F–N plots, showing linearity with a transition point at high current density. (**d**) Signal noise comparison for imaging the acceleration electrode in the SEM objective lens using the LaB_6 NW emitter and W emitter, respectively [68]. Copyright 2016, Springer Nature.

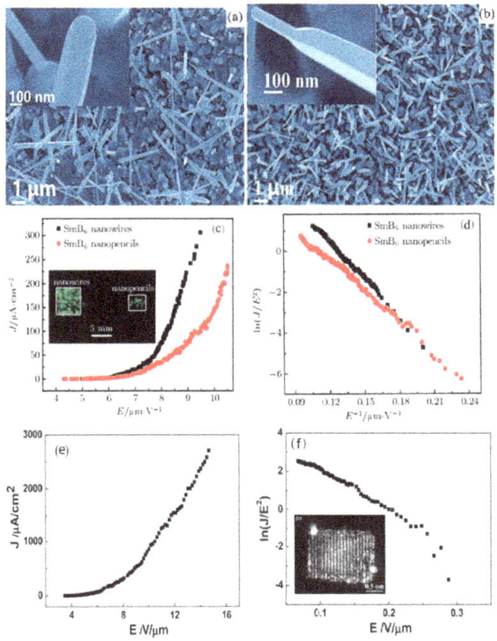

Figure 11. (**a**,**b**) Top-view SEM images of the SmB_6 nanowires and nanopencils, respectively. The insets are their corresponding high-magnification cross-sectional images. (**c**,**d**) J–E curves and FN plots of the SmB_6 nanowires and nanopencils. Their emission images are given in the inset [45]. Copyright 2017, IOP Publishing. (**e**,**f**) Typical J–E curves and FN plots of large-area boron nanowire patterns, and the inset is their corresponding emission image [51]. Copyright 2014, Wiley.

Similarly, the BNWs exhibited good emission properties [51]. The turn-on and threshold fields of large-area BNW patterns are respectively 4.3 V μm^{-1} and 10.4 V μm^{-1}, as provided in Figure 11e,f. In addition, they exhibited a very uniform emission image (inset), in which the distribution uniformity of the emission patterns was 81.8%, and their brightness distribution uniformity was over 88.9%. The FE behaviors of boron-based nanostructures can compare favorably to those of many nanomaterials with excellent emission performance, which suggests that they have potential applications in cold cathode electron sources (Table 3).

3.1.2. Capacitance Characteristics

Inorganic boron-based nanostructures also have great potential in supercapacitors. Zhi and Liu et al. [73] stated that elemental boron-based supercapacitors exhibited excellent performances whether they were placed in all alkaline, neutral, or acidic electrolytes. Boron nanowire–carbon fiber cloth (BNWs–CFC) electrodes achieve a capacitance up to 42.8 mF cm^{-2} at a scan rate of 5 mV s^{-1} and 60.2 mF cm^{-2} at a current density of 0.2 mA cm^{-2} in acidic electrolyte (Figure 12), respectively. In addition, the BNWs–CFC electrodes still retain high performance even after being bent 1000 times, revealing that they possess excellent mechanical properties. Subsequently, the LaB_6 nanowires electrode had an areal capacitance as high as 17.34 mF cm^{-2} in 1.0 M of Na_2SO_4 solution at a current density of 0.1 mA cm^{-2} and 16.03 mF cm^{-2} at a scan rate of 5 mV s^{-1}. Moreover, the LaB_6 nanowires electrodes displayed outstanding cycling stability after 10,000 charging/discharging cycles, which suggests that they are highly effective electrode materials for supercapacitors [74]. Soon afterwards, Teo et al. [62] used the exfoliated few-layer B sheets as the electrode materials of a supercapacitor. The supercapacitor using B-sheet electrodes exhibited impressive electrochemical performance with a wide potential window up to 3.0 V. Moreover, the energy density of the supercapacitor can reach as high as 46.1 W kg^{-1} at a power density of 478.5 W kg^{-1}, and its cycling stability was 88.7% after 6000 cycle operations. The excellent performance of inorganic boron-based supercapacitors suggests that they should be ideal choices for the flexible anode materials of high-performance batteries and supercapacitors.

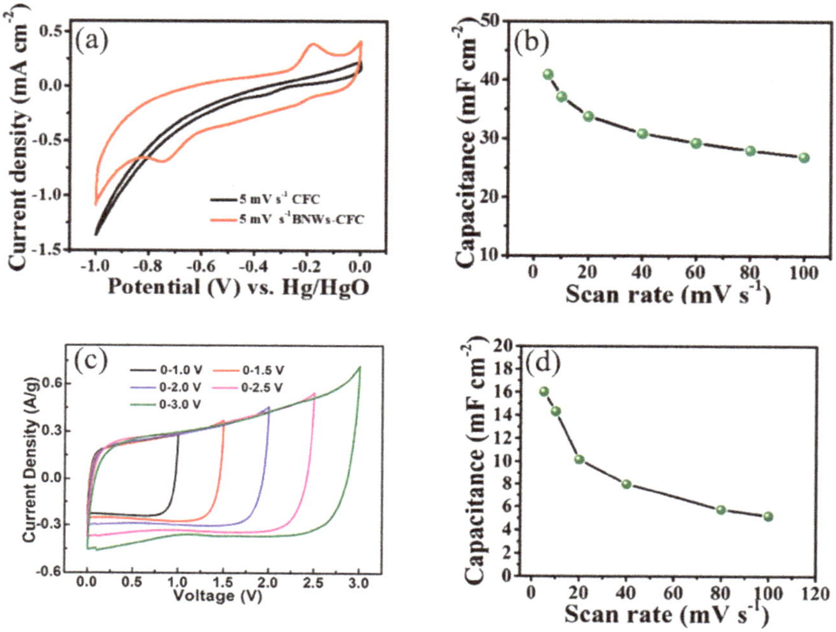

Figure 12. (**a**) Comparison of CV curves of carbon fiber cloth (CFC) and boron nanowire–carbon fiber cloth (BNWs–CFC) at a scan rate of 5 mV s^{-1}. (**b**) Areal capacitances of BNWs–CFC at different scan rates [73]. Copyright 2018, Wiley. (**c**) CV curves collected under various voltage windows at a scan rate of 10 mV s^{-1} of B sheets [62]. Copyright 2018, American Chemical Society. (**d**) Plots of scan rates against the areal capacitances of LaB_6–CFC electrodes [74]. Copyright 2018, Elsevier.

3.1.3. Surface Electrical Transport Property

The surface states of topological insulators are always concealed by the dominant bulk conduction due to the existence of massive dopants or defects. In this situation, the topological Kondo insulator (TKI) becomes very attractive because the strongly correlated electron system can assure that the surface conduction dominates over their electrical transport. As typical TKI materials, rare-earth hexaborides (such as SmB_6 and YbB_6) have been the research goals for theoretical calculations and experiments [22,75,76]. Wirth et al. [77] observed that the surface states of SmB_6 dominated the electron conduction when the temperature was below about 7 K due to a suppressed Kondo effect at the surface. Furthermore, Wirth et al. [78] elucidated the effect of magnetic or non-magnetic impurities on the topological surface states of SmB_6 by the combination of local STS and macroscopic transport measurements, unveiling that the local shielding lengths of the surface states can change with the magnetic properties of the substitutes. Simultaneously, Yu, He, and Liu et al. [22,79] respectively investigated the local and nonlocal magnetotransport properties of individual SmB_6 nanowires (NWs). Yu et al. observed an obvious transition from negative to positive magnetoresistance (MR) as the bias current increased in a single SmB_6 nanowire, as shown in Figure 13. He and Liu et al. found that the hysteretic magnetoresistance effect will emerge in TKI SmB_6 nanowires with diameters less than 58 nm.

A non-monotonically temperature-dependent positive magnetoresistance is observed at intermediate temperatures because of the strong magnetism on the narrow nanowire's surface [22], suggesting that impurity band conduction may exist in the SmB_6 nanowire (Figure 14). Meanwhile, YbB_6 is also predicted to be a TKI. Shi and Feng et al. [80,81] investigated the surface and bulk gap structures of YbB_6 by angle-resolved photoemission spectroscopy (ARPES). Shi disclosed that the *f*-orbitals of YbB_6 are fully occupied, and Yb exists in bivalence states, which is different from mixed-valence Sm in SmB_6. Furthermore, the metallic surface states of YbB_6 are believed to be topological surface states, which are spin-polarized in plane and locked to the crystal momentum. Feng et al. [81] directly observed the bands around the time-reversal invariant momenta exhibiting an obvious linearly dispersive relationship. The in-gap states possess the chirality of orbital angular momentum, which is attributed to the chiral spin texture. This unveils that YbB_6 is a moderately correlated topological insulator.

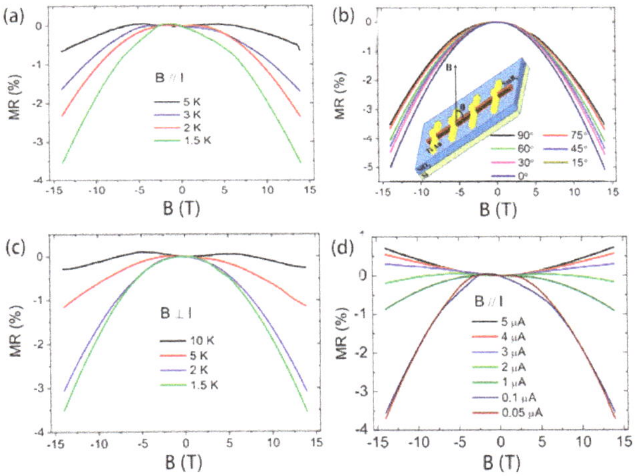

Figure 13. (**a**) The magnetoresistance (MR) for parallel magnetic field at different temperatures. (**b**) The MR measured at 1.5 K and under the magnetic field with different directions. The inset schematically shows the angle θ between the direction of the magnetic field and the longitudinal direction of the nanowire. (**c**) The MR for perpendicular magnetic fields at different temperatures. (**d**) The MR at 1.5 K for different bias currents ranging from 0.05 μA to 5 μA [79]. Copyright 2017, American Physical Society.

In addition, 2D boron-based nanomaterials (borophene and boronphane) were predicted to be ideal Dirac materials, which exhibit clear linear energy dispersion characteristics at the Fermi level and have the Dirac cones in the band diagram based on the first-principle calculations [63–65]. For monolayer β_{12} borophene, Ezawa [82] proposed the existence of triplet fermions at the high-symmetry points, and there should be no loop encircling the triple-band bonding point without touching the Fermi surface. Furthermore, bilayer borophene is also a Dirac material, while few-layered borophene retains robust metallic characteristics owing to its multiple band interactions [83]. Compared to monolayer borophene with a high Fermi velocity close to graphene (8.2×10^5 m/s) [84], borophane was calculated to possess a Fermi velocity (3.5×10^6 m/s) that was two to four times larger than graphene, implying that they should be promising systems for high-speed electronic or optoelectronic devices. However, to date, all of the research studies on the surface topological behaviors of 2D boron-based nanomaterials have focused on the theoretical predictions, which are indeed short of the experimental evidence.

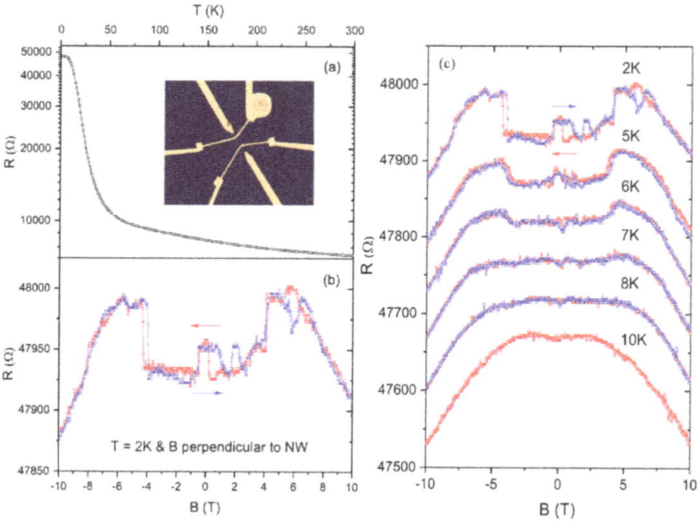

Figure 14. (**a**) The temperature dependence of resistance of a single SmB_6 nanowire with a D value of 45 nm. The inset shows the fabricated four-probe device. (**b**) Double-sweep magnetoresistance curve with B perpendicular to the NW and T = 2 K. The field sweeping directions are indicated by arrows. (**c**) Double-sweep MR curves of a SmB_6 NW in perpendicular magnetic fields at different temperatures. The NW diameter is 45 nm, and the field-sweeping directions are indicated by arrows. Curves are offset vertically for comparison. At T = 10 K, the MR curve was only measured in the sweep-down direction [22]. Copyright 2018, Wiley.

3.2. *Optical Properties*

3.2.1. Optical Absorption

Rare-earth hexaborides usually have high absorbance in the visible and near-infrared (NIR) wavelength ranges [85–87], which can vary with their lattice constants (Figure 15a). Moreover, LaB_6 nanoparticles with high conductivity create a promising metal-like plasmonic material that resembles Au or Ag nanoparticles. In this situation, $LaB_6@SiO_2$ (core/shell) nanoparticles were used as the photothermal catalysts in the reduction of 4-nitrophenol [87]. By the surface decoration of Au nanoparticles, $LaB_6@SiO_2$/Au composite nanoparticles (Figure 15b) exhibited a better photothermal conversion capacity, because the existence of Au nanoparticles offered more active spots on the catalyst surface and enhanced the temperature of the reaction process [88]. As observed in Figure 15, Sani et al. found that the absorbance coefficient of LaB_6 (0.7) is comparable to the advanced solar absorber materials, such as SiC (0.8) and HfB_2 (0.5) [89]. The absorption valleys of REB_6 (CeB_6, PrB_6, and NdB_6) nanostructures (Figure 15d) are respectively located at the visible regions of 670 nm, 785 nm, and 800 nm, suggesting that the rare-earth hexaborides may have potential applications as a sunlight absorber in the visible and NIR ranges [90].

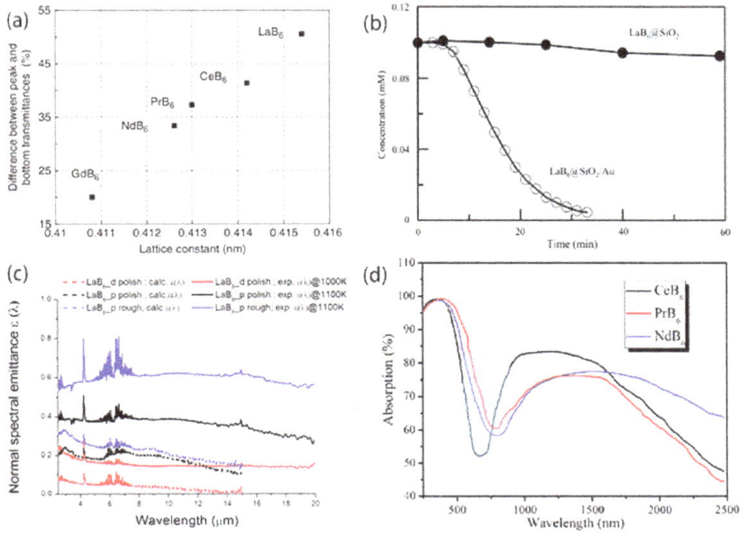

Figure 15. (**a**) Near-infrared absorption of rare-earth hexaboride nanoparticles with regard to their lattice constants [85]. Copyright 2008, Wiley. (**b**) Variation of 4-nitrophenol concentration with time by $LaB_6@SiO_2$ and $LaB_6@SiO_2/Au$ composite nanoparticles [88]. Copyright 2013, Elsevier. (**c**) Calculated (dashed lines) and experimental emittance (solid lines) of LaB_6 samples [89]. Copyright 2017, Springer Nature. (**d**) The optical absorption of REB_6 nanostructures at visible and near-infrared ranges [90]. Copyright 2017, Elsevier.

3.2.2. Photosensitive Properties

Boron and rare-earth boride nanostructures generally have small bandgaps and high refractive indices. Recently, our group observed the strong anisotropic light scattering behaviors and photocurrent response of tetragonal single crystalline boron nanowires (BNWs) in the visible region [48]. As indicated in Figure 16a,b, second harmonic generation (SHG) effects were discovered in a single boron nanowire under femtosecond laser irradiations. The individual BNW device has high device sensitivity (20), large responsivity (12.12 A W^{-1}), and a fast on–off response (18 ms) (Figure 16c,d). Also, Yu et al. [91] found that a single SmB_6 nanowire exhibited self-powered photodetector performances, in which the photovoltaic coming from the scanning photocurrent microscopy was responsible for the photocurrent. Moreover, the SmB_6 nanowire device has a broadband response from 488 nm to 10.6 μm at room temperature, and an on/off ratio of about 100, as seen in Figure 17. Their responsibility and specific detectivity were respectively 1.99 mA W^{-1} and 2.5×10^7 Jones. As observed in Table 4, the boron-based nanostructures have exhibited comparable photosensitive behaviors with many other nanostructures with excellent working performances, suggesting that they are promising candidates for future high-performance photodetectors.

Table 4. The comparable table of the working performances of various nanodevices. UV: ultraviolet, MIR: middle infrared.

Nanostructures	Operation Voltage [V]	Device Sensitivity (I_P/I_D)	Photoresponsivity (R_λ) [A W^{-1}]	Detection Range	Response Time	Ref.
ZnO nanowire	2	8	N/A	UV	50 s	[92]
AlN nanowire	40	20	2.7×10^6	UV-Visible	10 ms	[93]
GaN nanowire	0	13	N/A	UV	0.53 s	[94]
B nanowire	10	20	12.12	Visible	18 ms	[48]
SmB_6 nanowire	0	100	1.99×10^{-3}	Visible-MIR	N/A	[91]

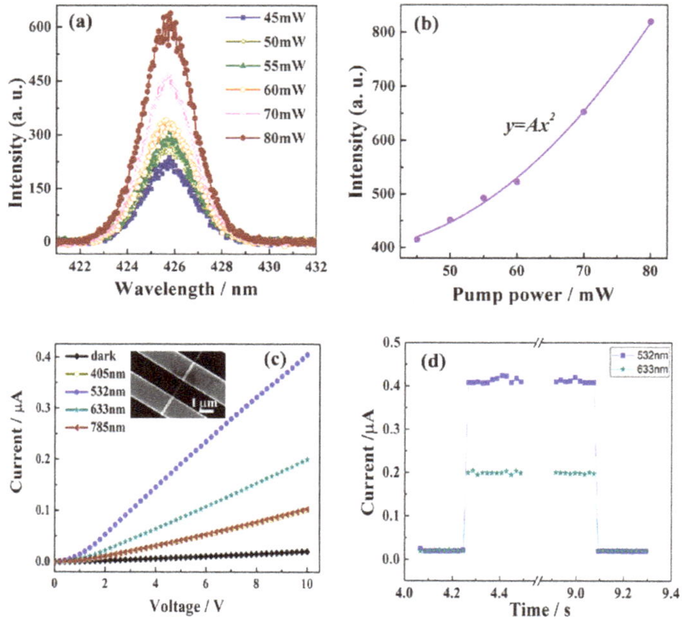

Figure 16. (**a**) Second-harmonic generation spectra of the BNW film under 850-nm illumination with different pump powers. (**b**) The curves of the second harmonic generation (SHG) peaks' intensity versus the pump powers. (**c**) The photocurrent-voltage curves of an individual BNW under different irradiations. The inset is the nanodevice structure. (**d**) The fast on–off curves of the BNW under 532-nm and 633-nm irradiations, respectively [50]. Copyright 2018, Wiley.

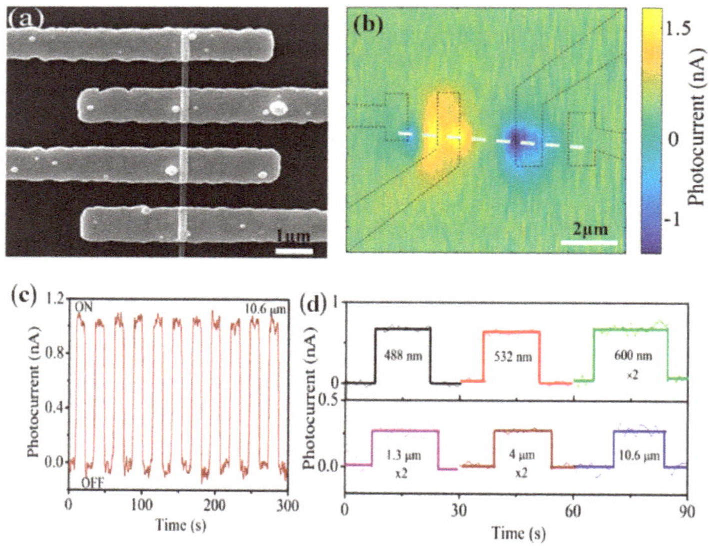

Figure 17. (**a**,**b**) SEM image and scanning photocurrent mapping of a single SmB_6 nanowire photodetector. (**c**) The time-dependent photocurrent measurement on the nanowire photodetector under 10.6-μm irradiation. (**d**) Room temperature photoresponse of the SmB_6 device under photoexcitation with different excitation wavelengths (as marked) [91]. Copyright 2018, American Institute of Physics.

4. Outlook and Conclusions

In conclusion, we have reviewed the recent progress of inorganic boron-based nanostructures (boron monoelement and rare-earth borides). As described above, CVD, laser ablation, magnetron sputtering, and thermal evaporation methods are the usual fabrication techniques of 1D boron-based nanostructures, whereas the MBE, CVD and liquid-phase exfoliation methods are more convenient for the synthesis of 2D boron-based nanostructures. The boron-based nanostructures have potential applications in field emission, supercapacitor, optical absorption, and photodetectors. Among them, 1D rare-earth boride nanostructures are promising as cold cathode electron sources, because they have a lower work function, higher melting point, metallic conductivity, and longer duration time, as well as a larger emission current. Since they belong to strongly correlated electron systems, SmB_6, CeB_6, and YbB_6 TKIs are ideal platforms for investigating the surface quantum behaviors in condensed matter physics and material sciences. For 2D borophene or borophane, they have distinct advantages in flexible energy conversion devices or high-speed electronic devices, because they have metallic behaviors, a larger specific surface area, a higher Young's modulus, and extremely high Fermi velocity. Also, 1D boron-based nanostructures exhibit strong light-absorption behaviors or rapid photocurrent responses, which suggests they are good candidates for flexible photodetectors at the visible and infrared ranges.

Although the research on the synthesis and optoelectronic properties of inorganic boron-based nanostructures has gained many achievements, there are still many essential and challenging issues that to be further explored. Designing and fabricating the LaB_6 or CeB_6 nanostructure-based cold cathode electron source remains a big challenge for the researchers. As a potential electron source, many important problems need to be solved in advance, such as a long-term and stable working performance at high current, long duration at severe conditions, a simple and cheap technique for a high-yield synthesis method, and low-power operation. For topological Kondo insulators, the fabrication of high-quality rare-earth boride nanostructure single crystals and exploration of the modulation techniques of their surface electrical transport behaviors are very important for their actual applications, which are still hard to solve. As for 2D boron-based nanostructures, the synthesis of a large-area single crystalline thin film with controllable layer thickness and chirality is a vital issue for their future applications. Another issue to explore and investigate is their optical and electrical properties in experiments, which are very essential because the current research studies mainly focus on the theoretical predictions. Although there are many challenges for the applications of inorganic boron-based nanostructures, they should have a very promising future, because they have exhibited enough fascinating properties in both experiments and theoretical calculations.

Author Contributions: Formal analysis, H.L. and Z.L.; writing—original draft preparation, Y.T., Z.G., T.Z.; writing—review and editing, Y.T., J.C., S.Z., F.L.; supervision, F.L.; funding acquisition, F.L.

Funding: This research was supported in part by the National Project for the Development of Key Scientific Apparatus of China (2013YQ12034506), National Natural Science Foundation of China (Grant Nos. 51872337, 51290271, 11474364 and 11574129), National Key Basic Research Program of China (Grant No.2013CB933601), the Fundamental Research Funds for the Central Universities of China, the Science and Technology Department of Guangdong Province and the Education Department of Guangdong Province.

Conflicts of Interest: The authors declare no conflict of interest.

References

1. Tsagareishvili, G.V.; Tavadze, F.N. Boron crystals: Preparation, structure and properties. *Prog. Cryst. Growth Charact.* **1988**, *16*, 341–365. [CrossRef]
2. Emin, D. Icosahedral boron-rich solids. *Phys. Today* **1987**, *40*, 55–62. [CrossRef]
3. Gaule, G.K. *Boron: Preparation, Properties, and Application*; Plenum: New York, NY, USA, 1965.
4. Ulas, A.; Kuo, K.K.; Gotzmer, C. Ignition and combustion of boron particles in fluorine-containing environments. *Combust. Flame* **2001**, *127*, 1935–1957. [CrossRef]

5. Davis, P.R.; Gesley, M.A.; Schwind, G.A.; Swanson, L.W.; Hutta, J.J. Comparison of thermionic cathode parameters of low index single crystal faces of LaB_6, CeB_6 and PrB_6. *Appl. Surf. Sci.* **1989**, *37*, 381–394. [CrossRef]
6. Otani, S.; Ishizawa, Y. Thermionic emission properties of boron-rich LaB_6 and CeB_6 crystal cathodes. *J. Alloys Compd.* **1996**, *1*, L18–L20. [CrossRef]
7. Sträter, N. William Nunn Lipscomb, Jr. (1919–2011). *Angew. Chem.* **2011**, *123*, 7874–7875.
8. Cui, Y.; Lieber, C.M. Functional nanoscale electronic devices assembled using silicon nanowire building blocks. *Science* **2001**, *291*, 851–853. [CrossRef]
9. Osada, M.; Sasaki, T. Two-dimensional dielectric nanosheets: Novel nanoelectronics from nanocrystal building blocks. *Adv. Mater.* **2012**, *24*, 210–228. [CrossRef]
10. Chen, C.H.; Aizawa, T.; Iyi, N.; Sato, A.; Otani, S. Structural refinement and thermal expansion of hexaborides. *J. Alloys Compd.* **2004**, *366*, L6–L8. [CrossRef]
11. Kimura, S.; Nanba, T.; Tomikawa, M.; Kunii, S.; Kasuya, T. Electronic structure of rare-earth hexaborides. *Phys. Rev. B* **1992**, *46*, 12196–12204. [CrossRef]
12. Peysson, Y.; Ayache, C.; Salce, B.; Rossat-Mignod, J.; Kunii, S.; Kasuya, T. Thermal properties of CeB_6 and LaB_6. *J. Magn. Magn. Mater.* **1985**, *47*, 63–65. [CrossRef]
13. Zhang, H.; Tang, J.; Yuan, J.S.; Ma, J.; Shinya, N.; Nakajima, K.; Murakami, H.; Ohkubo, T.; Qin, L.C. Nanostructured LaB_6 field emitter with lowest apical work function. *Nano Lett.* **2010**, *10*, 3539–3544. [CrossRef]
14. Zhang, H.; Zhang, Q.; Tang, J.; Qin, L.C. Single-crystalline LaB_6 nanowires. *J. Am. Chem. Soc.* **2005**, *127*, 2862–2863. [CrossRef]
15. Ordanyan, S.S.; Vikhman, S.V.; Nesmelov, D.D.; Danilovich, D.P.; Panteleev, I.B. Nonoxide high-melting point compounds as materials for extreme conditions. *Adv. Sci. Tech.* **2014**, *89*, 47–56. [CrossRef]
16. Zhang, H.; Tang, J.; Zhang, L.; An, B.; Qin, L.C. Atomic force microscopy measurement of the Young's modulus and hardness of single LaB_6 nanowires. *Appl. Phys. Lett.* **2008**, *92*, 173121. [CrossRef]
17. Moore, J.E. The birth of topological insulators. *Nature* **2010**, *464*, 194–198. [CrossRef]
18. Dzero, M.; Sun, K.; Galitski, V.; Coleman, P. Topological Kondo insulators. *Phys. Rev. Lett.* **2010**, *104*, 2909–2915. [CrossRef]
19. Jiang, J.; Li, S.; Zhang, T.; Sun, Z.; Chen, F.; Ye, Z.R.; Xu, M.; Ge, Q.Q.; Tan, S.Y.; Niu, X.H.; et al. Observation of possible topological in-gap surface states in the Kondo insulator SmB_6 by photoemission. *Nat. Commun.* **2013**, *4*, 3010. [CrossRef]
20. Li, G.; Xiang, Z.; Yu, F.; Asaba, T.; Lawson, B.; Cai, P.; Tinsman, C.; Berkley, A. Two-dimensional Fermi surfaces in Kondo insulator SmB_6. *Science* **2014**, *346*, 1208–1212. [CrossRef]
21. Kim, D.J.; Xia, J.; Fisk, Z. Topological surface state in the Kondo insulator samarium hexaboride. *Nat. Mater.* **2014**, *13*, 466–470. [CrossRef]
22. He, X.S.; Gan, H.B.; Du, Z.Z.; Ye, B.C.; Zhou, L.; Tian, Y.; Deng, S.Z.; Guo, G.P.; Lu, H.Z.; Liu, F.; He, H.T. Magnetoresistance anomaly in topological Kondo insulator SmB_6 nanowires with strong surface magnetism. *Adv. Sci.* **2018**, *5*, 1700753. [CrossRef]
23. Tai, G.; Hu, T.S.; Zhou, Y.G.; Wang, X.F.; Kong, J.Z.; Zeng, T.; You, Y.C.; Wang, Q. Synthesis of atomically thin boron films on copper foils. *Angew. Chem. Int. Ed.* **2015**, *127*, 15693–15697. [CrossRef]
24. Mannix, A.J.; Zhou, X.F.; Kiraly, B.; Wood, D.; Alducin, D.; Myers, B.D.; Liu, X.L.; Fisher, B.L.; Santiago, U.; Guest, J.R.; et al. Synthesis of borophenes: Anisotropic, two-dimensional boron polymorphs. *Science* **2015**, *350*, 1513–1516. [CrossRef] [PubMed]
25. Feng, B.J.; Zhang, J.; Zhong, Q.; Li, W.B.; Li, S.; Li, H.; Cheng, P.; Meng, S.; Chen, L.; Wu, K.H. Experimental realization of two-dimensional boron sheets. *Nat. Chem.* **2016**, *8*, 563–568. [CrossRef]
26. Jiang, H.R.; Lu, Z.; Wu, M.C.; Ciucci, F.; Zhao, T.S. Borophene: A promising anode material offering high specific capacity and high rate capability for lithium-ion batteries. *Nano Energy* **2016**, *23*, 97–104. [CrossRef]
27. Li, W.L.; Chen, Q.; Tian, W.J.; Bai, H.; Zhao, Y.F.; Hu, H.S.; Li, J.; Zhai, H.J.; Li, S.D.; Wang, L.S. The B_{35} cluster with a double-hexagonal vacancy: A new and more flexible structural motif for borophene. *J. Am. Chem. Soc.* **2014**, *136*, 12257–12260. [CrossRef]
28. Zhang, X.M.; Hu, J.; Cheng, Y.; Yang, H.Y.; Yao, Y.; Yang, S.A. Borophene as an extremely high capacity electrode material for Li-ion and Na-ion batteries. *Nanoscale* **2016**, *8*, 15340–15347. [CrossRef]

29. Xia, Y.; Yang, P.; Sun, Y.; Wu, Y.; Mayers, B.; Gates, B.; Yin, Y.; Kim, F.; Yan, H. One-dimensional nanostructures: Synthesis, characterization, and applications. *Adv. Mater.* **2003**, *15*, 353–389. [CrossRef]
30. Kauffman, D.R.; Star, A. Single-walled carbon-nanotube spectroscopic and electronic field-effect transistor measurements: A combined approach. *Small* **2007**, *3*, 1324–1329. [CrossRef]
31. Li, H.; Wang, X.; Xu, J.; Zhang, Q.; Bando, Y.; Golberg, D.; Ma, Y.; Zhai, T.Y. One-Dimensional CdS Nanostructures: A promising candidate for optoelectronics. *Adv. Mater.* **2013**, *25*, 3017–3037. [CrossRef] [PubMed]
32. Wang, Z.X.; Safdar, M.; Jiang, C.; He, J. High-performance UV–Visible–NIR broad spectral photodetectors based on one-dimensional In_2Te_3 nanostructures. *Nano Lett.* **2012**, *12*, 4715–4721. [CrossRef]
33. Ahn, S.H.; Kim, D.J.; Chi, W.S.; Kim, J.H. One-dimensional hierarchical nanostructures of TiO_2 nanosheets on SnO_2 Nanotubes for high efficiency solid-state dye-sensitized solar Cells. *Adv. Mater.* **2013**, *25*, 4893–4897. [CrossRef]
34. Xu, J.Q.; Hou, G.H.; Li, H.Q.; Zhai, T.Y.; Dong, B.P.; Yan, H.L.; Wang, Y.R.; Yu, B.H.; Bando, Y.; Golberg, D. Fabrication of vertically aligned single-crystalline lanthanum hexaboride nanowire arrays and investigation of their field emission. *NPG Asia Mater.* **2013**, *5*, e53. [CrossRef]
35. Zou, X.M.; Wang, J.L.; Liu, X.Q.; Wang, C.L.; Jiang, Y.; Wang, Y.; Xiao, X.H.; Ho, J.C.; Li, J.C.; Jiang, C.Z.; et al. Rational design of sub-parts per million specific gas sensors array based on metal nanoparticles decorated nanowire enhancement-mode transistors. *Nano Lett.* **2013**, *13*, 3287–3292. [CrossRef]
36. Zhao, X.; Cai, B.; Tang, Q.; Tong, Y.; Liu, Y. One-dimensional nanostructure field-effect sensors for gas detection. *Sensors* **2014**, *14*, 13999–14020. [CrossRef]
37. Tian, J.; Xu, Z.C.; Shen, C.M.; Liu, F.; Xu, N.S.; Gao, H.J. One-dimensional boron nanostructures: Prediction, synthesis, characterizations, and applications. *Nanoscale* **2010**, *2*, 1375–1389. [CrossRef]
38. Chi, M.; Zhao, Y.; Fan, Q.; Han, W. The synthesis of PrB_6 nanowires and nanotubes by the self-catalyzed method. *Ceram. Int.* **2014**, *40*, 8921–8924. [CrossRef]
39. Han, W.; Qiu, Y.; Zhao, Y.M.; Zhang, H.; Chen, J.; Sun, S.; Lan, L.F.; Fan, Q.H.; Li, Q.D. Low-temperature synthesis and electronic transport of topological insulator SmB_6 nanowires. *CrystEngComm.* **2016**, *18*, 7934. [CrossRef]
40. Fan, Q.; Zhang, Q.; Zhao, Y.; Ding, Q.W. Field emission from one-dimensional single-crystalline NdB_6 nanowires. *J. Rare Earths.* **2013**, *31*, 145–148. [CrossRef]
41. Ji, X.H.; Zhang, Q.Y.; Xu, J.Q.; Zhao, Y.M. Rare-earth hexaborides nanostructures: Recent advances in materials, characterization and investigations of physical properties. *Prog. Solid State Chem.* **2011**, *39*, 51–69. [CrossRef]
42. Zhang, H.; Zhang, Q.; Zhao, G.; Tang, J.; Zhou, O.; Qin, L.C. Single-crystalline GdB_6 nanowire field emitters. *J. Am. Chem. Soc.* **2005**, *127*, 13120–13121. [CrossRef]
43. Wang, X.J.; Jiang, Y.D.; Lin, Z.L.; Qi, K.C.; Wang, B.L. Field Emission Characteristics of Single Crystal LaB_6 Field Emitters Fabricated by Electrochemical Etching Method. *J. Phys. D Appl. Phys.* **2009**, *42*, 055409. [CrossRef]
44. Gan, H.B.; Peng, L.X.; Yang, X.; Tian, Y.; Xu, N.S.; Chen, J.; Liu, F.; Deng, S.Z. A moderate synthesis route of 5.6 mA-current LaB_6 nanowire film with recoverable emission performance towards cold cathode electron source applications. *RSC Adv.* **2017**, *7*, 24848–24855. [CrossRef]
45. Yang, X.; Gan, H.B.; Tian, Y.; Xu, N.S.; Deng, S.Z.; Chen, J.; Chen, H.J.; Liang, S.D.; Liu, F. An easy way to controllably synthesize one-dimensional SmB_6 topological insulator nanostructures and exploration of their field emission applications. *Chin. Phys. B.* **2017**, *26*, 118103. [CrossRef]
46. Cao, L.M.; Zhang, Z.; Sun, L.L.; Gao, C.X.; He, M.; Wang, Y.Q.; Li, Y.C.; Zhang, G.L.; Wang, W.K. Well aligned boron nanowire arrays. *Adv. Mater.* **2001**, *13*, 1701–1704. [CrossRef]
47. Meng, X.M.; Hu, J.Q.; Jiang, Y.; Lee, C.S.; Lee, S.T. Boron nanowires synthesized by laser ablation at high temperature. *Chem. Phys. Lett.* **2003**, *370*, 825. [CrossRef]
48. Liu, F.; Tian, J.F.; Bao, L.H.; Yang, T.Z.; Shen, C.M.; Lai, X.Y.; Xiao, Z.M.; Deng, S.Z.; Chen, J.; She, J.C.; Xu, N.S.; Gao, H.J. Fabrication of vertically aligned single crystalline boron nanowire arrays and investigation of their field emission behavior. *Adv. Mater.* **2008**, *20*, 2609–2615. [CrossRef]
49. Patel, R.B.; Chou, T.; Iqbal, Z. Synthesis of boron nanowires, nanotubes, and nanosheets. *J. NanoMater.* **2015**, *16*, 14. [CrossRef]

50. Peng, L.X.; Wen, J.X.; Chen, H.J.; Zheng, Z.B.; Xu, N.S.; Chen, J.; Deng, S.Z.; Liu, F. Tetragonal single crystalline boron nanowires with strong anisotropic light scattering behaviors and photocurrent response in visible light regime. *Small* **2018**, *14*, 1704135. [CrossRef] [PubMed]
51. Liu, F.; Gan, H.B.; Tang, D.M.; Cao, Y.Z.; Mo, X.S.; Chen, J.; Deng, S.Z.; Xu, N.S.; Golberg, D.; Bando, Y. Growth of large-scale boron nanowire patterns with identical Base-Up Mode and in Situ Field Emission Studies of Individual Boron Nanowire. *Small* **2014**, *10*, 685–693. [CrossRef]
52. Ciuparu, D.; Klie, R.F.; Zhu, Y.M.; Pfefferle, L. Synthesis of pure boron single-wall nanotubes. *J. Phys. Chem. B* **2004**, *108*, 3967–3969. [CrossRef]
53. Liu, F.; Shen, C.M.; Su, Z.J.; Ding, X.L.; Deng, S.Z.; Chen, J.; Xu, N.S.; Gao, H.J. Metal-like single crystalline boron nanotubes: Synthesis and in situ study on electric transport and field emission properties. *J. Mater. Chem.* **2010**, *20*, 2197–2205. [CrossRef]
54. Novoselov, K.S.; Geim, A.K.; Morozov, S.V.; Jiang, D.; Zhang, Y.; Dubonos, S.V.; Dubonos, S.V.; Grigorieva, I.V.; Firsov, A.A. Electric field effect in atomically thin carbon films. *Science* **2004**, *306*, 666–669. [CrossRef]
55. Sun, H.; Li, Q.; Wan, X.G. First-principles study of thermal properties of borophene. *Phys. Chem. Chem. Phys.* **2016**, *18*, 14927–14932. [CrossRef]
56. Peng, B.; Zhang, H.; Shao, H.Z.; Xu, Y.F.; Zhang, R.J.; Zhu, H.Y. The electronic, optical, and thermodynamic properties of borophene from first-principles calculations. *J. Mater. Chem. C* **2016**, *4*, 3592–3598. [CrossRef]
57. Wang, V.; Geng, W.T. Lattice defects and the mechanical anisotropy of borophene. *J. Phys. Chem. C* **2017**, *121*, 10224–10232. [CrossRef]
58. Campbell, G.P.; Mannix, A.J.; Emery, J.D.; Emery, J.D.; Lee, T.L.; Guisinger, N.P.; Hersam, M.C.; Bedzyk, M.J. Resolving the chemically discrete structure of synthetic borophene polymorphs. *Nano Lett.* **2018**, *18*, 2816–2821. [CrossRef]
59. Li, W.B.; Kong, L.J.; Chen, C.Y.; Gou, J.; Sheng, S.X.; Zhang, W.F.; Li, H.; Chen, L.; Cheng, P.; Wu, K.H. Experimental realization of honeycomb borophene. *Sci. Bull.* **2018**, *63*, 282–286. [CrossRef]
60. Wu, R.T.; Drozdov, I.K.; Eltinge, S.; Zahl, P.; Ismail-Beigi, S.; Božović, I.; Gozar, A. Large-area single-crystal sheets of borophene on Cu (111) surfaces. *Nat. Nanotechnol.* **2018**, *1*, 44–49. [CrossRef]
61. Tsai, H.S.; Hsiao, C.H.; Lin, Y.P.; Chen, C.W.; Ouyang, H.; Liang, J.H. Fabrication of multilayer borophene on insulator structure. *Small* **2016**, *12*, 5251–5255. [CrossRef]
62. Li, H.L.; Jing, L.; Liu, W.W.; Lin, J.J.; Tay, R.Y.; Tsang, S.H.; Teo, H.T. Scalable production of few-layer boron sheets by liquid-phase exfoliation and their superior supercapacitive performance. *ACS Nano* **2018**, *12*, 1262–1272. [CrossRef]
63. Xu, L.C.; Du, A.; Kou, L. Hydrogenated borophene as a stable two-dimensional Dirac material with an ultrahigh Fermi velocity. *Phys. Chem. Chem. Phys.* **2016**, *18*, 27284–27289. [CrossRef]
64. Nakhaee, M.; Ketabi, S.A.; Peeters, F.M. Tight-binding model for borophene and borophane. *Phys. Rev. B* **2018**, *97*, 125424. [CrossRef]
65. Zhou, X.F.; Dong, X.; Oganov, A.R.; Zhu, Q.; Tian, Y.J.; Wang, H.T. Semimetallic two-dimensional boron allotrope with massless Dirac fermions. *Phys. Rev. B* **2014**, *112*, 085502. [CrossRef]
66. Zhang, R.; Li, Z.; Yang, J. Two-dimensional stoichiometric boron oxides as a versatile platform for electronic structure engineering. *J. Phys. Chem. Lett.* **2017**, *8*, 4347–4353. [CrossRef]
67. Zhong, C.Y.; Wu, W.K.; He, J.J.; Ding, G.Q.; Liu, Y.; Li, D.F.; Yang, S.Y.; Zhang, G. Two-dimensional honeycomb borophene oxide: Strong anisotropy and nodal loop transformation. *Nanoscale* **2019**, *11*, 2468. [CrossRef]
68. Zhang, H.; Tang, J.; Yuan, J.S.; Yamauchi, Y.; Suzuki, T.T.; Shinya, N.; Nakajima, K.; Qin, L.C. An ultrabright and monochromatic electron point source made of a LaB_6 nanowire. *Nat. Nanotechnol.* **2016**, *11*, 273–279. [CrossRef]
69. Shen, Y.; Xu, N.S.; Deng, S.Z.; Zhang, Y.; Liu, F.; Chen, J. A Mo nanoscrew formed by crystalline Mo grains with high conductivity and excellent field emission properties. *Nanoscale* **2014**, *6*, 4659–4668. [CrossRef]
70. Sun, Y.N.; Yun, K.N.; Leti, G.; Lee, S.H.; Song, Y.H.; Lee, C.J. High-performance field emission of carbon nanotube paste emitters fabricated using graphite nanopowder filler. *Nanotechnology* **2017**, *28*, 065201. [CrossRef]

71. Xing, G.Z.; Fang, X.S.; Zhang, H.; Wu, T. Ultrathin Single-Crystal Zno Nanobelts: Ag-catalyzed growth and field emission property. *Nanotechnology* **2010**, *21*, 255701. [CrossRef]
72. Pan, Z.W.; Wang, N.; Lee, C.S.; Xie, S.S. Oriented silicon carbide nanowires: Synthesis and field emission properties. *Adv. Mater.* **2000**, *12*, 1186–1190. [CrossRef]
73. Xue, Q.; Gan, H.B.; Huang, Y.; Zhu, M.S.; Pei, Z.X.; Li, H.F.; Deng, S.Z.; Liu, F.; Zhi, C.Y. Boron element nanowires electrode for supercapacitors. *Adv. Energy Mater.* **2018**, 1703117. [CrossRef]
74. Xue, Q.; Tian, Y.; Deng, S.Z.; Huang, Y.; Zhu, M.S.; Pei, Z.X.; Li, H.F.; Liu, F.; Zhi, C.Y. LaB_6 nanowires for supercapacitors. *Mater. Today Energy* **2018**, *10*, 28–33. [CrossRef]
75. Dzero, M.; Sun, K.; Coleman, P.; Galitski, V. Theory of topological Kondo insulators. *Phys. Rev. B* **2012**, *85*, 045130. [CrossRef]
76. Neupane, M.; Alidoust, N.; Xu, S.Y.; Kondo, T.; Ishida, Y.; Kim, D.J.; Liu, C.; Belopolski, I.; Jo, Y.J.; Chang, T.R.; et al. Surface electronic structure of the topological Kondo-insulator candidate correlated electron system SmB_6. *Nat. Commun.* **2013**, *4*, 2991. [CrossRef]
77. Jiao, L.; Rößler, S.; Kim, D.J.; Tjeng, L.H.; Fisk, Z.; Steglich, F.; Wirth, S. Additional energy scale in SmB_6 at low-temperature. *Nat. Commun.* **2016**, *7*, 13762. [CrossRef]
78. Jiao, L.; Rößler, S.; Kim, D.J.; Tjeng, L.H.; Fisk, Z.; Steglich, F.; Wirth, S. Magnetic and defect probes of the SmB_6 surface state. *Sci. Adv.* **2018**, *4*, eaau4886. [CrossRef]
79. Kong, L.J.; Zhou, Y.; Liu, S.; Lin, Z.; Zhang, L.; Lin, F.; Tang, D.S.; Wu, H.C.; Liu, J.F.; Lu, H.Z.; et al. Spin-polarized surface state transport in a topological Kondo insulator SmB_6 nanowire. *Phys. Rev. B* **2017**, *95*, 235410. [CrossRef]
80. Xu, N.; Matt, C.E.; Pomjakushina, E.; Dil, J.H.; Landolt, G.; Ma, J.Z.; Shi, X.; Dhaka, R.S.; Plumb, N.S.; Radovic, M.; et al. Surface vs bulk electronic structures of a moderately correlated topological insulator YbB_6 revealed by ARPES. *arXiv*, 2014; arXiv:1405.0165.
81. Xia, M.; Jiang, J.; Ye, Z.R.; Wang, Y.H.; Zhang, Y.; Chen, S.D.; Niu, X.H.; Xu, D.F.; Chen, F.; Chen, X.H.; et al. Angle-resolved Photoemission Spectroscopy Study on the Surface States of the Correlated Topological Insulator YbB_6. *Sci. Rep.* **2014**, *4*, 5999. [CrossRef]
82. Ezawa, M. Triplet fermions and Dirac fermions in borophene. *Phys. Rev. B* **2017**, *96*, 035425. [CrossRef]
83. Zhong, H.X.; Huang, K.X.; Yu, G.D.; Yuan, S.J. Electronic and mechanical properties of few-layer borophene. *Phys. Rev. B* **2018**, *98*, 054104. [CrossRef]
84. Yi, W.C.; Liu, W.; Botana, J.; Zhao, L.; Liu, Z.; Liu, J.Y.; Miao, M.S. Honeycomb boron allotropes with Dirac cones: A true analogue to graphene. *J. Phys. Chem. Lett.* **2017**, *8*, 2647–2653. [CrossRef]
85. Takeda, H.; Kuno, H.; Adachi, K. Solar control dispersions and coatings with rare-earth hexaboride nanoparticles. *J. Am. Ceram. Soc.* **2008**, *91*, 2897–2902. [CrossRef]
86. Adachi, K.; Miratsu, M.; Asahi, T. Absorption and scattering of near-infrared light by dispersed lanthanum hexaboride nanoparticles for solar control filters. *J. Mater. Res.* **2010**, *25*, 510–521. [CrossRef]
87. Chen, C.J.; Chen, D.H. Preparation of LaB_6 nanoparticles as a novel and effective near-infrared photothermal conversion material. *Chem. Eng. J.* **2012**, *180*, 337–342. [CrossRef]
88. Lai, B.H.; Lin, Y.R.; Chen, D.H. Fabrication of LaB_6@ SiO_2/Au composite nanoparticles as a catalyst with near infrared photothermally enhanced activity. *Chem. Eng. J.* **2013**, *223*, 418–424. [CrossRef]
89. Sani, E.; Mercatelli, L.; Meucci, M.; Zoli, L.; Sciti, D. Lanthanum hexaboride for solar energy applications. *Sci. Rep.* **2017**, *7*, 718. [CrossRef] [PubMed]
90. Li, Q.; Zhao, Y.; Fan, Q.; Han, W. Synthesis of one-dimensional rare earth hexaborides nanostructures and their optical absorption properties. *Ceram. Int.* **2017**, *43*, 10715–10719. [CrossRef]
91. Zhou, Y.; Lai, J.W.; Kong, L.J.; Ma, J.C.; Lin, Z.L.; Lin, F.; Zhu, R.; Xu, J.; Huang, S.M.; Tang, D.S.; et al. Single crystalline SmB_6 nanowires for self-powered, broadband photodetectors covering mid-infrared. *Appl. Phys. Lett.* **2018**, *112*, 162106. [CrossRef]
92. Liao, Z.M.; Xu, J.; Zhang, J.M.; Yu, D.P. Photovoltaic effect and charge storage in single ZnO nanowires. *Appl. Phys. Lett.* **2008**, *93*, 023111. [CrossRef]

93. Liu, F.; Li, L.F.; Guo, T.Y.; Gan, H.B.; Mo, X.S.; Chen, J.; Deng, S.Z.; Xu, N.S. Investigation on the photoconductive behaviors of an individual AlN nanowire under different excited lights. *Nanoscale Res. Lett.* **2012**, *7*, 454. [CrossRef]
94. Zhang, M.X.; Liu, Y.; Yang, M.Q.; Zhang, W.; Zhou, J.Y.; Zhang, Z.X.; Xie, E.Q.; Pan, X.J.; Li, S.B. High performance self-powered ultraviolet photodetectors based on electrospun gallium nitride nanowires. *Appl. Surf. Sci.* **2018**, *452*, 43–48. [CrossRef]

nanomaterials

MDPI

Review

High Sensitivity Resists for EUV Lithography: A Review of Material Design Strategies and Performance Results

Theodore Manouras [1,2,*] and Panagiotis Argitis [3,*]

1 Department of Materials Science and Technology, University of Crete, 70013 Heraklion, Greece
2 Institute of Electronic Structure and Laser, Foundation for Research and Technology-Hellas, 70013 Heraklion, Greece
3 Institute of Nanoscience and Nanotechnology, NCSR "Demokritos", 15310 Athens, Greece
* Correspondence: tman@iesl.forth.gr (T.M.); p.argitis@inn.demokritos.gr (P.A.)

Received: 6 July 2020; Accepted: 12 August 2020; Published: 14 August 2020

Abstract: The need for decreasing semiconductor device critical dimensions at feature sizes below the 20 nm resolution limit has led the semiconductor industry to adopt extreme ultra violet (EUV) lithography with exposure at 13.5 nm as the main next generation lithographic technology. The broad consensus on this direction has triggered a dramatic increase of interest on resist materials of high sensitivity especially designed for use in the EUV spectral region in order to meet the strict requirements needed for overcoming the source brightness issues and securing the cost efficiency of the technology. To this direction both fundamental studies on the radiation induced chemistry in this spectral area and a plethora of new ideas targeting at the design of new highly sensitive and top performing resists have been proposed. Besides the traditional areas of acid-catalyzed chemically amplified resists and the resists based on polymer backbone breaking new unconventional ideas have been proposed based on the insertion of metal compounds or compounds of other highly absorbing at EUV atoms in the resist formulations. These last developments are reviewed here. Since the effort targets to a new understanding of electron-induced chemical reactions that dominate the resist performance in this region these last developments may lead to unprecedented changes in lithographic technology but can also strongly affect other scientific areas where electron-induced chemistry plays a critical role.

Keywords: EUV lithography; chemically amplified resists; inorganic resists; main chain scission resists; electron-induced chemistry

1. Introduction—Nanostructure Formation in Semiconductor Lithography

The standard methodology for fabricating miniaturized devices with critical dimensions in the micro- and nano- domain in the semiconductor industry has been photolithography, using photosensitive polymeric mainly materials, known as resists, for imaging, accompanied with pattern transfer to the substrate of interest with plasma etching. The uninterrupted patterning technology evolution from the 1960s to the last decade allowed the miniaturization of critical device dimensions from a few microns to the sub 25 nm domain allowing the semiconductor technology progress according to the well-known Moore's law [1–3]. Although this impressive technology development has been mainly achieved by adopting reduced wavelength of the imaging radiation, in the last two decades additional technology breakthroughs such as immersion and double exposure patterning allowed the shrinkage of device dimensions, while keeping the wavelength at 193 nm [4]. In addition, options like directed self-assembly [5–8], e-beam lithography [9–12], maskless techniques [13,14], and nanoimprint technology [15,16] have been heavily explored. At this period, it seems that the implementation of exposure at reduced wavelength, and in particular at 13.5 nm, is the choice of the big semiconductor industries for the future device technology. Extreme ultraviolet (EUV) lithography using 13.5 nm wavelength exposure is expected to be the main industrial option for pushing further the resolution limit in sub 20 nm region. However, EUV sources have limited power making the improvement in resist sensitivity a high-importance issue to fulfill the throughput requirements for high volume manufacturing while maintaining pattern fidelity and uniformity.

Reviews on EUV resists have been published in 2017 summarizing important developments [17,18]. Nevertheless, dramatic changes are under way in the field after consensus in the semiconductor industry community has been reached to accept EUV as the main next generation lithographic option. New unconventional materials are introduced and fundamental studies are under way enabled by improvements in the scientific understanding of the EUV radiation interactions with the resist materials. In this context a very recent review [19], focuses especially on inorganic photoresists covering materials that have been introduced to address the challenges encountered in EUV lithography.

In the present review we will discuss primarily the main factors that influence the resist sensitivity at EUV and we will give priority to promising materials for achieving the high sensitivity demanded. We will focus mostly on the developments during the last years aiming at new generations of materials and on the scientific support from fundamental studies toward technology optimization. Following this approach we discuss first in two separate sections the main characteristics of EUV technology and the issues related to materials absorption at this spectral region and then we review the chemical approaches for high sensitivity EUV resists and the results reported so far on resist materials of advanced performance.

2. EUV Introduction and Main Technical Challenges

The critical dimension (CD) of a projection imaging system such as a lithographic scanner is given by the fundamental equation governing the resolution: $CD = k\lambda/NA$, where λ is the wavelength of the light source and NA is the numerical aperture of the imaging optics. The value of κ is depending on the quality of the imaging optics, the way the mask is illuminated, and the mask pattern itself. As it is obvious from the latter equation, the critical dimension value is proportional to the wavelength of the radiation applied to the photosensitive materials.

The choice of 13.5 nm, or EUV as it is commonly referred to, as the imaging radiation wavelength after the 193 nm has a long history of scientific investigations, technology breakthroughs and controversies that lasted more than two decades [20]. A review of the first years of EUV lithography has been presented in [21], where also the successful demonstration of printing for the first time 19 nm dense lines/spaces with optical lithography using an interferometric Lloyd's mirror maskless set-up is discussed [22]. The current patterning challenges at EUV aiming at a viable industrial technology have been also presented in [21] and more recently in [23,24]. One attractive characteristic of the EUV radiation at 13.5 nm is that it can be combined with imaging technology based on reflective optics, a technology that has reached an acceptable maturity level. This imaging technology, based on multilayer mirrors composed of technologically acceptable materials, Mo and Si, allows imaging of the mask pattern to the wafer. The reflectivity of such multilayers is discussed for instance in [25], and a characteristic graph from this paper is depicted in Figure 1a, where also emission spectra of different alternative sources are given. This graph is also based on data from reference [26]. The reflective optics used in EUV technology is apparently more complicated than the refractive optics used at 193 nm and longer wavelengths, see for instance a scheme of a projection system depicted in Figure 1b, adopted by ref [17], which as stated there is based on a previous presentation [27]. In addition, new technological challenges in mask fabrication had to be addressed, but the technology nowadays has been proved capable for effective radiation imaging. The successful development of this imaging technology for 13.5 nm radiation has played a very important role for the choice of this particular wavelength for next generation lithography.

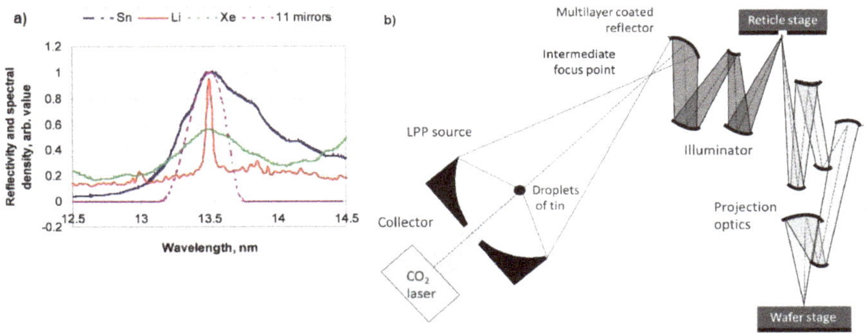

Figure 1. Extreme ultra violet (EUV) technology characteristics: (**a**) Emission of different sources based on Sn, Xe, and Li at the EUV spectral region and calculated near normal incidence reflectivity of a 11 mirror system in the same area. Adapted from [25], with permission from IOP Publishing, 2020 (**b**) A scheme of a EUV lithography system where the different parts, including source, illuminator, reticle stage (mask), and projection optics are depicted. Adapted from [17], with permission from De Gruyter, 2020.

One very challenging issue for the development of industrially competitive EUV lithography technology has been the brightness of the sources that could be used for generation of radiation at 13.5 nm. Plasma discharge approaches and approaches based on laser induced radiation through focusing on specific material targets were explored by many groups toward the development of efficient sources. Today the approach of choice seems to be laser-induced radiation generation by focusing on liquid Sn droplets. Sources based on this approach have been incorporated in the commercial EUV lithography exposure tools available today, as is also shown in the scheme of Figure 1b. The brightness of these sources has been greatly improved over the past decade and played a significant role for the decision to adopt EUV lithography as the next production technology made by a number of big semiconductor manufacturers [28]. Nevertheless, the throughput of the exposure tool remains relatively low compared with older lithographic technologies and for this reason there is an urgent demand for high sensitivity resists that can help the technology to achieve the desired cost efficiency. In addition, we should always keep in mind that the EUV resists must also provide advanced performance characteristics to be suitable for sub 20 nm patterning with industrial standards, related primarily to pattern quality and process demands.

This demand for high sensitivity EUV resists caused a real awakening of the research in the resists field. At this point it should be emphasized that the research effort in the resist development area has been rather slow for a period of more than 10 years (schematically extending from about 2005 to 2015). Indeed, after the abandonment of the option for 157 nm lithography, which had provoked an intense resist development effort, the resist material research was not intense since the 193 nm resists were already mature materials and the modifications needed for immersion lithography were rather minor. Only few groups worldwide were active in the resist field and often the efforts were diverted in different directions. On the other hand a significant research direction for lithographic materials followed by many polymer groups was toward self-assembled block copolymers for directed self-assembling (DSA) lithography [1–4]. The momentum seems to have changed during the last five years after the choice of EUV as the next industrial technology became clear in the lithography community along with the now well-recognized demand for high sensitivity EUV resists.

In the following we will try to make clear why the demand for high sensitivity resists causes a mobilization of the lithography researchers and brings again the resist development effort at a spot of increased interest by many scientists or engineers new to the field.

3. Material Absorption at EUV

The absorption of resist materials at EUV is an issue that only lately has started to attract the attention it deserves [29,30]. From an historical point of view, it should be mentioned that the resist absorbance at EUV was first considered at late nineties and it was recognized that the absorption of resist materials at EUV is mainly determined by their atomic composition. Indeed, the photon energy at 13.5 nm is close to 92 eV which is well above the ionization potential of the atoms that constitute the resist material. The photoemission cross sections of different atoms that can be potentially included in a resist composition are shown in the graph below, Figure 2, image adopted from ref [30] based on data from ref [31].

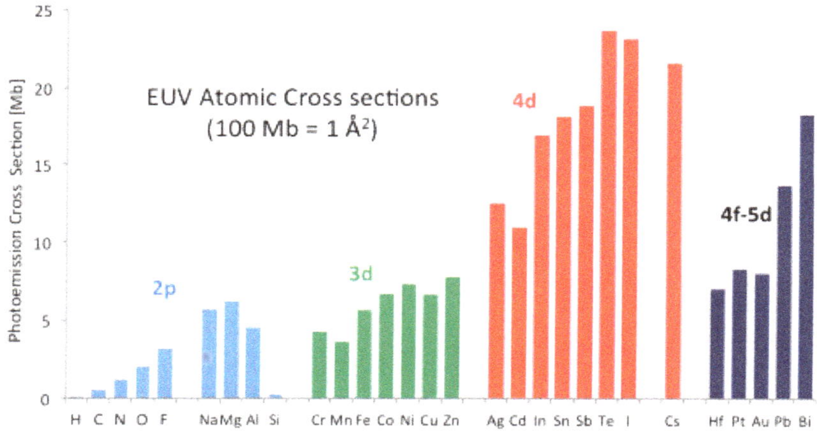

Figure 2. Photoemission cross sections at 92 eV calculated for selected atoms in Mb (Megabarn), where 1 Mb = 10^{-22} m^2 in SI units. Figure adapted from [30], with permission from Elsevier, 2020.

The atoms encountered in a typical resist composition are carbon, oxygen, and hydrogen. As shown in the graph of Figure 2, oxygen has a much higher photoemission cross section at EUV than C, thus increased the ratio of oxygen atoms in the composition is expected to increase the material absorption of 92 eV radiation resulting to enhanced electron emission [30]. On the other hand, one can further control the material absorption at EUV by incorporating molecules containing other atoms such as Hf, Zn, Sn, some of which are considered for the first time, in the resist composition.

In the first years of investigating options for resist materials suitable for exposure at 13.5 nm it was realized that high material absorption was not an issue of concern at EUV as was for 193 nm and 157 nm lithography [32]. On the contrary, it was soon noticed that the typical organic resist materials had rather low absorbance at the thicknesses of interest and this low absorbance would also result in low sensitivity. Thus, the incorporation of atoms in the resist composition that could substantially increase the materials absorbance was proposed among other strategies to increase sensitivity [17,18]. According to the data presented in Figure 2 such atoms could be Sn and I, compounds of which are indeed under investigation as the main components of highly sensitive resists and they will be discussed below.

In a deeper examination of the facts related to the increase of the resists absorbance one should consider the probabilities of absorption exhibited by the different atoms and correlate them with data for the absorption cross section for different wavelengths of ionization radiation. Relevant data can be also found in ref [33] where the different atom cross sections at a range of radiation energies are presented and correlated with specific atomic energy levels. By examining the tables for different atoms, it is clear that the main tendency is that the ionizing radiation interacts preferentially with electrons lying at deeper levels if they are at a distance smaller than 92 eV from the vacuum level. Thus, although for the case of C electrons can be released only from the 2s and 2p levels, for heavier elements there are core electrons at deeper levels which interact preferentially with the radiation. In ref [29] a recent study is reported where molecules of similar composition are compared (see Figure 3). There, it is shown that the substitution of H with e.g., halogen atoms has a dramatic influence on the formation of photoelectrons. The most interesting case is the one of iodine-substituted compound where the photoelectron spectrum intensity is much higher than that of the corresponding molecules containing lower atomic number atoms. This high intensity is due to the existence of core levels at favorable energy position and results in the formation of photoelectrons with rather low electron energies. The understanding of the impact of low energy electrons is a significant issue in designing efficient EUV resists. This point is under investigation in different resist systems, see for instance the case presented below in Figure 9.

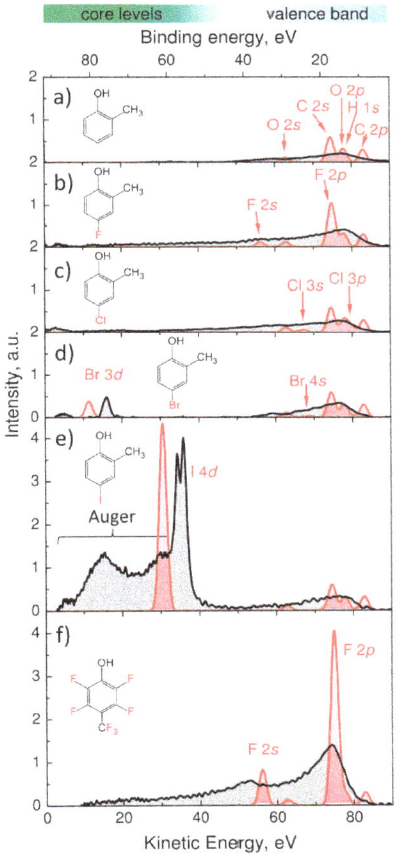

Figure 3. Photoelectron spectra of gas-phase molecules measured using 13.5 nm (92 eV) EUV radiation. Adapted from [29], with permission from AIP publishing, 2020. The kinetic energy of photoelectrons is shown in x axis. The black line corresponds to experimental data, and the red line corresponds to the model. (**a**) 2-methylphenol, (**b**) 4-fluoro-2-methylphenol, (**c**) 4-chloro-2-methylphenol, (**d**) 4-bromo-2-methylphenol, (**e**) 4-iodo-2-methylphenol, and (**f**) 2,3,5,6-tetrafluoro-4-(trifluoromethyl)phenol. The presented data clarify the importance of the inclusion of certain atoms in the resist composition. For instance, from the data presented it is clear that the inclusion of I in the resist composition is expected to greatly enhance the material absorption at EUV.

One more remark that should be made here is that according to the discussion in ref [30] and the tables of references [31] and [33] it appears that it is quite improbable to have interaction of the incoming 92 eV photon with the electrons participating in the chemical bonds since these are substantially closer to the vacuum level. This remark rationalizes the hypothesis that the absorbance at EUV is mainly controlled by the atomic composition of the materials and not by the specific molecules used. Nevertheless, a deeper examination of the issue still has to be undertaken since the above considerations are usually based on energy position arguments and a detailed examination of the different electron states and their possible interactions with electromagnetic radiation has been undertaken only in very few cases.

4. Chemical Directions for Highly Sensitive Resists at EUV

General Considerations

Although the photochemistry involved in 248 and 193 nm is fully studied and the basic events occurring in those wavelengths are well-known, little work has been performed regarding organic material exposure at 13.5 nm. Actually, the chemical roots activated by the ionizing electromagnetic radiation have been so far considered mostly in very general terms not only for EUV but also for other wavelength ranges of ionizing electromagnetic radiation and even for different types of ionizing radiations including electron beams. The main reason for this fact is the difficulty encountered to identify and follow the plethora of possible events that are possible in specific organic chemical systems. Indeed, the ionizing electromagnetic radiation mainly gives rise to the formation of free electrons that can transfer different amounts of energy to the molecules through a number of interactions with the bound electrons. Nevertheless, in many cases the prevailing chemical outcomes are due to the availability of efficient chemical routes in the material, triggered by the formation of active intermediates like certain radicals, acids, or bases that act as initiators or catalysts. Some examples of such routes are the polymer degradation encountered in poly(meth) acrylates and related back-bone breakable polymers or the acid catalyzed deprotection or crosslinking reactions encountered in chemical systems best known as chemically amplified resists.

Such chemical reactions were discussed in X-ray lithography resists investigated until early 2000s when the EUV lithography appeared as an attractive option [34]. X-ray lithography based on radiation with wavelengths in the 0.1–10 nm range had been broadly investigated since these small wavelengths were considered attractive for high resolution applications. Nevertheless, the lack of good imaging technologies for such radiation had led to the adoption of the proximity exposure option which limited the possible gain from using small wavelength. The resists mainly explored during this period had been adaptations of UV or e-beam resists. The chemistries for solubility change observed were similar to the ones encountered in UV region especially for the case of poly(meth)acylate-based polymer resists such as PMMA (poly methyl methacrylate) and also for the acid-catalyzed systems (chemically amplified resists). The sensitivity of such resists in these wavelengths was not investigated in depth and the primary events after the absorption of electromagnetic radiation were not elucidated. The above approach was also adopted in the first resist efforts in EUV (13.5 nm) and the behavior of resists was compared with results obtained with different types of ionizing radiation as in ref [34]. A review of 2010 presents in a detailed manner radiation chemistry issues especially for the case of chemically amplified resists [35].

During the past decade the need for increased sensitivity which proved to be a significant requirement for the EUV lithography to become a competent industrial option ignited the research effort for especially designed sensitive and high-performing materials. The development and optimization of such high sensitivity resists pushed the scientific community to investigate in depth the radiation induced events and initiated a very demanding research area for new types of lithographic resists and for understanding fundamental issues [36]. In this respect, since photoelectrons and secondary electrons resulting from of ionizing radiation were recognized to play the main role in solubility changing reactions [37], attention was drawn to the fundamentals of electron-induced chemistry known from other scientific fields as described for instance in [38,39]. Recently a study on the role of low energy electrons in a tin containing resist was published [40] and revealed their important role in resist chemistry. Similar studies are expected to dramatically improve the understanding of the resist chemistries at EUV.

In the following sections the different material options under investigation are discussed. These options include the introduction of unconventional materials such as the metal oxide containing resists and the in-depth understanding and optimization of the radiation chemistry encountered in more traditional resists as the back bone breakable polymers and the chemically amplified systems.

4.1. Chemically Amplified Resists

The imaging chemistry of the industrial resists used in 248 nm and 193 nm is based on an acid-catalyzed mechanism known as "Chemical Amplification." The chemical amplified resists (CARs) are mainly consisted of a main polymeric or molecular component as a matrix, photoacid generator molecules, and base quencher molecules. Upon irradiation, the photo-acid generator molecules (PAGs) interact with the light to generate acids; then the change in dissolution rate of the matrix begins during a subsequent post-exposure bake (PEB) step. During this step reactions of specific groups of the matrix are catalyzed by the photogenerated acid molecules changing hydrophilicity, or inducing crosslinking or back bone scission, and hence alter solubility, usually in aqueous base developers. The design of the resists used for 248 nm and 193 nm relied on the tune of the absorptivity of the matrix, the type and size of the molecular components in the matrix, the expected etch resistance during pattern transfer as well as on several additional properties which controlled the performance of the resist [4]. Nevertheless, no such absorption selectivity of the main components of the resist materials can be expected when an ionizing radiation such as EUV (13.5 nm) is used. In fact, traditional CAR materials are very transparent in this wavelength, a fact that constitutes a major problem since it reduces the sensitivity of these materials. For this reason, new materials consisting of atoms with enhanced absorptivity in EUV regime have been proposed as alternative to traditional CARs. As discussed above the interaction of EUV photons with matter is mediated by photoelectrons, as well as secondary electrons. The broad energy distribution of these electrons that induce chemical changes in the material results in a variety of chemical events. The processes involved are quite complex and hard to disentangle, as photon illumination initiates often an entire electron cascade and the possibility of discerning the role of electrons with different energies is in many cases practically impossible. Consequently, it is not surprising that also in the case of CARs, as in other material cases, the interaction of low energy electrons (<15 eV) with soft matter is not well understood.

Toward improving our understanding on the behavior of CAR materials at EUV new experimental approaches have been proposed. Pollentier et al. in their paper [41] proposed an experimental approach based on residual gas analysis (RGA) to distinguish between photogenerated acid related reactions and direct scission reactions in model resist systems. In a more recent paper [42], also by Polentier et al., the residual gas analysis (RGA) was optimized and this method was successfully used to quantify the photogenerated acid yield and the reactions leading to insolubility as a function of EUV dose for a number of CAR formulations related to a commercial material.

On the other hand, new PAGs are introduced and are expected to perform more effectively at EUV compared to traditional PAGs. In a characteristic paper by Torti et al. [41], new fluorinated aryl sulfonates were investigated as initiators for EUV lithography and compared with traditional PAGs in epoxy-based hybrid materials. In Figure 4 the new proposed fluorinated PAGs are shown (upper row) along with the traditional PAGs which were used for comparison. As expected according to the atomic cross sections presented in Figure 2, the use of fluorinated PAGs resulted in resist formulations of enhanced sensitivity.

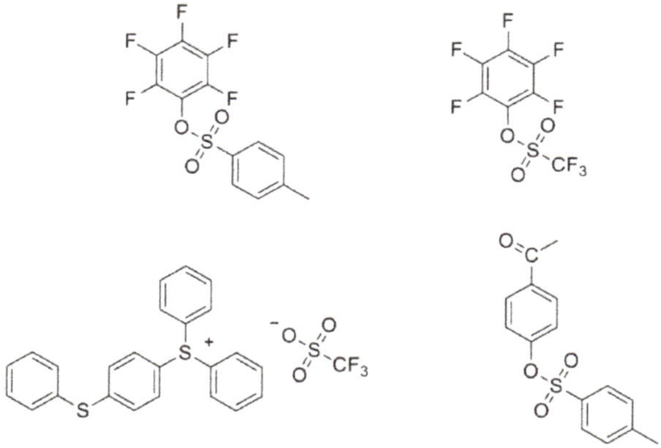

Figure 4. Fluorinated PAGs especially designed for EUV (**upper row**) and traditional PAGs (**lower row**) that were used for comparison in epoxy-based resist formulations. Adapted from [43], with permission from John Wiley & Sons, 2020.

In the next section, we review the CAR materials used nowadays in EUVL as well as the proposed ways to improve their performance.

4.1.1. Polymeric Systems Based on Hydrophilicity Change

Polymeric chemically amplified resists have been deeply investigated and look most established at this point. Traditional 248 nm and 193 nm photoresists based on acid catalyzed hydrophilicity change of polymer pendant groups were first evaluated in EUV lithography. Typically, photoacid generator (PAG) molecules are applied as a source of acid catalyst triggered by EUV exposure. The PAG components can exist in the resist formulation as individual moieties or as incorporated groups in the resins (polymer bound PAGs). In this section, recent progress in the design of these materials following various strategies is described.

In an early study, Yamamoto et al. used polyhydroxystyrene and partially protected polyhydroxystyrene with tert-butyl and adamantly groups to examine the dependence of acid generation efficiency on the protection ratio of hydroxy groups in chemically amplified EUV resists. Their study showed that protection of hydroxy groups clearly affects the acid generation process. The incorporation of t-butyl groups decreased the acid generation efficiency while adamanthoxyethyl groups increased this efficiency [44]. As has been mentioned above the primitive efforts for development of resists for EUV lithography followed the deep knowledge that had been acquired from the mature resists used in longer wavelengths. The first evaluated resists in EUV lithography were simply extensions of the commercially available previously developed environmentally stable chemical amplification photoresists (ESCAP) consisting of poly(p-hydroxy)styrene (PHS)/styrene/t-butyl acrylate copolymers as they are presented in Figure 5.

Fedynyshyn et al. reported a study on ESCAP resist using an EUV illumination source which revealed the effect of the matrix on the acid generation efficiency. They found out that while the absorbance was considered to be the important parameter, other polymer properties also strongly influenced the acid generation efficiency of the used PAG. The nature of the polymeric matrix, i.e., the different atoms and chemical groups that are present, plays an important role in the acid efficiency and has to be taken into account in order to achieve effective sensitization on the PAG. It should be noticed that these specific resists have well-defined processes and therefore they were extensively used in the early development of exposure tools. They showed capability for 30 nm half-pitch (hp) resolution with the required sensitivity, but it became clear that new designed resist materials optimized for the 13.5 nm wavelength exposure were needed for further performance improvement [45].

Tamaoki et al. compared polymer-bound PAG and polymer-blended PAG type resist materials in terms of blur, swelling properties, and lithographic performance. They found out that the polymer with bound PAG acquired very small blur with higher sensitivity and suppressed swelling very well [46]. Tarutani et al. investigated the effect of the hydrophobicity on the ultimate resolution of a photoresist material. Accordingly, they synthesized a series of polymers in which their hydrophobicity was changed by utilizing polymers having a different chemical structure and protection ratio. The polymer with the higher hydrophobicity can resolve 16 nm hp line/spacer using an EUV tool [47]. In another work of the same group, CARs with different sensitivities were synthesized in order to examine the impact of sensitivity on 15 nm hp resolution. Their results suggested that there was a strong relation between the exposure dose and the quality of the created structures. The resist with lower sensitivity (>30 mJ/cm^2) could resolve 15 nm hp because of the low impact of photon shot noise [48]. Other issues concerning EUV resists include Out of Band (OoB) radiation which deteriorates the resist performance. OoB is estimated to be about 4% of the radiation from the EUV tool. PAGs with selectivity to EUV radiation have been designed and synthesized to minimize the effect of OoB in resist performance. These PAGs exhibited decreased Deep UV (DUV) absorption by the incorporation of insensitive cations and they were considered in resist formulations based both on blended PAG and polymer-bound PAG. The concept and merit of OoB insensitivity was confirmed by the obtained DUV and EUV sensitivity [49–52].

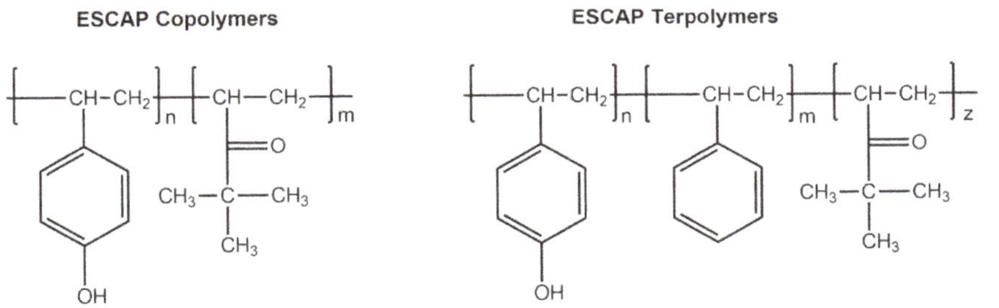

Figure 5. Polymers based on the environmentally stable chemical amplification photoresist (ESCAP) approach that are discussed in EUV resist formulations in ref [43]. The copolymer (**left**) consisted of poly(p-hydroxy styrene)-r-poly(t-butyl acrylate), the terpolymer (**right**) consisted of poly(p-hydroxy styrene)-r-poly(styrene)-r-poly(t-butyl acrylate).

Liu et al. found out a chemical way to introduce PAG moieties in polymeric chains. This direct modification of polymeric chains was based on the introduction of sulfonium chloride onto the benzene ring of PHS by a convenient direct reaction at a high rate and then the anion was exchanged into perfluroalkyl sulfonate [53]. Narasimhan et al. studied the interaction between electrons having similar energies to secondary electrons produced during the EUV exposures and investigated resist materials using both experimentation and modeling [54]. JSR corporation developed new CAR EUV resist formulations showing short acid diffusion length as well as new sensitizers with higher EUV photo absorption atoms. These resist formulations showed capability to resolve 13 nm hp. Furthermore, addition of a new sensitizer to conventional CARs can improve the sensitivity about 9–16% with no affection on resolution and line width roughness (LWR) [55,56]. Krysak et al. established a method of pattern collapse mitigation in CARs using a dry develop rise material. This method is able to extend the resolution limit of chemical amplified resist susceptible to pattern collapse, resolving 24 nm pitch features [57]. Thackeray et al. found out that resists with low activation energies of deprotection can achieve superior process window and exposure latitude in the 35 nm resolution regime. In addition, they use photo-destroyable quenchers to minimize the loss of the photogenerated protons [58,59]. Fujii et al. improved resist sensitivity by increasing the proton source content in the polymer and applying an electron withdrawing group on PAG cation. They fabricated 13 nm hp line/spaces using newly developed chemical amplified resist materials combining the aforementioned items for enhancing the acid generation efficiency and suppressing the acid diffusion length [60]. Yamamoto et al. improved the sensitivity of chemically amplified resist by adding a metal sensitizer. The improvement in sensitivity is not a result of higher EUV photon absorption but of higher acid yield and electron efficiency. They achieved 43% improvement in sensitivity as well as reduction in LWR [61]. Fallica et al. measured the rate of bleaching by tracking the change in absorptivity of CARs during exposure to EUV light. They found out that the bleaching speed depended on the PAG-polymer interaction. This is an important effort in resist design and development because of the fact that Dill C parameter can be tuned in a variety of ways [62]. Lee et al. developed a multiscale model for EUV patterning of CARs. This model gives insight information about the chemical reactions (diffusion, quenching, deprotection etc.,) taking place during the structuring. Furthermore, it can predict the polymer loss during PEB as well as LER performance [63].

Additional ways to improve chemical amplified resist performance have also been proposed. Brainard et al. developed and evaluated eleven acid amplifiers for use in EUV photoresist. Acid amplifier (AA) is a compound that decomposed rapidly after the influence of an acid to generate more acid. They used an ESCAP photoresist measuring the performance after the addition of an AA. They found out that AA producing fluorinated sulfonic acids shows great promise in helping EUV resists, simultaneously improving the resolution, LER, and sensitivity [64]. Sekigushi et al. studied how the addition of metal into an ESCAP resist influences the sensitivity. They performed transmittance measurements and sensitivity evaluation of an ESCAP type resist doped with ZrO_2 and TeO_2 nanoparticles which have low and high absorptivity in EUV radiation respectively. ZrO_2 nanoparticles caused no change in absorption and only slight sensitivity enhancement, whereas TeO_2 nanoparticles enhanced both absorption and sensitivity [65]. Jiang et al. compared the impact of metal salt sensitizers and halogenated sensitizers on EUV sensitivity. Metal sensitizers improve both EUV photon absorption and electron yield resulting in higher sensitivity. Fluorine and iodine sensitizers also improved electron generation with their higher absorption but the chemical environment where these halogens are bonded influences heavily the sensitivity [66]. Nagahara et al. introduced Flood exposure Assisted Chemical gradient Enhancement Technology (FACET) to improve the resolution, process control, roughness, patterning failure, and sensitivity in EUV resist. Their concept is based on the increase in UV absorption after the influence of a EUV generated acid. After that the UV flood exposure induces acid production in these areas [67]. Okamoto et al. investigated the effect of the addition of diphenyl sulfones into EUV CARs. They confirmed that the addition of sulfones significantly increases the acid yield leading to the increase of the sensitivity of the CARs [68].

4.1.2. Polymeric Systems Based on Acid-Catalyzed Main Chain Scission

Chemically amplified resists that can undergo acid catalyzed chain scission have been proposed as an alternative to resists based on hydrophilicity change in order to achieve high sensitivity at EUV. In these resists the imaging chemistry is based on acid-catalyzed back bone breaking instead of the deprotection of a pendant group. The removable units are attached at low molecular weight monomeric units instead of polymeric chains. Cardineau et al. synthesized polymers containing either tertiary aliphatic or tertiary benzylic cleavable ethers. But, further development would be made to overcome damage drawbacks such as serpentine pattern deformation and bridging [69]. In a similar approach, Manouras et al. designed and synthesized a random copolymer containing acid-cleavable bonds along the main chain. As it is presented in Figure 6, the random copolymer consisted of three different monomers corresponding at percentages of approximately 85%, 10%, and 5%, connected with acid labile bonds. Each monomer introduces or tunes a specific property of the random copolymer such as etch resistance, EUV absorptivity, Tg etc. Obviously, the bond strength of the monomer at high percentage dominated the polymer back bone breaking. Resist films based on the synthesized polymer have shown satisfactory etch resistance, due to the high aromatic moiety content. The sensitivity of this polymer is very high and the high values of Tg maintain the polymeric thin film untouched in the exposure step. The main chain of the polymer steadily chopped in the post exposure bake step in which the photogenerated acid can easily penetrate into the polymeric matrix. Resist formulations based on the synthesized random copolymer were exposed to EUV radiation exhibiting a high potential for industrial applications. A characteristic contrast curve shows that the resists are ultra-high sensitive (~0.5 mJ/cm^2 using 5% PAG and 0.25% quencher) with satisfactory contrast, whereas they also showed very good etch resistance (1/10 selectivity to SiO_2). Imaging experiments using EUV lithography have demonstrated capability for 20 nm lines with ultra-low doses <4 mJ/cm^2 using 2.5% PAG and 0.15% quencher [70].

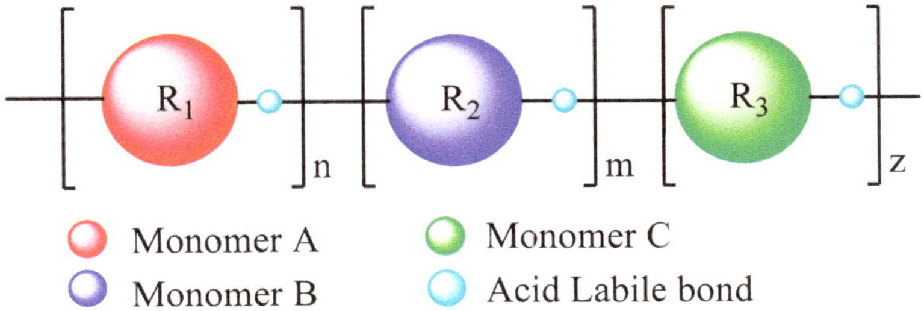

Figure 6. Chemical structure of the backbone breakable random copolymer. The copolymer is consisted of three different monomers connected by an acid labile bond. Each monomer introduces or tunes a specific property of the polymer. Adapted from [70], with permission from SPIE and the author Theodore Manouras, 2020.

4.1.3. Molecular Chemically Amplified Systems

Molecular type resists have attracted much attention for years since the small size of basic matrix material and its well-defined molecular structure are expected to be beneficial to better resolution and lower LWR. Past developments in the field of positive molecular resists including EUV materials had been reviewed in 2016 [71]. Currently, material's robustness is often discussed like Tg, modulus, and adhesion to substrate along with efforts for high sensitivity. Echigo et al. developed a new molecular photoresist based on calixarene chemistry (Figure 7b). This resist has excellent solubility in conventional resist solvents and can be developed with standard alkaline developer TMAH. EUV patterning results showed resolution capability of 45 nm line and space as an EUV dose of 10.3 mJ/cm^2 [72]. Figure 7 contains the precursor molecules of Noria photoresists (Figure 7a) and calixarene-based photoresists (Figure 7b). The hydroxyl groups contained in the aforementioned molecules can easily be modified producing photoresists with improved properties. Owada et al. prepared cyclic low molecular (CLM) weight resists with different protecting number. CLM resist achieved resolution of sub 30 nm hp patterns with high sensitivity [73]. Kudo et al. synthesized Noria derivatives with pendant adalantyl ester groups. They created 25 nm resolution pattern using EUV lithography. These patterns were obtained with less than 10 mJ/cm^2 irradiation dose [74]. Kulshreshtha et al. synthesized a negative tone chemically amplified molecular resist based on modified Noria molecule having oxetane crosslinking moieties. Optimization of crosslinking can improve the balance between sensitivity, LER, and swelling. They have patterned 1:1 line structures with 20 nm resolution and 3.2 nm LER [75]. Another work of the same group demonstrated a blended resist system with higher performance by combining enthalpic and entropic contributions to solubility contrast. These resists have shown significant advancements in resolution, LER, and processability [76]. Dow electronics designed and synthesized several molecular glasses (MG) resists based on calixarene cores as alternatives to polymeric resists. They studied the relationship between the structure and the properties of MG resists in order to improve the lithographic performance. They created patterns with 28 nm 1:1 space lines using EUV exposure [77]. Frommhold et al. developed a new molecular resist system that showed high resolution capability. They optimized the performance of this system at 14 nm hp by 50% using a new quencher. Furthermore, dose improvements up to 60% was observed using metals as additives [78,79]. Irresistible materials Ltd. developed negative resist materials based on a multi-trigger concept. In a multi-trigger material, a catalytic process is utilized following the resist exposure in a similar manner to a chemical-amplified photoresist. However, in multi-trigger resist, multiple photoacids activate multiple acid sensitive molecules, which then react with each other to cause a single resist event. Figure 7c depicts the basic components of the multi-trigger resist formulations which are a molecular resin and an epoxy crosslinker molecule. Instead of a photoacid causing a single resist chemistry event as occurred in traditional CARs, in the case of multi-trigger resist concept, the photogenerated acid is being regenerated. This concept enables a high sensitivity solubility change above a certain dose threshold, but turns the resist respond off at lower dosages. This behavior is expected to lead to sharper lines and lower LER. Several studies have been performed to tune the quencher loading, metal addition, and resist design. Improvements on the design of these materials led to some excellent and very promising results in resolution, LWR-LER, and sensitivity. Resist formulations based on the multi-trigger concept showed capability to resolve 13.3 nm lines on 28 nm pitch, with 2.97 nm LWR and dose of 26 mJ/cm^2 as well as 14.7 nm lines on 30 nm pitch, with 2.72 nm LWR and dose of 34 mJ/cm^2 [80–84].

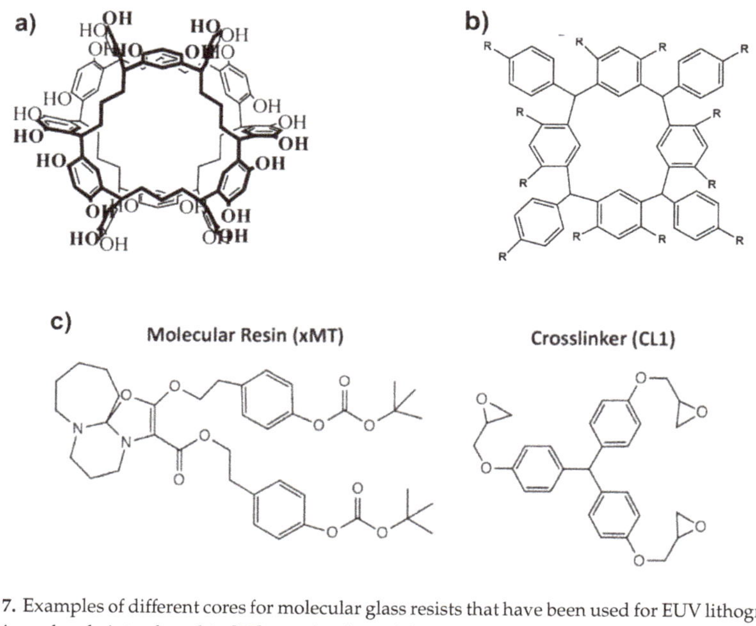

Figure 7. Examples of different cores for molecular glass resists that have been used for EUV lithography. (**a**) Noria molecule introduced in [85], graph adapted from [85], with permission from John Wiley & Sons, 2020, (**b**) a generic calixarene structure that has been the basis of a number of resist formulations and (**c**) basic components of a multi-triggered resist i.e., a molecular resin and a crosslinker, adapted from [83], with permission from The Society of Photopolymer Science and Technology, 2020.

4.2. Non Chemically Amplified Photoresists

Development of non-chemically amplified resists has been proposed as an alternative to chemically amplified because of acid diffusion which may limit resolution improvement and worsen LWR for 22 nm HP and beyond. A key example is the excellent resolution and LER performance of the chain-scission polymethylmethacrylate (PMMA) resists. The chain scission reactions in these systems are well-known from the early days of Semiconductor Lithography [86], and the relevant reactions have been studied extensively. The backbone of the PMMA resist is cleaved under UV, ionizing radiation and electron beam exposures as shown in Figure 8, in the scheme adopted from ref [87]. Despite some excellent characteristics, PMMA resists have a number of drawbacks that have prevented their widespread use, such as the need for organic solvent development, high outgassing, poor etch resistance, and poor sensitivity. Different approaches are proposed to address these drawbacks and even recently it was a reported a new approach for enhancing PMMA pattern transfer performance through an infiltration process resulting in the formation of an AlOx film on top of the PMMA resist film [88]. This approach has been demonstrated so far with e-beam exposure. On the other hand in an increasing number of papers the performance of PMMA and related materials at EUV is investigated. Fallica et al. compared the performance of three high resolution lithographic tools. They used EUV interference lithography (EUV-IL), electron beam lithography, and He ion beam lithography tools to evaluate PMMA and hydrogen silsesquioxane (HSQ) resists under the same conditions. EUV-IL is a technique capable to pattern large areas of dense features with good resolution. Electron beam lithography is effective to fabricate high resolution arbitrary patterns and He ion beam is a promising technique to create both isolated and dense patterns because of almost negligible backscattered electrons [89]. ZEP resin has been extensively studied as an electron beam lithography resist material and has capability to provide high resolution patterns. The imaging enabling degradation mechanism is well studied and understood [90,91]. As it is shown in the Figure 8, PMMA and its derivatives undergo main chain scission when it is irradiated with proper radiation. ZEP520A was evaluated as a EUV resist using EUV-IL and exhibited better sensitivity in EUV radiation compared to PMMA. It yielded excellent dense arrays of 50 nm hp resolution and down to 25 nm hp with acceptable LER [92]. Sharma et al. developed a non-chemically amplified photoresist consisting of 4-(methacryloyloxy)phenyl dimethyl sulfonium triflate-r-isopropyl methacrylate. The resist has shown sensitivity of about 11.3 mJ/cm^2 avoiding acid diffusion and blurring of resist pattern [93]. Oyama et al. introduced an easy method to predict EUV sensitivity by using electron beam sources. Considering that e-beam and EUV can induce the same chemical reactions, the required expose doses for the e-beam and EUV are expected to be related [94]. In a study at EUV [95] the influence of the PMMA molecular weight and processing parameters was examined. A characteristic EUV dose to clear was found to be 25 mJ/cm^2 whereas crosslinking was observed at a dose of 600 mJ/cm^2. The higher Mw material (950 K) gave the best printing results for 50 nm lines/spaces.

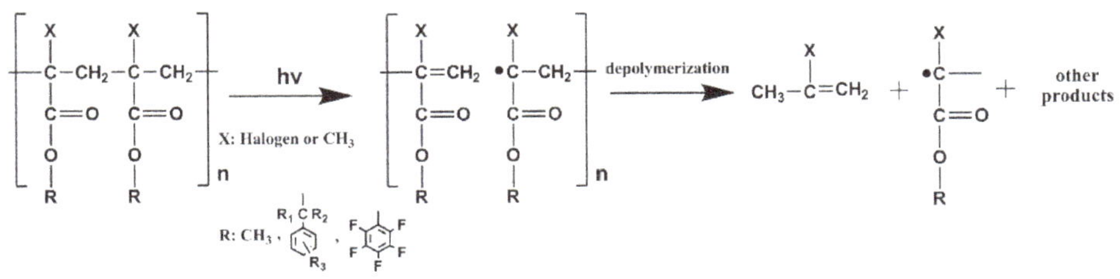

Figure 8. Example of a photodegradation mechanism for poly(methyl methacrylate) (PMMA) and its derivatives. Different atoms or groups strongly influence the properties of the corresponding resists including sensitivity and etch resistance. Irradiation of the polymer leads to the breaking of the main chain resulting to its depolymerization Adapted from [87], with permission from John Wiley & Sons, 2020.

4.3. Inorganic Resists

In the first steps of EUV resist development, in an analogous way to the design of KrF or ArF resists, materials containing silicon attracted attention despite its low absorbing character to EUV light. However, realizing the importance of enhancing resist film absorbance to EUV light due to the low light source output at 13.5 nm, the silicon materials investigation as candidate EUV resists was substantially reduced. Alternatively, because of higher absorptivity in EUV radiation, resist materials containing metals were gaining interest. These materials are expected to have excellent sensitivity, better robustness and good etch resistance. Furthermore, it is expected to have improved properties concerning the relation among resolution, line edge roughness (in general, pattern quality), and sensitivity.

Most of the promising materials proposed are based on the inclusion of Hf, Zr, Zn, and Sn atoms in the resist formulation which are expected to enhance photoelectron production upon EUV radiation as expected from the data in Figure 2. In the following section we will review the promising materials that have been proposed trying to keep a chronological order, starting from the older ones. Since the imaging chemistries encountered are still under intense investigation, the recent papers focusing mostly on mechanistic issues will be discussed at the end of this topic on inorganic resists.

Ober's group introduced metal oxide nanoparticles (NP) as next generation photoresist materials. The synthesis of these nanoparticles includes the controlled hydrolysis of zirconium or hafnium alkoxides in an excess of carboxylic acid followed by precipitation treatments to give ZrO_2-NP or HfO_2-NP with organic ligands. The size of the NP was controlled below 3 nm which is suitable for sub-20 nm lithography. Using either photoradical initiator or PAG, these NP gave both positive and negative tone patterns. These materials have shown high etch resistance as well as thermal and chemical stability. They have capability to resolve 26 nm lines using only 4.2 mJ/cm^2 EUV dose. Furthermore, they made several efforts to understand the imaging mechanism of these hybrid photoresists [96–98].

Cardineau et al. studied the photolysis of tin clusters of the type $[(RSn)_{12}O_4(OH)_6]X_2$ using EUV radiation and explored these clusters as novel high-resolution photoresist materials. The photolysis of the organic ligand after the EUV irradiation activates the cluster leading to agglomeration and results in the observed negative-tone imaging. They have resolved dense line patterns with 18 nm dimension [99]. Passarelli et al. developed organometallic carboxylate compounds $[R_nM(O_2(R')_2]$ as negative-tone EUV resists candidates. The imaging chemistry of such a resist is based on the polymerization of its acrylic substituents. This system has demonstrated exceptional sensitivity printing 35 nm dense lines with 5.6 mJ/cm^2. Furthermore, they found out that among antimony, bismuth, tin and tellurium containing materials, the antimony incorporation provides the more sensitive resist while tellurium the least [100]. Sortland et al. investigated the photoreactivity of platinum and palladium mononuclear complexes. Despite the fact that many platinum and palladium complexes show little or no EUV sensitivity, they have found that metal carbonates ($L_2M(CO_3)$) and metal oxalates ($L_2M(C_2O_4)$) (M is either Pt or Pd) are sensitive to EUV radiation. They demonstrated that the use of palladium as a core metal offers faster resists than the use of platinum [101]. Fugimori et al. developed a metal containing non chemically amplified resist material showing ultra-high sensitivity and capability to resolve 17 nm resolution features with 7 mJ/cm^2 [102]. Li et al. prepared Hf-based photoresist materials with three different organic ligands by a sol-gel method. These resists have shown high sensitivity in EUV radiation as well as capability to create high resolution patterns [103]. Inpria corporation developed directly patternable, metal oxide hardmasks as robust, high resolution photoresists for EUV lithography. They have achieved 13 nm half-pitch at 35 mJ/cm^2 and 11 nm hp with 1.7 nm LWR [104]. On the other hand, Hinsberg et al. proposed a numeric model describing the chemical and physical mechanisms governing pattern formation in metal oxide (MO_x) EUV photoresists. They used experimental measurements to develop a quantitative representation of the chemical and physical state of the MO_x resist film at each step of the lithographic process [105]. Xu et al. prepared discrete nanometer scale zinc-based clusters and used them as resist materials for EUV lithography. These materials have shown capability to resolve 15 nm features [106]. Zang et al. reported the dual tone property of the tin-oxo cage $[(BuSn)_{12}O_{14}(OH)_6](OH)_2$ photoresist. This resist has shown a positive tone behavior when it is irradiated with low dose of EUV or E-beam and a negative tone behavior when it is irradiated at higher dosages [107]. Sitterly et al. investigated the photoreactivity of six organometallic complexes of the type ph_nMX_2 containing bismuth, antimony, and tellurium as metals and acetate or pivalate as ligands. They monitored the photodecomposition using mass spectroscopy when they were irradiated with EUV. They found that both the metal center and the carboxylate ligands have significant influence on the EUV photoreactivity of these compounds [108]. Rantala et al. developed novel EUV resists based on organohydrogensilsesquioxane. These materials worked as negative tone resists and have shown excellent etch selectivity and ability to form patterns by using industry standard TMAH development process. Furthermore, they exhibited low LWR (<2 nm) with sufficient sensitivity (40–60 mJ/cm^2) [109,110]. Thakur et al. prepared Zn-based oxoclusters having trifluoroacetate (TFA) and methacrylate (MA) ligands. The Zn(MA)TFA photoresist displays appreciable sensitivity toward EUV radiation [111].

The mechanisms in all the above materials are still not adequately understood. Recently a study on hybrid HfO containing resists was reported by Mattson et al. [112]. In this study EUV-induced reactions were studied by using in situ IR spectroscopy of films irradiated by a variable energy electron gun and insights on the solubility changing mechanisms were obtained. Another study for HfO and ZrO materials was published by Wu et al. [113]. They applied different spectroscopic techniques and confirmed the higher sensitivity for a Hf-based material, as expected from its higher absorptivity. The role of carboxylate ligands in the resist crosslinking was also confirmed.

Lately, a lot of interest has been devoted to tin oxo cage materials. Haitjema et al. studied the chemical behavior observed in tin oxo cage materials under UV exposure to get clues that could also be of use for determining the EUV imaging chemistry [114]. Further insights on tin oxo cages chemistry at EUV was provided in the recent work by Bespalov where the influence of electron energy was studied [40]. The main results from this study are depicted in Figure 9. It was found that when the electron energy was under 2 eV incomplete material crosslinking was achieved. On the other hand for energies above 2 eV the material was crosslinked providing a denser final film.

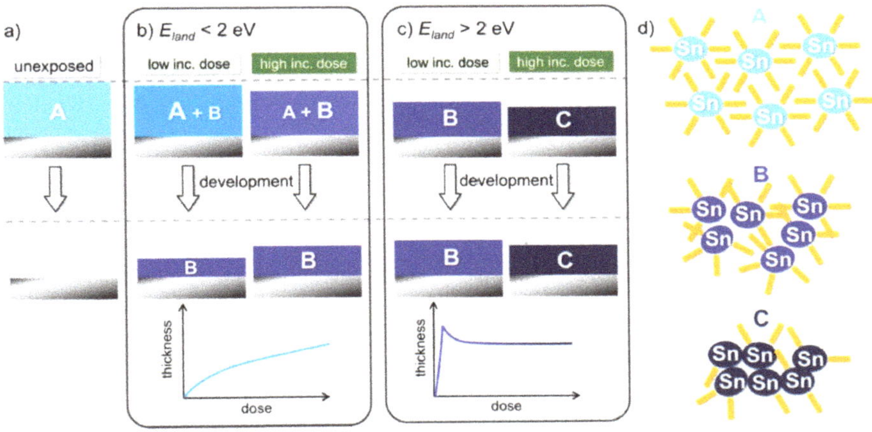

Figure 9. Scheme representing the patterning of the Tin-OH material. (**a**) The unexposed resist is removed completely during the development. (**b**) For electron exposure with E < 2 eV, only a small fraction of electrons impinging on the surface reach the material and only low conversion is attained. (**c**) For E > 2 eV, as the incident dose increases, consecutive reactions lead to the insoluble products B (denser than A) and C (denser than B). (**d**) Schematic representation of the initial Tin-OH molecular material A and of the two insoluble networks B and C. Blue ellipses represent the Sn-based inorganic core and orange bars the butyl chain. Adapted from [40] (https://pubs.acs.org/doi/10.1021/acsami.9b19004), with permission from American Chemical Society, 2020.

The above mechanistic studies are now starting to help the elucidation of imaging mechanisms in the novel inorganic or hybrid component containing materials that are proposed as EUV resists. The continuation of this effort is expected to lead in designing high sensitivity EUV resists of advanced performance in the near future.

5. Summary and Outlook

The developments on the evolution of EUV resist technology during the past decade reviewed in the previous chapters, which have been actually intensified during the last few years, deserve the attention not only of the lithographic community but also of the broader nanomaterials community.

In the Table 1 below we summarize the performance results of the most promising resist materials reviewed. The materials are listed according to the design principle and the type of the main component in the resist composition. The performance indicators include the best resolution reported along with the corresponding dose, whereas data on the pattern quality i.e., LER-LWR (line edge roughness or line width roughness are provided, where available).

The main directions of the EUV resist technology evolution are also depicted in the graph of Figure 10.

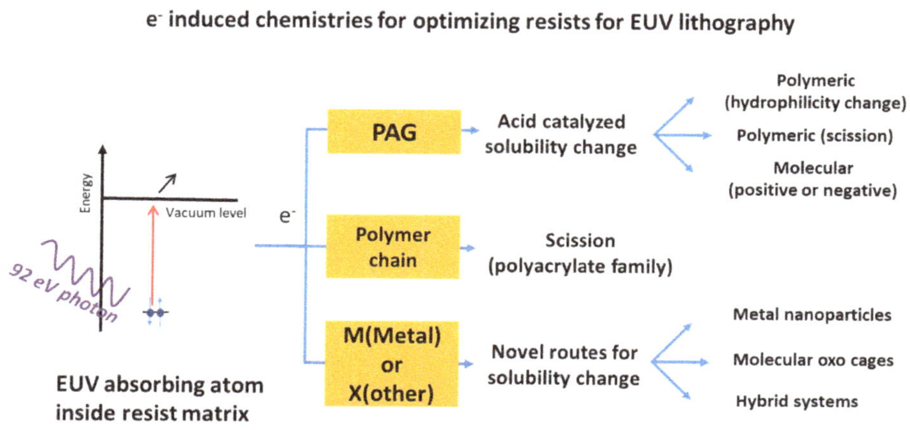

Figure 10. A summary of the most promising research directions that are currently explored toward high sensitivity and high performance EUV resists.

Based on the reviewed ongoing efforts towards new resist materials we could emphasize the following points:

First, the demand for high performing resists according to industry standards for sub 20 nm nanostructures and also the demand of high sensitivity for securing acceptable technology throughput led to both, revisiting the fundamental issues and proposing new non-traditional material solutions. The ongoing resist field effort is expected to enable the successful adoption of EUV technology by the semiconductor industry.

Second, the new materials that have been proposed as resist candidates have incorporated many novelties. Especially interesting is the case of resists incorporating metal nanoclusters and related compounds. These compounds were selected because of their high absorbance of ionizing radiation, in particular EUV, but it is conceivable that they could absorb in other spectral regions as well and can be of use in a plethora of other radiation-related applications. The fact that well-defined nanostructures can be formed with such materials can also have important implications.

Table 1. Performance characteristics (see text) of the most promising resists materials reviewed based on different design approaches.

Design Principle	Materials	Resolution	Sensitivity (Dose to Size)	Pattern Quality (LER-LWR)
CAR [45]	Polymeric	30 nm	<20 mJ/cm^2	-
CAR [46]	Polymer bound PAG	24 nm	14 mJ/cm^2	5.3 nm
CAR [47]	Polymer bound PAG-increased Hydrophobicity	16 nm	24 mJ/cm^2	3 nm
CAR [48]	Polymeric	15 nm	25–30 mJ/cm^2	6 nm
CAR [56]	Polymeric with different PAGs	13 nm	35.5 mJ/cm^2	
CAR [59]	Polymeric	20 nm	31 mJ/cm^2	-
CAR [60]	Polymeric	14 nm	43 mJ/cm^2	5.8 nm
CAR [64]	Polymeric with Acid Amplifier (AA)	60 nm	1.9 mJ/cm^2	7.9 nm
CAR-Multi-triggered resist [80–83]	Molecular	12.7 nm	53 mJ/cm^2	4.2 nm
CAR [69]	Polymeric-main chain scission	20 nm	4 mJ/cm^2	-
CAR [72]	Molecular	45 nm	10.3 mJ/cm^2	-
CAR [73]	Molecular	45 nm	9.5 mJ/cm^2	6.2 nm
CAR [74]	Molecular	26 nm	14.5 mJ/cm^2	-
CAR [76]	Molecular	20 nm	40.5 mJ/cm^2	3.2 nm
CAR [77]	Molecular	28 nm	22 mJ/cm^2	3.7 nm
CAR [78,79]	Molecular	14 nm	36.1 mJ/cm^2	3.26 nm
Non-CAR [93]	Polymeric	22 nm	78 mJ/cm^2	<6 nm
Non-CAR [94]	Polymeric	20 nm	26.6 mJ/cm^2	-
Non-CAR [95]	polymeric	50 nm	52 mJ/cm^2	4.1 nm
Inorganic [97,98]	Nanoparticles	26 nm	4.2 mJ/cm^2	-
Inorganic [101]	Clusters	18 nm	350 mJ/cm^2	-
Organometallic [102]	Molecular	35 nm	5.6 mJ/cm^2	-
Organometallic [103]	Complexes	30 nm	90 mJ/cm^2	5.5 nm
Metal [104]	-	17 nm	7 mJ/cm^2	5.6 nm
Metal oxide [105]	-	13 nm	35 mJ/cm^2	-
Metal-organic [107]	Clusters	13 nm	35 mJ/cm^2	-
Metal [108]	Complexes	50 nm	53.5 mJ/cm^2	-
Organohydrogen silsesquioxane [110]	Molecule	22 nm	65.4 mJ/cm^2	1.4 nm
Metal oxide [112]	Clusters	25 nm	37 mJ/cm^2	-

Third, electron-induced chemical changes that can be controlled and directed toward specific results are encountered in other fields of chemistry. Thus knowledge that can be acquired in the resist development effort can be eventually utilized for devising new routes in chemical synthesis. It can be also of great help in biology-related studies where very often the results of ionizing radiation should be investigated in depth.

It should be finally noticed here that open questions related to resist chemistry optimization remain and that the last developments can be viewed as the beginning rather than the end in the effort toward devising ionizing radiation sensitive materials of high performance.

Author Contributions: P.A. and T.M. conceived and wrote the review. All authors have read and agreed to the published version of the manuscript.

Funding: We acknowledge support from "National Infrastructure in Nanotechnology, Advanced Materials and Micro-/Nanoelectronics" (MIS 5002772) which is implemented under the Action "Reinforcement of the Research and Innovation Infrastructure", funded by the Operational Programme "Competitiveness, Entrepreneurship and Innovation" (NSRF 2014- 2020) and co-financed by Greece and the European Union (European Regional Development Fund).

Conflicts of Interest: The authors declare no conflict of interest.

References

1. Pease, R.F.; Chou, S.Y. Lithography and Other Patterning Techniques for Future Electronics. *Proc. IEEE* **2008**, *96*, 248–270. [CrossRef]
2. Pimpin, A.; Srituravanich, W. Review on Micro- and Nanolithography Techniques and Their Applications. *Eng. J.* **2011**, *16*, 37. [CrossRef]
3. Seisyan, R.P. Nanolithography in microelectronics: A review. *Tech. Phys.* **2011**, *56*, 1061. [CrossRef]
4. Lawson, R.A.; Robinson, A.P.G. Chapter 1—Overview of materials and processes for lithography. In *Frontiers of Nanoscience*; Robinson, A., Lawson, R., Eds.; Elsevier: Amsterdam, The Netherlands, 2016; Volume 11, pp. 1–90.
5. Jeong, S.-J.; Kim, J.Y.; Kim, B.H.; Moon, H.-S.; Kim, S.O. Directed self-assembly of block copolymers for next generation nanolithography. *Mater. Today* **2013**, *16*, 468–476. [CrossRef]
6. Kim, B.H.; Kim, J.Y.; Kim, S.O. Directed self-assembly of block copolymers for universal nanopatterning. *Soft Matter* **2013**, *9*, 2780–2786. [CrossRef]
7. Stoykovich, M.P.; Nealey, P.F. Block copolymers and conventional lithography. *Mater. Today* **2006**, *9*, 20–29. [CrossRef]
8. Yang, G.-W.; Wu, G.-P.; Chen, X.; Xiong, S.; Arges, C.G.; Ji, S.; Nealey, P.F.; Lu, X.-B.; Darensbourg, D.J.; Xu, Z.-K. Directed Self-Assembly of Polystyrene-b-poly(propylene carbonate) on Chemical Patterns via Thermal Annealing for Next Generation Lithography. *Nano Lett.* **2017**, *17*, 1233–1239. [CrossRef]
9. Chen, Y. Nanofabrication by electron beam lithography and its applications: A review. *Microelectron. Eng.* **2015**, *135*, 57–72. [CrossRef]
10. Okazaki, S. High resolution optical lithography or high throughput electron beam lithography: The technical struggle from the micro to the nano-fabrication evolution. *Microelectron. Eng.* **2015**, *133*, 23–35. [CrossRef]
11. Tseng, A.A.; Kuan, C.; Chen, C.D.; Ma, K.J. Electron beam lithography in nanoscale fabrication: Recent development. *IEEE Trans. Electron. Packag. Manuf.* **2003**, *26*, 141–149. [CrossRef]
12. Vieu, C.; Carcenac, F.; Pépin, A.; Chen, Y.; Mejias, M.; Lebib, A.; Manin-Ferlazzo, L.; Couraud, L.; Launois, H. Electron beam lithography: Resolution limits and applications. *Appl. Surf. Sci.* **2000**, *164*, 111–117. [CrossRef]
13. Öktem, B.; Pavlov, I.; Ilday, S.; Kalaycıoğlu, H.; Rybak, A.; Yavaş, S.; Erdoğan, M.; Ilday, F.Ö. Nonlinear laser lithography for indefinitely large-area nanostructuring with femtosecond pulses. *Nat. Photonics* **2013**, *7*, 897–901. [CrossRef]
14. Walsh, M.E.; Zhang, F.; Menon, R.; Smith, H.I. 6—Maskless photolithography. In *Nanolithography*; Feldman, M., Ed.; Woodhead Publishing: Cambridge, UK, 2014; pp. 179–193. [CrossRef]
15. Douglas, J.R.; Jin, C. A review of nanoimprint lithography for high-volume semiconductor device manufacturing. *Adv. Opt. Technol.* **2017**, *6*, 229–241. [CrossRef]
16. Jan van, S.; Helmut, S. Next-generation lithography—An outlook on EUV projection and nanoimprint. *Adv. Opt. Technol.* **2017**, *6*, 159–162. [CrossRef]

17. Danilo De, S.; Yannick, V.; Geert, V. Photoresists in extreme ultraviolet lithography (EUVL). *Adv. Opt. Technol.* **2017**, *6*, 163–172. [CrossRef]
18. Li, L.; Liu, X.; Pal, S.; Wang, S.; Ober, C.K.; Giannelis, E.P. Extreme ultraviolet resist materials for sub-7 nm patterning. *Chem. Soc. Rev.* **2017**, *46*, 4855–4866. [CrossRef]
19. Luo, C.; Xu, C.; Lv, L.; Li, H.; Huang, X.; Liu, W. Review of recent advances in inorganic photoresists. *RSC Adv.* **2020**, *10*, 8385–8395. [CrossRef]
20. Wu, B.; Kumar, A. Extreme ultraviolet lithography: A review. *J. Vac. Sci. Technol. B* **2007**, *25*, 1743–1761. [CrossRef]
21. Buitrago, E.; Kulmala, T.S.; Fallica, R.; Ekinci, Y. Chapter 4—EUV lithography process challenges. In *Frontiers of Nanoscience*; Robinson, A., Lawson, R., Eds.; Elsevier: Amsterdam, The Netherlands, 2016; Volume 11, pp. 135–176.
22. Solak, H.H.; He, D.; Li, W.; Singh-Gasson, S.; Cerrina, F.; Sohn, B.H.; Yang, X.M.; Nealey, P. Exposure of 38 nm period grating patterns with extreme ultraviolet interferometric lithography. *Appl. Phys. Lett.* **1999**, *75*, 2328–2330. [CrossRef]
23. Bhattarai, S.; Neureuther, A.R.; Naulleau, P.P. Study of shot noise in photoresists for extreme ultraviolet lithography through comparative analysis of line edge roughness in electron beam and extreme ultraviolet lithography. *J. Vac. Sci. Technol. B* **2017**, *35*, 061602. [CrossRef]
24. Naulleau, P. Chapter 5—EUV lithography patterning challenges. In *Frontiers of Nanoscience*; Robinson, A., Lawson, R., Eds.; Elsevier: Amsterdam, The Netherlands, 2016; Volume 11, pp. 177–192.
25. Banine, V.; Moors, R. Plasma sources for EUV lithography exposure tools. *J. Phys. D Appl. Phys.* **2004**, *37*, 3207–3212. [CrossRef]
26. Banine, V.; Benschop, J.; Leenders, M.; Moors, R. Relationship between an EUV source and the performance of an EUV lithographic system. In *Emerging Lithographic Technologies IV*; SPIE: Bellingham, WA, USA, 2000; Volume 3997, pp. 126–135.
27. Photonic Frontiers: EUV Lithography: EUV Lithography Has Yet to Find Its Way into the Fab. Available online: https://www.laserfocusworld.com/lasers-sources/article/16557008/photonic-frontiers-euv-lithography-euv-lithography-has-yet-to-find-its-way-into-the-fab (accessed on 2 August 2020).
28. Wood Ii, O.R. EUVL: Challenges to Manufacturing Insertion. *J. Photopolym. Sci. Technol.* **2017**, *30*, 599–604. [CrossRef]
29. Kostko, O.; Xu, B.; Ahmed, M.; Slaughter, D.S.; Ogletree, D.F.; Closser, K.D.; Prendergast, D.G.; Naulleau, P.; Olynick, D.L.; Ashby, P.D.; et al. Fundamental understanding of chemical processes in extreme ultraviolet resist materials. *J. Chem. Phys.* **2018**, *149*, 154305. [CrossRef]
30. Ogletree, D.F. Chapter 2—Molecular excitation and relaxation of extreme ultraviolet lithography photoresists. In *Frontiers of Nanoscience*; Robinson, A., Lawson, R., Eds.; Elsevier: Amsterdam, The Netherlands, 2016; Volume 11, pp. 91–113.
31. Henke, B.L.; Gullikson, E.M.; Davis, J.C. X-Ray Interactions: Photoabsorption, Scattering, Transmission, and Reflection at E = 50–30,000 eV, Z = 1–92. *At. Data Nucl. Data Tables* **1993**, *54*, 181–342. [CrossRef]
32. Irie, S.; Endo, M.; Sasago, M.; Kandaka, N.; Kondo, H.; Murakami, K. Study of Transmittance of Polymers and Influence of Photoacid Generator on Resist Transmittance at Extreme Ultraviolet Wavelength. *Jpn. J. Appl. Phys.* **2002**, *41*, 5864–5867. [CrossRef]
33. Yeh, J.J.; Lindau, I. Atomic subshell photoionization cross sections and asymmetry parameters: $1 \leq Z \leq 103$. *At. Data Nucl. Data Tables* **1985**, *32*, 1–155. [CrossRef]
34. He, D.; Solak, H.; Li, W.; Cerrina, F. Extreme ultraviolet and x-ray resist: Comparison study. *J. Vac. Sci. Technol. B* **1999**, *17*, 3379–3383. [CrossRef]
35. Kozawa, T.; Tagawa, S. Radiation Chemistry in Chemically Amplified Resists. *Jpn. J. Appl. Phys.* **2010**, *49*, 030001. [CrossRef]
36. Narasimhan, A.; Wisehart, L.; Grzeskowiak, S.; Ocola, L.E.; Denbeaux, G.; Brainard, R.L. What We Don't Know About EUV Exposure Mechanisms. *J. Photopolym. Sci. Technol.* **2017**, *30*, 113–120. [CrossRef]
37. Torok, J.; Re, R.D.; Herbol, H.; Das, S.; Bocharova, I.; Paolucci, A.; Ocola, L.E.; Ventrice, C., Jr.; Lifshin, E.; Denbeaux, G.; et al. Secondary Electrons in EUV Lithography. *J. Photopolym. Sci. Technol.* **2013**, *26*, 625–634. [CrossRef]
38. Böhler, E.; Warneke, J.; Swiderek, P. Control of chemical reactions and synthesis by low-energy electrons. *Chem. Soc. Rev.* **2013**, *42*, 9219–9231. [CrossRef]

39. van Dorp, W.F. Chapter 3—Theory: Electron-induced chemistry. In *Frontiers of Nanoscience*; Robinson, A., Lawson, R., Eds.; Elsevier: Amsterdam, The Netherlands, 2016; Volume 11, pp. 115–133.
40. Bespalov, I.; Zhang, Y.; Haitjema, J.; Tromp, R.M.; van der Molen, S.J.; Brouwer, A.M.; Jobst, J.; Castellanos, S. Key Role of Very Low Energy Electrons in Tin-Based Molecular Resists for Extreme Ultraviolet Nanolithography. *ACS Appl. Mater. Interfaces* **2020**, *12*, 9881–9889. [CrossRef]
41. Pollentier, I.; Vesters, Y.; Petersen, J.; Vanelderen, P.; Rathore, A.; de Simone, D.; Vandenberghe, G. Unraveling the role of photons and electrons upon their chemical interaction with photoresist during EUV exposure. In *Advances in Patterning Materials and Processes XXXV*; SPIE: Bellingham, WA, USA, 2018; Volume 10586, p. 105860C.
42. Pollentier, I.; Petersen, J.; De Bisschop, P.; De Simone, D.; Vandenberghe, G. Unraveling the EUV photoresist reactions: Which, how much, and how do they relate to printing performance. In *Extreme Ultraviolet (EUV) Lithography X*; SPIE: Bellingham, WA, USA, 2019; Volume 10957, p. 109570I.
43. Torti, E.; Protti, S.; Bollanti, S.; Flora, F.; Torre, A.; Brusatin, G.; Gerardino, A.; Businaro, L.; Fagnoni, M.; Della Giustina, G.; et al. Aryl Sulfonates as Initiators for Extreme Ultraviolet Lithography: Applications in Epoxy-Based Hybrid Materials. *ChemPhotoChem* **2018**, *2*, 425–432. [CrossRef]
44. Yamamoto, H.; Kozawa, T.; Nakano, A.; Okamoto, K.; Tagawa, S.; Ando, T.; Sato, M.; Komano, H. Dependence of acid generation efficiency on the protection ratio of hydroxyl groups in chemically amplified electron beam, x-ray and EUV resists. *J. Vac. Sci. Technol. B* **2004**, *22*, 3522–3524. [CrossRef]
45. Fedynyshyn, T.; Goodman, R.; Roberts, J. Polymer matrix effects on acid generation. In *Advances in Resist Materials and Processing Technology XXV*; SPIE: Bellingham, WA, USA, 2008; Volume 6923, p. 692319.
46. Tamaoki, H.; Tarutani, S.; Tsubaki, H.; Takahashi, T.; Inoue, N.; Tsuchihashi, T.; Takizawa, H.; Takahashi, H. Characterizing polymer bound PAG-type EUV resist. In *Advances in Resist Materials and Processing Technology XXVIII*; SPIE: Bellingham, WA, USA, 2011; Volume 7972, p. 79720A.
47. Tarutani, S.; Tsubaki, H.; Takizawa, H.; Goto, T. EUV Resist Materials for 16 nm and below Half Pitch Applications. *J. Photopolym. Sci. Technol.* **2012**, *25*, 597–602. [CrossRef]
48. Tsubaki, H.; Tarutani, S.; Inoue, N.; Takizawa, H.; Goto, T. EUV Resist Materials Design for 15 nm Half Pitch and Below. *J. Photopolym. Sci. Technol.* **2013**, *26*, 649–657. [CrossRef]
49. Iwashita, J.; Hirayama, T.; Matsuzawa, K.; Utsumi, Y.; Ohmori, K. Out-of-band insensitive polymer-bound PAG for EUV resist. In *Extreme Ultraviolet (EUV) Lithography III*; SPIE: Bellingham, WA, USA, 2012; Volume 8322, p. 83220Y.
50. Jain, V.; Coley, S.; Lee, J.J.; Christianson, M.; Arriola, D.; LaBeaume, P.; Danis, M.; Ortiz, N.; Kang, S.-J.; Wagner, M.; et al. Impact of polymerization process on OOB on lithographic performance of a EUV resist. In *Extreme Ultraviolet (EUV) Lithography II*; SPIE: Bellingham, WA, USA, 2011; Volume 7969, p. 796912.
51. Park, C.-M.; Kim, I.; Kim, S.-H.; Kim, D.-W.; Hwang, M.-S.; Kang, S.-N.; Park, C.; Kim, H.-W.; Yeo, J.-H.; Kim, S.-S. Prospects of DUV OoB suppression techniques in EUV lithography. In *Extreme Ultraviolet (EUV) Lithography V*; SPIE: Bellingham, WA, USA, 2014; Volume 9048, p. 90480S.
52. Roberts, J.; Bristol, R.; Younkin, T.; Fedynyshyn, T.; Astolfi, D.; Cabral, A. Sensitivity of EUV resists to out-of-band radiation. In *Advances in Resist Materials and Processing Technology XXVI*; SPIE: Bellingham, WA, USA, 2009; Volume 7273, p. 72731W.
53. Liu, J.; Qiao, Y.; Wang, L. Novel polymeric sulfonium photoacid generator and its application for chemically amplified photoresists. In *Advances in Patterning Materials and Processes XXXI*; SPIE: Bellingham, WA, USA, 2014; Volume 9051, p. 905124.
54. Narasimhan, A.; Grzeskowiak, S.; Srivats, B.; Herbol, H.; Wisehart, L.; Kelly, C.; Earley, W.; Ocola, L.; Neisser, M.; Denbeaux, G.; et al. Studying secondary electron behavior in EUV resists using experimentation and modeling. In *Extreme Ultraviolet (EUV) Lithography VI*; SPIE: Bellingham, WA, USA, 2015; Volume 9422, p. 942208.
55. Fujiwara, K. Novel EUV resist development for sub-14 nm half pitch. In Proceedings of the 2017 China Semiconductor Technology International Conference (CSTIC), Shangai, China, 12–13 March 2017; pp. 1–2.
56. Kimoto, T.; Naruoka, T.; Nakagawa, H.; Fujisawa, T.; Shiratani, M.; Nagai, T.; Ayothi, R.; Hishiro, Y.; Hori, M.; Hoshiko, K.; et al. Novel EUV Resist Development for Sub-14nm Half Pitch. *J. Photopolym. Sci. Technol.* **2015**, *28*, 519–523. [CrossRef]
57. Krysak, M.; Leeson, M.; Han, E.; Blackwell, J.; Harlson, S. Extending resolution limits of EUV resist materials. In *Extreme Ultraviolet (EUV) Lithography VI*; SPIE: Bellingham, WA, USA, 2015; Volume 9422, p. 942205.

58. Thackeray, J.; Cameron, J.; Jain, V.; LaBeaume, P.; Coley, S.; Ongayi, O.; Wagner, M.; Rachford, A.; Biafore, J. Progress in resolution, sensitivity, and critical dimensional uniformity of EUV chemically amplified resists. In *Advances in Resist Materials and Processing Technology XXX*; SPIE: Bellingham, WA, USA, 2013; Volume 8682, p. 868213.
59. Thackeray, J.W.; Nassar, R.A.; Brainard, R.L.; Goldfarb, D.L.; Wallow, T.; Wei, Y.; Mackey, J.; Naulleau, P.P.; Pierson, B.; Solak, H.H. Chemically amplified resists resolving 25 nm 1:1 line: Space features with EUV lithography. In *Emerging Lithographic Technologies XI*; SPIE: Bellingham, WA, USA, 2007; Volume 6517, p. 651719.
60. Fujii, T.; Matsumaru, S.; Yamada, T.; Komuro, Y.; Kawana, D.; Ohmori, K. Patterning performance of chemically amplified resist in EUV lithography. In *Extreme Ultraviolet (EUV) Lithography VII*; SPIE: Bellingham, WA, USA, 2016; Volume 9776, p. 97760Y.
61. Yamamoto, H.; Vesters, Y.; Jiang, J.; Simone, D.D.; Vandenberghe, G.; Kozawa, T. Role of Metal Sensitizers for Sensitivity Improvement in EUV Chemically Amplified Resist. *J. Photopolym. Sci. Technol.* **2018**, *31*, 747–751. [CrossRef]
62. Fallica, R.; Ekinci, Y. Photoacid generator-polymer interaction on the quantum yield of chemically amplified resists for extreme ultraviolet lithography. *J. Mater. Chem. C* **2018**, *6*, 7267–7273. [CrossRef]
63. Lee, H.; Kim, M.; Moon, J.; Park, S.; Lee, B.; Jeong, C.; Cho, M. Multiscale approach for modeling EUV patterning of chemically amplified resist. In *Advances in Patterning Materials and Processes XXXVI*; SPIE: Bellingham, WA, USA, 2019; Volume 10960, p. 1096008.
64. Brainard, R.; Kruger, S.; Higgins, C.; Revuru, S.; Gibbons, S.; Freedman, D.; Yueh, W.; Younkin, T. Kinetics, Chemical Modeling and Lithography of Novel Acid Amplifiers for Use in EUV Photoresists. *J. Photopolym. Sci. Technol.* **2009**, *22*, 43–50. [CrossRef]
65. Sekiguchi, A.; Harada, T.; Watanabe, T. A study on enhancing EUV resist sensitivity. In *Extreme Ultraviolet (EUV) Lithography VIII*; SPIE: Bellingham, WA, USA, 2017; Volume 10143, p. 1014322.
66. Jiang, J.; Giordano, G.; Fallica, R.; DeSimone, D.; Vandenberghe, G. Sensitizer for EUV Chemically Amplified Resist: Metal versus Halogen. *J. Photopolym. Sci. Technol.* **2019**, *32*, 21–25. [CrossRef]
67. Nagahara, S.; Dinh, C.Q.; Yoshida, K.; Shiraishi, G.; Kondo, Y.; Yoshihara, K.; Nafus, K.; Petersen, J.; De Simone, D.; Foubert, P.; et al. EUV resist chemical gradient enhancement by UV flood exposure for improvement in EUV resist resolution, process control, roughness, sensitivity, and stochastic defectivity. In *Advances in Patterning Materials and Processes XXXVII*; SPIE: Bellingham, WA, USA, 2020; Volume 11326, p. 113260A.
68. Okamoto, K.; Kawai, S.; Kozawa, T. Sensitivity enhancement of chemically amplified EUV resist by adding diphenyl sulfone derivatives. In *Advances in Patterning Materials and Processes XXXVII*; SPIE: Bellingham, WA, USA, 2020; Volume 11326, p. 113260D.
69. Cardineau, B.; Garczynski, P.; Earley, W.; Brainard, R.L. Chain-Scission Polyethers for EUV Lithography. *J. Photopolym. Sci. Technol.* **2013**, *26*, 665–671. [CrossRef]
70. Manouras, T.; Kazazis, D.; Koufakis, E.; Ekinci, Y.; Vamvakaki, M.; Argitis, P. Ultra-sensitive EUV resists based on acid-catalyzed polymer backbone breaking. In *Extreme Ultraviolet (EUV) Lithography IX*; SPIE: Bellingham, WA, USA, 2018; Volume 10583, p. 105831R.
71. Argitis, P.; Vidali, V.P.; Niakoula, D. Chapter 9—Positive molecular resists. In *Frontiers of Nanoscience*; Robinson, A., Lawson, R., Eds.; Elsevier: Amsterdam, The Netherlands, 2016; Volume 11, pp. 319–348.
72. Echigo, M.; Oguro, D. Development of new phenylcalix [4] resorcinarene: Its application to positive-tone molecular resist for EB and EUV lithography. In *Advances in Resist Materials and Processing Technology XXVI*; SPIE: Bellingham, WA, USA, 2009; Volume 7273, p. 72732Q.
73. Owada, T.; Yomogita, A.; Kashiwamura, T.; Kusaba, T.; Miyamoto, S.; Takeya, T. Development of novel positive-tone resists for EUVL. In *Advances in Resist Materials and Processing Technology XXVI*; SPIE: Bellingham, WA, USA, 2009; Volume 7273, p. 72732R.
74. Kudo, H.; Nina, N.; Sato, T.; Oizumi, H.; Itani, T.; Miura, T.; Watanabe, T.; Kinoshita, H. Extreme Ultraviolet (EUV)-Resist Material Based on Noria (Water Wheel-like Macrocycle) Derivatives with Pendant Alkoxyl and Adamantyl Ester Groups. *J. Photopolym. Sci. Technol.* **2012**, *25*, 587–592. [CrossRef]
75. Kulshreshtha, P.; Maruyama, K.; Kiani, S.; Dhuey, S.; Perera, P.; Blackwell, J.; Olynick, D.; Ashby, P. Sub-20nm lithography negative tone chemically amplified resists using cross-linker additives. In *Advances in Resist Materials and Processing Technology XXX*; SPIE: Bellingham, WA, USA, 2013; Volume 8682, p. 86820N.

76. Kulshreshtha, P.K.; Maruyama, K.; Kiani, S.; Blackwell, J.; Olynick, D.L.; Ashby, P.D. Harnessing entropic and enthalpic contributions to create a negative tone chemically amplified molecular resist for high-resolution lithography. *Nanotechnology* **2014**, *25*, 315301. [CrossRef]
77. Green, D.P.; Jain, V.; Bailey, B.; Wagner, M.; Clark, M.; Valeri, D.; Lakso, S. Development of molecular resist derivatives for EUV lithography. In *Extreme Ultraviolet (EUV) Lithography IV*; SPIE: Bellingham, WA, USA, 2013; Volume 8679, p. 867912.
78. Frommhold, A.; McClelland, A.; Roth, J.; Fallica, R.; Ekinci, Y.; Robinson, A.P. Optimization and sensitivity enhancement of high-resolution molecular resist for EUV lithography. In *Extreme Ultraviolet (EUV) Lithography VII*; SPIE: Bellingham, WA, USA, 2016; Volume 9776, p. 977614.
79. Frommhold, A.; Yang, D.; McClelland, A.; Roth, J.; Xue, X.; Rosamund, M.C.; Linfield, E.H.; Robinson, A.P.G. Novel Molecular Resist for EUV and Electron Beam Lithography. *J. Photopolym. Sci. Technol.* **2015**, *28*, 537–540. [CrossRef]
80. O'Callaghan, G.; Popescu, C.; McClelland, A.; Kazazis, D.; Roth, J.; Theis, W.; Ekinci, Y.; Robinson, A.P. Multi-trigger resist: Novel synthesis improvements for high resolution EUV lithography. In *Advances in Patterning Materials and Processes XXXVI*; SPIE: Bellingham, WA, USA, 2019; Volume 10960, p. 109600C.
81. Popescu, C.; Frommhold, A.; McClelland, A.; Roth, J.; Ekinci, Y.; Robinson, A.P. Sensitivity enhancement of the high-resolution xMT multi-trigger resist for EUV lithography. In *Extreme Ultraviolet (EUV) Lithography VIII*; SPIE: Bellingham, WA, USA, 2017; Volume 10143, p. 101430V.
82. Popescu, C.; Kazazis, D.; McClelland, A.; Dawson, G.; Roth, J.; Theis, W.; Ekinci, Y.; Robinson, A.P. High-resolution EUV lithography using a multi-trigger resist. In *Extreme Ultraviolet (EUV) Lithography IX*; SPIE: Bellingham, WA, USA, 2018; Volume 10583, p. 105831L.
83. Popescu, C.; Vesters, Y.; McClelland, A.; De Simone, D.; Dawson, G.; Roth, J.; Theis, W.; Vandenberghe, G.; Robinson, A.P.G. Multi Trigger Resist for EUV Lithography. *J. Photopolym. Sci. Technol.* **2018**, *31*, 227–232. [CrossRef]
84. Popescu, C.; O'Callaghan, G.; McClelland, A.; Roth, J.; Lada, T.; Robinson, A.P. Performance enhancements with high opacity multi-trigger resist. In *Advances in Patterning Materials and Processes XXXVII*; SPIE: Bellingham, WA, USA, 2020; Volume 11326, p. 1132611.
85. Kudo, H.; Hayashi, R.; Mitani, K.; Yokozawa, T.; Kasuga, N.C.; Nishikubo, T. Molecular waterwheel (Noria) from a simple condensation of resorcinol and an alkanedial. *Angew. Chem. Int. Ed.* **2006**, *45*, 7948–7952. [CrossRef]
86. Moreau, W.M. *Semiconductor Lithography, Principles, Particles and Materials*; Plenum Press: New York, NY, USA, 1988.
87. Pasparakis, G.; Manouras, T.; Argitis, P.; Vamvakaki, M. Photodegradable Polymers for Biotechnological Applications. *Macromol. Rapid Commun.* **2012**, *33*, 183–198. [CrossRef]
88. Tiwale, N.; Subramanian, A.; Kisslinger, K.; Lu, M.; Kim, J.; Stein, A.; Nam, C.-Y. Advancing next generation nanolithography with infiltration synthesis of hybrid nanocomposite resists. *J. Mater. Chem. C* **2019**, *7*, 8803–8812. [CrossRef]
89. Fallica, R.; Kirchner, R.; Ekinci, Y.; Mailly, D. Comparative study of resists and lithographic tools using the Lumped Parameter Model. *J. Vac. Sci. Technol. B* **2016**, *34*, 06K702. [CrossRef]
90. Hosaka, Y.; Oyama, T.G.; Oshima, A.; Enomoto, S.; Washio, M.; Tagawa, S. Pulse Radiolysis Study on a Highly Sensitive Chlorinated Resist ZEP520A. *J. Photopolym. Sci. Technol.* **2013**, *26*, 745–750. [CrossRef]
91. Oyama, T.G.; Enomoto, K.; Hosaka, Y.; Oshima, A.; Washio, M.; Tagawa, S. Electron-Beam-Induced Decomposition Mechanisms of High-Sensitivity Chlorinated Resist ZEP520A. *Appl. Phys. Express* **2012**, *5*, 036501. [CrossRef]
92. Fallica, R.; Kazazis, D.; Kirchner, R.; Voigt, A.; Mochi, I.; Schift, H.; Ekinci, Y. Lithographic performance of ZEP520A and mr-PosEBR resists exposed by electron beam and extreme ultraviolet lithography. *J. Vac. Sci. Technol. B* **2017**, *35*, 061603. [CrossRef]
93. Sharma, S.K.; Pal, S.P.; Reddy, P.G.; Kumar, P.; Ghosh, S.; Gonsalves, K.E. Design and development of low activation energy based nonchemically amplified resists (n-CARs) for next generation EUV lithography. *Microelectron. Eng.* **2016**, *164*, 115–122. [CrossRef]
94. Oyama, T.G.; Oshima, A.; Tagawa, S. Estimation of resist sensitivity for extreme ultraviolet lithography using an electron beam. *AIP Adv.* **2016**, *6*, 085210. [CrossRef]

95. Rathore, A.; Pollentier, I.; Singh, H.; Fallica, R.; De Simone, D.; De Gendt, S. Effect of molecular weight on the EUV-printability of main chain scission type polymers. *J. Mater. Chem. C* **2020**, *8*, 5958–5966. [CrossRef]
96. Trikeriotis, M.; Bae, W.J.; Schwartz, E.; Krysak, M.; Lafferty, N.; Xie, P.; Smith, B.; Zimmerman, P.; Ober, C.; Giannelis, E. Development of an inorganic photoresist for DUV, EUV, and electron beam imaging. In *Advances in Resist Materials and Processing Technology XXVII*; SPIE: Bellingham, WA, USA, 2010; Volume 7639, p. 76390E.
97. Trikeriotis, M.; Krysak, M.; Chung, Y.S.; Ouyang, C.; Cardineau, B.; Brainard, R.; Ober, C.; Giannelis, E.; Cho, K. A new inorganic EUV resist with high-etch resistance. In *Extreme Ultraviolet (EUV) Lithography III*; SPIE: Bellingham, WA, USA, 2012; Volume 8322, p. 83220U.
98. Kosma, V.; Kasahara, K.; Xu, H.; Odent, J.; Ober, C.; Giannelis, E. Elucidating the patterning mechanism of zirconium-based hybrid photoresists. *J. Micro Nanolithogr. MEMS MOEMS* **2017**, *16*, 041007. [CrossRef]
99. Cardineau, B.; Del Re, R.; Marnell, M.; Al-Mashat, H.; Vockenhuber, M.; Ekinci, Y.; Sarma, C.; Freedman, D.A.; Brainard, R.L. Photolithographic properties of tin-oxo clusters using extreme ultraviolet light (13.5 nm). *Microelectron. Eng.* **2014**, *127*, 44–50. [CrossRef]
100. Passarelli, J.; Murphy, M.; Del Re, R.; Sortland, M.; Hotalen, J.; Dousharm, L.; Fallica, R.; Ekinci, Y.; Neisser, M.; Freedman, D.; et al. Organometallic carboxylate resists for extreme ultraviolet with high sensitivity. *J. Micro Nanolithogr. MEMS MOEMS* **2015**, *14*, 043503. [CrossRef]
101. Sortland, M.; Del Re, R.; Passarelli, J.; Hotalen, J.; Vockenhuber, M.; Ekinci, Y.; Neisser, M.; Freedman, D.; Brainard, R. Positive-tone EUV resists: Complexes of platinum and palladium. In *Extreme Ultraviolet (EUV) Lithography VI*; SPIE: Bellingham, WA, USA, 2015; Volume 9422.
102. Fujimori, T.; Tsuchihashi, T.; Minegishi, S.; Kamizono, T.; Itani, T. Novel ultra-high sensitive 'metal resist' for EUV lithography. In *Extreme Ultraviolet (EUV) Lithography VII*; SPIE: Bellingham, WA, USA, 2016; Volume 9776, p. 977605.
103. Li, L.; Chakrabarty, S.; Spyrou, K.; Ober, C.K.; Giannelis, E.P. Studying the Mechanism of Hybrid Nanoparticle Photoresists: Effect of Particle Size on Photopatterning. *Chem. Mater.* **2015**, *27*, 5027–5031. [CrossRef]
104. Grenville, A.; Anderson, J.; Clark, B.; De Schepper, P.; Edson, J.; Greer, M.; Jiang, K.; Kocsis, M.; Meyers, S.; Stowers, J.; et al. Integrated fab process for metal oxide EUV photoresist. In *Advances in Patterning Materials and Processes XXXII*; SPIE: Bellingham, WA, USA, 2015; Volume 9425, p. 94250S.
105. Hinsberg, W.; Meyers, S. A numeric model for the imaging mechanism of metal oxide EUV resists. In *Advances in Patterning Materials and Processes XXXIV*; SPIE: Bellingham, WA, USA, 2017; Volume 10146, p. 1014604.
106. Xu, H.; Sakai, K.; Kasahara, K.; Kosma, V.; Yang, K.; Herbol, H.C.; Odent, J.; Clancy, P.; Giannelis, E.P.; Ober, C.K. Metal-Organic Framework-Inspired Metal-Containing Clusters for High-Resolution Patterning. *Chem. Mater.* **2018**, *30*, 4124–4133. [CrossRef]
107. Zhang, Y.; Haitjema, J.; Baljozovic, M.; Vockenhuber, M.; Kazazis, D.; Jung, T.A.; Ekinci, Y.; Brouwer, A.M. Dual-tone Application of a Tin-Oxo Cage Photoresist Under E-beam and EUV Exposure. *J. Photopolym. Sci. Technol.* **2018**, *31*, 249–255. [CrossRef]
108. Sitterly, J.; Murphy, M.; Grzeskowiak, S.; Denbeaux, G.; Brainard, R. Molecular organometallic resists for EUV (MORE): Reactivity as a function of metal center (Bi, Sb, Te and Sn). In *Advances in Patterning Materials and Processes XXXV*; SPIE: Bellingham, WA, USA, 2018; Volume 10586, p. 105861P.
109. Rantala, J.; Gädda, T.; Laukkanen, M.; Dang, L.N.; Karaste, K.; Kazizis, D.; Ekinci, Y. New resist and underlayer approaches toward EUV lithography. In Proceedings of the International Conference on Extreme Ultraviolet Lithography, Monterey, CA, USA, 17–20 September 2018; SPIE: Bellingham, WA, USA, 2018; Volume 10809, p. 108090X.
110. Gädda, T.; Luong, N.D.; Laukkanen, M.; Karaste, K.; Kähkönen, O.; Kauppi, E.; Kazazis, D.; Ekinci, Y.; Rantala, J. Advanced EUV negative tone resist and underlayer approaches exhibiting sub-20 nm half-pitch resolution. In *Advances in Patterning Materials and Processes XXXVI*; SPIE: Bellingham, WA, USA, 2019; Volume 10960, p. 109600B.
111. Thakur, N.; Tseng, L.-T.; Vockenhuber, M.; Ekinci, Y.; Castellanos, S. Stability studies on a sensitive EUV photoresist based on zinc metal oxoclusters. *J. Micro Nanolithogr. MEMS MOEMS* **2019**, *18*, 043504. [CrossRef]
112. Mattson, E.C.; Cabrera, Y.; Rupich, S.M.; Wang, Y.X.; Oyekan, K.A.; Mustard, T.J.; Halls, M.D.; Bechtel, H.A.; Martin, M.C.; Chabal, Y.J. Chemical Modification Mechanisms in Hybrid Hafnium Oxo-methacrylate Nanocluster Photoresists for Extreme Ultraviolet Patterning. *Chem. Mater.* **2018**, *30*, 6192–6206. [CrossRef]

113. Wu, L.; Baljozovic, M.; Portale, G.; Kazazis, D.; Vockenhuber, M.; Jung, T.; Ekinci, Y.; Castellanos Ortega, S. Mechanistic insights in Zr- and Hf-based molecular hybrid EUV photoresists. *J. Micro Nanolithogr. MEMS MOEMS* **2019**, *18*, 013504. [CrossRef]
114. Haitjema, J.; Zhang, Y.; Ottosson, N.; Brouwer, A.M. Photoreactions of Tin Oxo Cages, Model EUV Photoresists. *J. Photopolym. Sci. Technol.* **2017**, *30*, 99–102. [CrossRef]

MDPI
St. Alban-Anlage 66
4052 Basel
Switzerland
Tel. +41 61 683 77 34
Fax +41 61 302 89 18
www.mdpi.com

Nanomaterials Editorial Office
E-mail: nanomaterials@mdpi.com
www.mdpi.com/journal/nanomaterials

www.ingramcontent.com/pod-product-compliance
Lightning Source LLC
Chambersburg PA
CBHW040759250926
51850CB00007BA/1038
9783039433285